NEUROMETHODS □ 8

Imaging and Correlative Physicochemical Techniques

NEUROMETHODS

Program Editors: Alan A. Boulton and Glen B. Baker

1. **General Neurochemical Techniques**
   Edited by ***Alan A. Boulton and Glen B. Baker,*** 1985
2. **Amines and Their Metabolites**
   Edited by ***Alan A. Boulton, Glen B. Baker, and Judith M. Baker,*** 1985
3. **Amino Acids**
   Edited by ***Alan A. Boulton, Glen B. Baker, and James D. Wood,*** 1985
4. **Receptor Binding Techniques**
   Edited by ***Alan A. Boulton, Glen B. Baker, and Pavel D. Hrdina,*** 1986
5. **Neurotransmitter Enzymes**
   Edited by ***Alan A. Boulton, Glen B. Baker, and Peter H. Yu,*** 1986
6. **Peptides**
   Edited by ***Alan A. Boulton, Glen B. Baker, and Quentin Pittman,*** 1987
7. **Lipids and Related Compounds**
   Edited by ***Alan A. Boulton, Glen B. Baker, and Lloyd A. Horrocks,*** 1988
8. **Imaging and Correlative Physicochemical Techniques**
   Edited by ***Alan A. Boulton, Glen B. Baker, and Donald P. Boisvert,*** 1988
9. **The Neuronal Microenvironment**
   Edited by ***Alan A. Boulton, Glen B. Baker, and Wolfgang Walz,*** 1988
10. **Analysis of Psychiatric Drugs**
    Edited by ***Alan A. Boulton, Glen B. Baker, and Ronald T. Coutts,*** 1988
11. **Carbohydrates and Energy Metabolism**
    Edited by ***Alan A. Boulton, Glen B. Baker, and Roger F. Butterworth***
12. **Drugs as Tools in Neurotransmitter Research**
    Edited by ***Alan A. Boulton, Glen B. Baker, and Augusto V. Juorio***
13. **Psychopharmacology**
    Edited by ***Alan A. Boulton, Glen B. Baker, and Andrew J. Greenshaw***
14. **Neurophysiology**
    Edited by ***Alan A. Boulton, Glen B. Baker, and Case H. Vanderwolf***
15. **Neuropsychology**
    Edited by ***Alan A. Boulton, Glen B. Baker, and Merril Hiscock***
16. **Molecular Neurobiology**
    Edited by ***Alan A. Boulton, Glen B. Baker, and Anthony T. Campagnoni***

NEUROMETHODS

Series 1: Neurochemistry

Program Editors: Alan A. Boulton and Glen B. Baker

Imaging and Correlative

Physicochemical Techniques

Edited by

Alan A. Boulton, Glen B. Baker,
and Donald P. J. Boisvert

Springer Science+Business Media, LLC

Library of Congress Cataloging in Publication Data

Main entry under title:

Imaging and correlative physiochemical techniques / edited by Alan A. Boulton, Glen B. Baker, and Donald P.J. Boisvert.
p. cm. — (Neuromethods; 8. Series I, Neurochemistry)
Includes bibliographies and index.

ISBN 978-1-4899-4111-4 ISBN 978-1-59259-613-3 (eBook)
DOI 10.1007/978-1-59259-613-3

1. Cerebral circulation—Measurement. 2. Brain—Blood-vessels--Imaging. I. Boulton, A. A. (Alan A.) II. Baker, Glen B., 1947– . III. Boisvert, Donald P. J. IV. Series: Neuromethods; 8. V. Series: Neuromethods. Series I, Neurochemistry.
[DNLM: 1. Brain—metabolism. 2. Brain—radiography. 3. Brain—-radionuclide imaging. 4. Cerebrovascular Circulation. 5. Neurochemistry—methods. W1 NE337G v. 8 / WL 300 I31]
QP108.5.C4I45 1988
612' .824—dc19
DNLM/DLC
for Library of Congress 88-9191
CIP

Originally published by The Humana Press Inc. in 1988
Softcover reprint of the hardcover 1st edition 1988

Preface to the Series

When the President of Humana Press first suggested that a series on methods in the neurosciences might be useful, one of us (AAB) was quite skeptical; only after discussions with GBB and some searching both of memory and library shelves did it seem that perhaps the publisher was right. Although some excellent methods books have recently appeared, notably in neuroanatomy, it is a fact that there is a dearth in this particular field, a fact attested to by the alacrity and enthusiasm with which most of the contributors to this series accepted our invitations and suggested additional topics and areas. After a somewhat hesitant start, essentially in the neurochemistry section, the series has grown and will encompass neurochemistry, neuropsychiatry, neurology, neuropathology, neurogenetics, neuroethology, molecular neurobiology, animal models of nervous disease, and no doubt many more "neuros." Although we have tried to include adequate methodological detail and in many cases detailed protocols, we have also tried to include wherever possible a short introductory review of the methods and/or related substances, comparisons with other methods, and the relationship of the substances being analyzed to neurological and psychiatric disorders. Recognizing our own limitations, we have invited a guest editor to join with us on most volumes in order to ensure complete coverage of the field and to add their specialized knowledge and competencies. We anticipate that this series will fill a gap; we can only hope that it will be filled appropriately and with the right amount of expertise with respect to each method, substance or group of substances, and area treated.

Alan A. Boulton
Glen B. Baker

Preface

Recent years have seen remarkable advances in the development of techniques that have direct applications in neurological research. In consequence, the circulatory and metabolic status of the brain can be measured and correlated with changes in structure and integrated function, often noninvasively, in the same experimental subject. This has stimulated an increased awareness of the complexity, under normal and pathological conditions, of the interdependence of these factors. Through the application of the methods described in this volume, however, these complexities can now be analyzed.

The chapters in this volume present methodological descriptions of some of the most powerful "physicochemical" methods for studying the brain. Multidisciplinary teams are required to develop some of these methods, which are extremely expensive in terms of capital equipment costs and technological personnel support. Thus, they will likely remain restricted to major medical research centers. Nevertheless, many recent concepts of brain responses to disease are a result of their application.

We have been fortunate in convincing active, leading scientists to contribute to this volume. The descriptions of the basic principles of each method, and its applications and limitations, are derived primarily from their personal experiences. The first two chapters (Rowan; Auer) deal with methods for assessing brain hemodynamics. The two subsequent chapters (Greenberg; Herscovitch) describe autoradiography and positron emission tomography techniques, which provide quantitative measurements of brain metabolism as well as blood flow. The chapters by Pritchard and Schulman and by Allen present more recent developments in methods for studying the brain, namely, those based on nuclear magnetic resonance research. Recently developed methods for measuring ion homeostasis and neurotransmitter activity are described in chapters by Branston and Harris and by Bruun and

Edvinsson. In the final chapter (Grundy), methodological considerations for measuring integrated neural function through brain-evoked potentials are presented.

Because of its scope and diversity, this volume will be of value to most clinical and laboratory neuroscientists.

Donald P. J. Boisvert

Contents

Preface to the Series v
Preface vii

Measurement of Cerebral Blood Flow Using Diffusible Gases
J. O. Rowan

1. Introduction 1
2. Basic Mathematical Principles of the Inert Gas Clearance Techniques 2
3. Determination of Cerebral Blood Flow by the Kety-Schmidt Technique 4
4. Methods Based on Monitoring the Clearance of Xenon-133 5
   4.1. Intracarotid Artery Injection Technique 5
   4.2. Inhalation Technique 8
   4.3. Intravenous Injection Technique 11
5. Sources of Error and Limitations of the Xenon-133 Clearance Methods Used for the Determination of Cerebral Blood Flow 12
6. Tomographic Techniques 17
   6.1. Xenon-133 Single Photon Tomography 18
   6.2. Short-Lived Radionuclides and Dual Photon Tomography 20
   6.3. Stable Xenon and X-Ray CT Scanning 21
7. Hydrogen Clearance Methods 23
   7.1. Principles Underlying Hydrogen Polarography 24
   7.2. Electrodes and Recording Circuit 25
   7.3. Hydrogen Administration 26
   7.4. Clearance Curves 28
   7.5. Advantages 30

7.6. Limitations and Sources of Error 30
References 32

Measurement of Pial Vessel Hemodynamics
Ludwig M. Auer

1. Introduction and Historical Survey 37
2. Recent Methods for Direct Investigation of Brain Vessels In Vivo 38
2.1. The Cranial Window Technique 38
2.2. Microscopes and Microscopical Techniques 43
2.3. Imaging of Pial Vessels 48
2.4. Analysis of Pial Vessel Caliber 51
2.5. Measurement of Flow Velocity 60
2.6. Calculation of Flow in Single Pial Vessels 72
2.7. Measurement or Calculation of Pial Blood Volume 72
3. Pial Vessels and Cerebral Blood Flow—Analysis of Relevance of Techniques 72
3.1. Relevance of Direct Pial Vessel Observation for the Definition of Normal CBF Autoregulation 74
3.2. Observations of Pathological Reactions 80
4. Direct Pharmacological Cerebrovascular Effects 84
5. Conclusions and Summary 91
References 97

Autoradiography and Cerebral Function
Joel H. Greenberg

1. Introduction 107
2. Local Cerebral Blood Flow 108
2.1. Theory 108
2.2. Experimental Technique 111
2.3. Diffusible Tracers 112
2.4. Compound Characterization and Technique Limitations 116
2.5. Indicator Fractionation Technique 122

2.6. Error Analysis of Blood Flow Measurement .. 124
3. Local Cerebral Glucose Utilization 131
3.1. Introduction 131
3.2. Theoretical Basis 133
3.3. Determination of Rate Constants 136
3.4. Determination of Lumped Constant 138
3.5. Determination of lCGU 141
3.6. Error Analysis 142
4. Resolution and Tritiated Autoradiography 146
4.1. Resolution 146
4.2. ^{3}H Autoradiography 151
5. Multiple Label Autoradiography 155
5.1. Carbon/Iodide Double Label 158
5.2. ^{14}C/^{3}H Double Label 159
5.3. ^{14}C/^{14}C Double Label 162
5.4. ^{18}F/^{14}C Double Label 165
5.5. Autoradiography with More Than Two Labels 165
6. Image Analysis 166
7. Applications 169
References 170

Measurement of Regional Cerebral Hemodynamics and Metabolism by Positron Emission Tomography
Peter Herscovitch

1. Introduction 179
2. PET Detection Systems 180
3. Radiotracers for PET 187
4. Tracer Techniques 191
5. Cerebral Blood Volume 193
6. Cerebral Blood Flow 196
6.1. Background 196
6.2. Steady-State Method 198
6.3. PET/Autoradiographic Method 200
6.4. Other rCBF Methods 202
7. Cerebral Oxygen Metabolism 203
7.1. Background 203
7.2. Steady-State Method 204

7.3. Brief Inhalation Method 205
8. Glucose Metabolism.................................... 206
8.1. Deoxyglucose Method............................. 206
8.2. Use of ^{11}C-Glucose 212
9. Analysis of PET Data 213
10. Applications of PET..................................... 217
References.. 222

NMR Spectroscopy of Brain Metabolism In Vivo
James W. Prichard and Robert G. Shulman

1. Introduction.. 233
2. Phosphorus (^{31}P) Studies 235
2.1. Animals 235
2.2. Magnetization Transfer Experiments........... 240
2.3. Humans 241
3. Carbon (^{13}C) Studies 242
4. Hydrogen (^{1}H) Studies 243
5. Editing Techniques 250
5.1. Homonuclear Decoupling.......................... 250
5.2. Heteronuclear Decoupling......................... 252
5.3. Selective Excitation................................. 255
6. Prospects.. 257
References 258

NMR Imaging of the Central Nervous System
Peter S. Allen

1. Introduction.. 265
2. Some Basic Principles of Nuclear Magnetic Resonance 266
3. The Encoding of Spatial Information................. 271
4. NMR Imaging 273
4.1. Slice Definition.. 273
4.2. Image Production 274
5. Nuclear Relaxation and Tissue Discrimination 275
5.1. Longitudinal Relaxation (T_1) 276
5.2. Transverse Relaxation (T_2)......................... 276
5.3. Nuclear Relaxation in Biological Systems 277

5.4. Some Methods of Encoding Images with Relaxation Information 278
5.5. Paramagnetic Agents to Enhance Tissue Contrast 283
6. NMR Imaging of the Central Nervous System..... 285
6.1. Introduction and the Normal Anatomy 285
6.2. Cerebrovascular Disease 288
6.3. Tumors of the CNS 289
6.4. Assessment of the State of Myelination........ 290
6.5. Craniovertebral Junction and the Spinal Cord 291
6.6. Imaging with the Sodium Nucleus, ^{23}Na 291
7. Summary.. 293
References 295

Measurement of Cerebral Ions
Neil M. Branston and Robert J. Harris

1. Introduction.. 299
1.1. Historical—Why Cerebral Ions Are Measured 299
1.2. Ions in Normal Cerebral Physiology 300
1.3. Ions in Cerebral Pathophysiology............... 302
2. General Methods 305
2.1. Introduction.. 305
2.2. Ion-Selective Electrodes (ISE) 305
2.3. Spectrophotometry.................................. 312
2.4. Electron Probe X-ray Microanalysis............. 312
2.5. Isotope, Fluorescence, and Other Methods for Measuring pH 313
2.6. Other Methods for Measuring Ca_e............... 315
3. Making and Calibrating a Common Type of Extracellular Ion-Selective Microelectrode 315
3.1. Manufacture 316
3.2. Calibration.. 319
3.3. Electrode Signal: Generation, Amplification, and Recording.. 321
4. Use of Ion-Selective Electrodes in Studies of Focal Ischemia 322

4.1. Introduction 322
4.2. Examples of Measurements 323
5. Conclusions 329
References 329

Immunocytochemical Procedures for the Demonstration of Putative Neurotransmitters in Cerebral Vessels, Cerebrum, and Retina
Anitha Bruun and Lars Edvinsson

1. Introduction 343
2. Immunological Basis 344
3. Types of Immunocytochemistry Methods 345
3.1. Indirect Fluorescence 345
3.2. Peroxidase–Antiperoxidase (PAP) Complex ... 345
3.3. Avidin-Biotin Procedure 346
3.4. Colloidal Gold 347
4. Indirect Immunofluorescence Method 347
4.1. Fixation 347
4.2. Sections and Whole Mounts 351
4.3 Immunolabeling 351
4.4. Microscopy 352
5. Immunocytochemistry of Selected Tissues 352
5.1. Neuropeptide Y 352
5.2. 5-Hydroxytryptamine 356
5.3. Substance P 361
5.4. Vasoactive Intestinal Peptide 363
6. Conclusion 365
Appendix I. Immunocytochemistry Procedures ... 366
Appendix II. Buffers, Fixatives, and Other Solutions 368
References 369

Sensory-Evoked Potentials
Betty L. Grundy

1. Introduction 375

2. Classification of Evoked Potentials 375
2.1. Types of Stimulation 375
2.2. Modes of Recording 381
2.3. Methods of Signal Processing 384
2.4. Poststimulus Latencies 390
2.5. Purported Neural Generators 391
3. Equipment and Techniques 396
4. Applications 400
4.1. Experimental 402
4.2. Clinical 403
5. Conclusions 405
References 426
Index 435

Contributors

PETER S. ALLEN • *Department of Applied Sciences in Medicine, University of Alberta, Edmonton, Alberta, Canada*

LUDWIG M. AUER • *Department of Neurosurgery, University of Graz, Graz, Austria*

GLEN B. BAKER • *Neurochemical Research Unit, Department of Psychiatry, University of Alberta, Edmonton, Alberta, Canada*

DONALD P. J. BOISVERT • *Division of Neurosurgery, University of Alberta, Edmonton, Alberta, Canada*

ALAN A. BOULTON • *Neuropsychiatric Research Unit, University of Saskatchewan, Saskatoon, Saskatchewan, Canada*

NEIL M. BRANSTON • *Department of Neurological Surgery, Institute of Neurology, London, UK*

ANITHA BRUUN • *Department of Opthalmology, University Hospital, Lund, Sweden*

LARS EDVINSSON • *Department of Clinical Pharmacology, University Hospital, Lund, Sweden*

JOEL H. GREENBERG • *Cerebrovascular Research Center, University of Pennsylvania, Philadelphia, Pennsylvania*

BETTY L. GRUNDY • *University of Florida College of Medicine and Verterans Administration Medical Center, Gainesville, Florida*

ROBERT J. HARRIS • *Department of Neurological Surgery, Institute of Neurology, London, UK*

PETER HERSCOVITCH • *National Institutes of Health, Bethesda, Maryland*

JAMES W. PRICHARD • *Department of Neurology, Yale Medical School, New Haven, Connecticut*

J. O. ROWAN • *West of Scotland Health Boards, Department of Clinical Physics and Bioengineering, Glasgow, Scotland*

ROBERT G. SCHULMAN • *Department of Molecular Biophysics and Biochemistry, Yale Medical School, New Haven, Connecticut*

Measurement of Cerebral Blood Flow Using Diffusible Gases

J. O. Rowan

1. Introduction

Certain gaseous tracers, characterized by being metabolically inert and possessing the capability of diffusing rapidly between blood and tissue, have been used for many years to derive estimates of cerebral blood flow in vivo, as well as in the animal laboratory. The work of Kety and Schmidt (1945), who developed the nitrous oxide technique for the determination of cerebral blood flow in humans, has provided the foundation for a range of techniques for the measurement of total and regional cerebral blood flow, many of them highly sophisticated and employing a wide spectrum of physical principles. As these techniques found their way into clinical practice, it was inevitable that pressure should build up to limit the invasive nature of the investigations. Thus from the xenon-133 intracarotid artery injection technique, we have seen the development of inhalation and intravenous (iv) injection methods. Furthermore, the development and rapid acceptance of X-ray CT scanning into the diagnostic armamentarium has led to blood flow determinations that rely on the use of stable xenon as an enhancement agent, whereas the introduction of positron emission tomography gave rise to three-dimensional assessment of regional cerebral blood flow and metabolism utilizing short-lived radionuclides such as oxygen-15. In these latter situations the radionuclide half-life is so short that quick and easy access to a local cyclotron installation is essential. Over the years strenuous efforts have been made to reconcile the conflicting requirements of high sensitivity, good spatial resolution, and negligible trauma. The hydrogen clearance technique comes very close to this ideal, but in general, ethical considerations prevent its widespread use in humans.

2. Basic Mathematical Principles of the Inert Gas Clearance Techniques

The Fick principle provides the basis for the methodology. This principle states that the quantity of a metabolically inert substance taken up by tissue in unit time is equal to the quantity brought to the tissue by arterial blood less the quantity carried away by venous blood. This is fundamentally a restatement of the law of conservation of mass.

In practice the methods rely on three basic assumptions:

1. The tracer employed is metabolically inert and diffuses rapidly between blood and tissue with no significant diffusion limitation.
2. Arterial flow is equal to venous flow.
3. The blood in brain tissue forms only a small fraction of the total brain tissue volume under consideration.

The Fick Principle can be stated in a mathematical form as follows:

$$\frac{dQ_i}{dt} = \mathrm{F_i}\,(C_a - C_v) \tag{1}$$

Where Q_i is the quantity of the tracer in brain tissue, F_i is the blood flow, and C_a and C_v are the tracer concentrations in arterial and venous blood, respectively.

If W_i denotes the weight of the tissue, then Eq. (1) can be rewritten to give the concentration of the tracer in tissue, C_i. Hence

$$\frac{dC_i}{dt} = \frac{F_i}{W_i}\,(C_a - C_v) \tag{2}$$

If we let λ_i denote the partition coefficient, i.e., the ratio of the solubilities of the diffusible tracer in tissue and blood, then

$$C_v = \frac{C_i}{\lambda_i}$$

and

$$\frac{dC_i}{dt} = \frac{F_i}{\lambda_i W_i}\,(\lambda_i C_a - C_i) \tag{3}$$

Let f_i denote the flow per unit weight of tissue, i.e.,

$$f_i = \frac{F_i}{W_i}$$

and putting

$$k_i = \frac{f_i}{\lambda_i}$$

then

$$\frac{dC_i}{dt} = k_i\,(\lambda_i\, C_a - C_i) \tag{4}$$

The undernoted solutions can be obtained for this linear differential equation.

1. When the arterial concentration varies with time, we have the general solution:

$$C_{i(T)} = \lambda_i k_i\, e^{-k_i T} \int_0^T C_{a(t)}\, e^{k_i t}\, dt \tag{5}$$

where T is a given time after the administration of the tracer, $C_{i(T)}$ is the concentration of the tracer in tissue at time T, and $C_{a(t)}$ is the concentration of the tracer in arterial blood at any time t.

2. If we consider the tissue saturation process and assume that the arterial tracer concentration rises sharply to a constant value, C_a, then

$$C_{i(t)} = \lambda_i C_a\,(1 - e^{-k_i t}) \tag{6}$$

3. If we now consider the tissue desaturation process, or clearance phase, and assume that the arterial tracer concentration is zero at the onset of tissue desaturation, then

$$C_{i(t)} = C_{i0}\, e^{-k_i t} \tag{7}$$

where C_{i0} is the tissue tracer concentration at the beginning of the desaturation period.

3. Determination of Cerebral Blood Flow by the Kety-Schmidt Technique

Kety and Schmidt introduced the nitrous oxide technique for the determination of cerebral blood flow in humans in 1945 and provided the impetus for the later developments that have led to the range of methods we see today. They administered low concentrations of the gas to their subjects by inhalation over a 10-min period. During tracer administration, blood samples were taken from an artery and from the internal jugular vein and analyzed for nitrous oxide content. Immediately upon completion of tracer administration, an equilibrium situation existed, with the nitrous oxide concentrations in arterial blood, tissue, and venous blood being approximately equal. Flow was then calculated using the integrated form of Eq. (2).

$$C_{i(T)} = \frac{F_i}{W_i} \int_0^T (C_a - C_v)\, dt \tag{8}$$

Thus

$$f_i = \frac{F_i}{W_i} = \frac{\lambda_i\, C_{v(T)}}{\int_0^T (C_a - C_v)\, dt} \tag{9}$$

The Kety-Schmidt method has been modified in a number of ways since 1945. These modifications have included continuous withdrawal of blood samples during the tracer administration period to provide integrated arterial and venous tracer concentrations and use of the radioactive gas xenon-133 instead of nitrous oxide so that tracer concentrations could be determined by radiation detection methods, thus avoiding the tedious nitrous oxide measurements involved in the original method. In order to overcome the need for precise control of tracer concentration during inhalation, the desaturation phase of the clearance curve has been used instead of the saturation phase.

The Kety-Schmidt technique provides an average value for cerebral hemisphere blood flow over the 10-min saturation or desaturation period involved. If the level of blood flow is low, tracer equilibrium between brain tissue and blood will be unlikely to occur in this 10-min period and, as a consequence, blood flow will be overestimated. Use of tracers with low solubility in blood will help to reduce this inaccuracy. The method depends critically on the assumption that blood samples taken from the internal jugular bulb are representative of the venous drainage from the brain. Except in rare occasions, it is found in practice that only 3% of the blood in the internal jugular bulb comes from extracerebral sources.

In addition to providing an average value of cerebral hemisphere blood flow per unit weight of brain tissue, the technique makes it possible to investigate cerebral metabolism utilizing measurements of arteriovenous differences of cerebral metabolites such as oxygen and glucose.

4. Methods Based on Monitoring the Clearance of Xenon-133

In essence the Kety-Schmidt technique measures blood flow in a cerebral hemisphere, providing an average value of blood flow per unit weight of tissue. A number of methods were developed subsequently with the aim of determining regional cerebral blood flow utilizing the administration of the radionuclide xenon-133 and monitoring the subsequent clearance of gamma radiation from the brain by means of a number of collimated radiation detectors, or gamma cameras, mounted around the head.

4.1. *Intracarotid Artery Injection Technique*

The intracarotid artery injection method was first introduced by Lassen et al. (1963), who employed krypton-85 as the radioactive tracer, and was later modified by Glass and Harper (1963), who used xenon-133, which emits 81 keV gamma radiation and has a physical half-life of 5.27 d.

Xenon-133, dissolved in saline, is injected, as a bolus, into one of the internal carotid arteries. The tracer is transported by the blood to brain tissue, where equilibrium between tissue and blood is rapidly established. Furthermore, since xenon is highly soluble in air, over 90% of the tracer reaching the lungs is excreted. The 10% or so remaining is then redistributed throughout the body as a whole, with the result that there is no significant recirculation of tracer to the brain. Once the injection has been completed, fresh arterial blood, free from xenon, clears the radioactive tracer from brain tissue, with the rate of clearance being dependent on the level of blood flow.

Since after injection the arterial blood is tracer-free, $C_a = 0$ and Eq. (7), which mathematically expresses a monoexponential decay, is the applicable solution of the Fick equation in this situation. In practice, with normal subjects, two exponential components, representing the levels of blood flow in gray and white matter, can be extracted from the clearance curves (Fig. 1). The slopes of these two components can be defined by their clearance half-times ($t_{½}$).

$$t_{1/2} = \frac{\log_e^2}{k_i} \tag{10}$$

Hence since

$$k_i = \frac{f_i}{\lambda_i}$$

$$f_i = \frac{\lambda_i \log_e^2}{t_{1/2}} \tag{11}$$

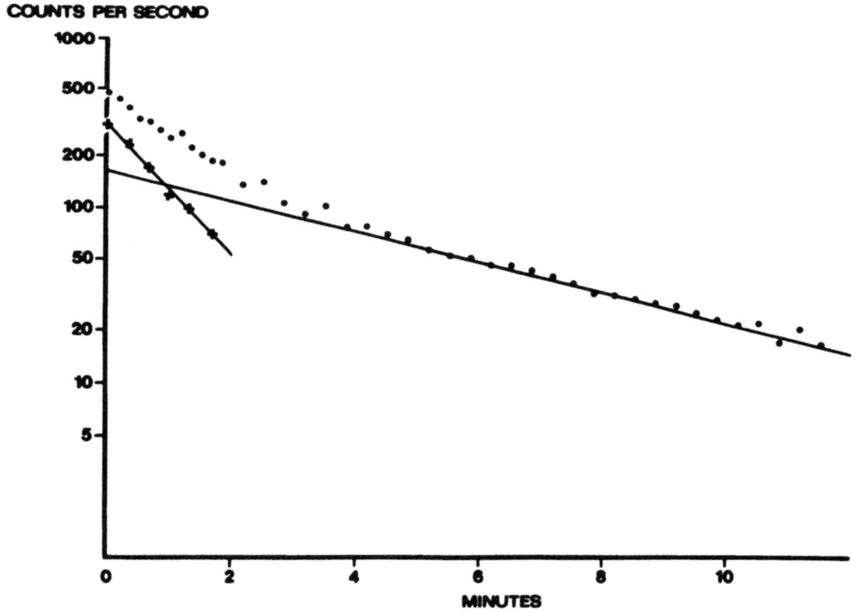

Fig. 1. Extraction of exponential components from a semilogarithmic plot of xenon-133 clearance from the brain following intracarotid artery injection.

Thus flow values in mL/100 g/min for gray and white matter can be found from the equation.

$$\text{Flow (mL/100 g/min)} = \frac{\lambda_i \log_e^2}{t_{1/2}} \times 100 \times 60 \qquad (12)$$

where $t_{1/2}$ is the clearance half-time in seconds.

The partition coefficient (λ_i) values for gray and white matter are 0.81 and 1.5, respectively.

A value for the mean cerebral blood flow through the region of the brain under consideration can be calculated from the ratio of the peak of the clearance curve (H_{max}) to the area under the curve (Fig. 2) (Zierler, 1965). Thus

$$\text{Mean flow (mL/100 g/min)} = \frac{\lambda_i H_{max}}{A} \times 100 \times 60 \qquad (13)$$

where H_{max} is measured in counts per s, and $\lambda_i = 1.1$, the mean value for the whole brain.

When a computer is not available and a typical 10-min investigation period is used, the equation for the mean flow can be modified to

$$\text{Mean flow (mL/100 g/min)} = \frac{\lambda_i (H_{max} - H_{10})}{A} \times 100 \times 60 \qquad (14)$$

where H_{10} is the value of the count rate at 10 min and A_{10} is the total number of counts recorded during the 10-min investigation period. An index based on the initial slope of the clearance curve can also be used to express mean flow (Olesen et al., 1971). This provides a significant advantage in practice when physiological stability throughout the 10-min investigation period cannot be guaranteed. The underlying assumption here is that the slope of the clearance curve during the first 1–2 min after the injection is proportional to mean flow.

$$\text{Mean flow} = \frac{\text{Empirical constant}}{t_{1/2}} \qquad (15)$$

Highly significant correlations have been demonstrated, over a wide range of flow values, between mean flow values calculated from initial slope and height/area measurements (Rowan et al., 1975a). At very high mean flow levels (greater than 100 mL/100 g min), the flow index calculated from the initial slope tends to be

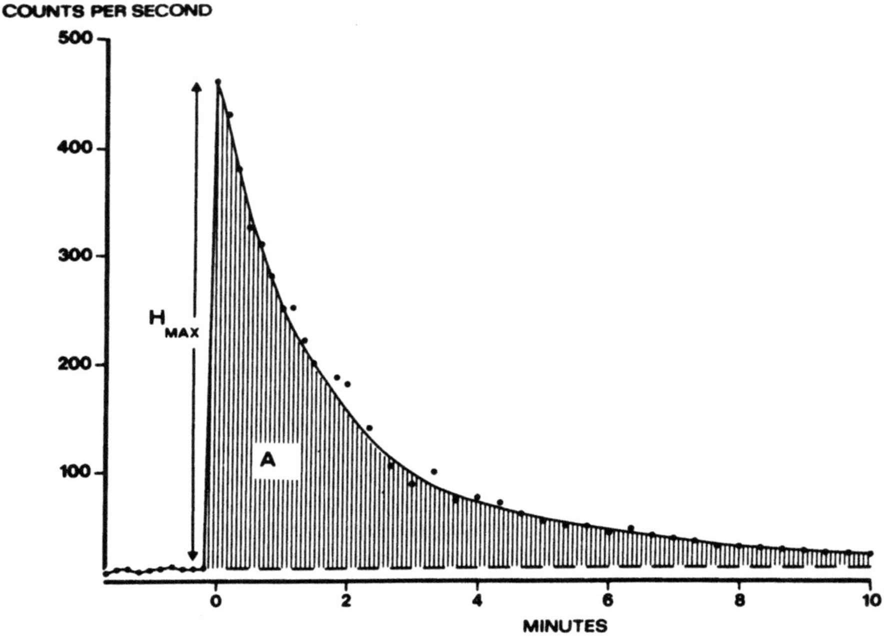

Fig. 2. Linear plot of xenon-133 clearance from the brain following intracarotid artery injection.

weighted toward gray matter flow values. The intracarotid artery injection technique has been widely used in physiological research investigations. Its use in routine clinical practice, however, has been limited because of the need to puncture a carotid artery, the associated hazard to the patient being far from insignificant. The result has been that most clinical centers have used this method only when the carotid artery bifurcation has been exposed in the operating theater, or when the patient has had to be investigated by carotid artery angiography.

4.2. Inhalation Technique

Because of the limited application of the xenon-133 intracarotid artery injection technique in clinical practice, many research workers put strenuous efforts into developing so-called atraumatic techniques for the measurement of cerebral blood flow that would be applicable in a wide range of clinical situations. The basic specification for the ideal cerebral blood flow measurement technique can be outlined as follows.

1. The technique should be completely atraumatic.
2. Blood flow data from well-defined regions of the brain should be provided by the technique, so that interregional comparisons can be made.
3. The investigation time should be as short as possible, so that the assumption that physiological stability is maintained during the investigation period is valid.
4. Repeated measurements of blood flow should be possible.
5. The precision of the measurements should be such that clinically significant changes in cerebral blood flow can be determined.

The xenon-133 inhalation technique initially devised by Mallet and Veall in 1963, and later modified by Obrist et al. in 1967 and 1970 and by Wyper et al. in 1976, was the first attempt to design a measurement technique that could come close to the ideal basic specification. The inhalation technique can certainly be considered atraumatic, but it introduces two major difficulties. During xenon inhalation a proportion of the tracer enters the extra cerebral vessels, with the result that the observed clearance curves are modified to some extent by the radioactivity detected from the scalp and other extracerebral tissues. Thus, in contrast to the intracarotid artery injection technique, in which clearance curves are observed to reach radiation background levels in approximately 20 min with normal patients, clearance curves with the inhalation technique may remain above background for an hour or more. To further complicate the situation, after inhalation, xenon-133 is taken up by tissues throughout the body, resulting in significant arterial recirculation. Thus, when xenon-133 administration ceases, the arterial tracer concentration is not negligible and, in addition, varies with time. Thus, during the period the clearance curves are being recorded, the xenon-133 input to the brain is varying. Therefore Eq. (5), the general solution of the Fick equation, must be used in this situation.

Extracerebral contamination and tracer recirculation tend to reduce the observed clearance rate of tracer from the head, and their significance has to be estimated when the inhalation technique is used. Correction for tracer recirculation is, in general, carried out by monitoring the radioactivity of end-expired air, which is assumed to reflect the arterial concentration of xenon-133 in patients with normal pulmonary function. Computer iterative methods are then used to find k_i.

Xenon-133 at a concentration of the order of 37 MBq (1 mCi)/L of air, is generally administered to the patient by means of a face mask or mouth-piece fitted with a one-way valve. Both closed and open looped breathing systems have been used. Mallet and Veall originally used a 5-min inhalation period, whereas later modified methods tended to limit the tracer administration period to 1–2 min. Longer inhalation times result in greater contributions from slowly clearing extracerebral tissues, whereas shorter periods increase the significance of tracer recirculation. Expired air at the mouth-piece is sampled utilizing a pump that is often mounted in a standard infrared carbon dioxide analyzer. The carbon dioxide analyzer determines the percentage of carbon dioxide in the expired air, and the expired air sample is then passed through a flow cell and the radioactivity monitored by a scintillation detector. As with the intracarotid artery injection method, the clearance of gamma radiation from the head is monitored by externally mounted radiation detectors.

Three-compartmental analysis can be carried out on the clearance curves obtained, so as to take into account the effect of the slow third exponential component caused by the presence of tracer in extracerebral tissues. However, this requires recording the clearance curves for 45–60 min. Even after that time it is not possible to discriminate between the second and third components with a high degree of efficiency. The use of exponential stripping procedures in the analyses of clearance curves, when a wide range of flow values is being considered, is severely limited because in reality there is a range of flows in each of the identifiable compartments, and the efficiency with which the different components can be resolved changes as the relative distributions change with the overall level of blood flow (Rowan, 1977).

In an attempt to overcome these problems, Obrist et al. (1970) devised a two-component analysis procedure involving the early part of the clearance curve. The first component is considered to represent gray matter flow, and the second component, flow in both white matter and extracerebral tissues. Both components are corrected for arterial recirculation of tracer, but only the first component is considered to provide meaningful values for cerebral blood flow. This modification to the inhalation technique reduces the investigation time from 40 (to 60) min to 12 min. The Glasgow 2-min slope xenon inhalation technique (Wyper et al., 1976) was devised to reduce this investigation time even further. With this method the patient inhales xenon-133 for 2 min (Fig. 3), but the

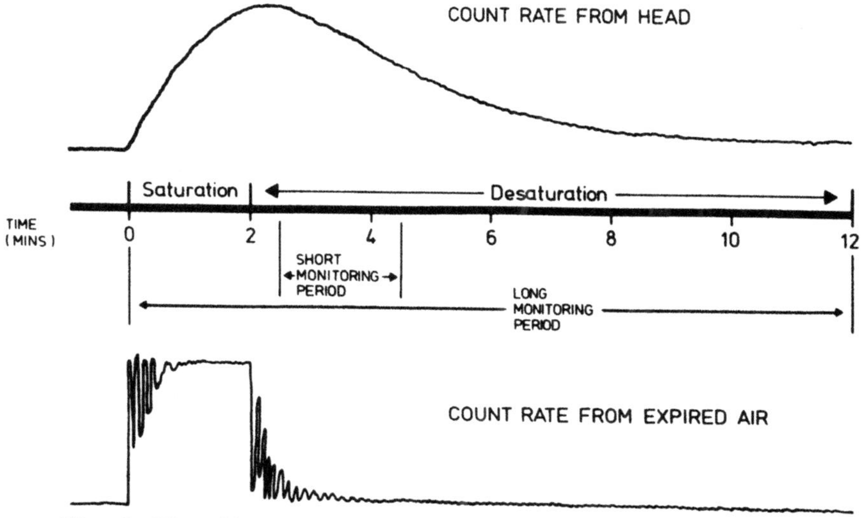

Fig. 3. The Glasgow 2-min slope xenon-133 inhalation technique (from Wyper et al., 1976).

clearance curves are only monitored for 2.5 min. An initial slope index corrected for arterial recirculation is estimated. This method provides indices of mean flow in contrast to the gray matter flow indices provided by the Obrist method. The development of computer-generated nomograms reduces the need for an on-line computer (Wyper and Rowan, 1976).

Compared to the intracarotid artery injection technique, inhalation methods tend to underestimate flow values by about 10–13%. On the other hand, high correlations (r approximately 0.90) are obtained between flow indices estimated by both techniques. Repeated measurements over hours, days, and weeks are possible.

4.3. Intravenous Injection Technique

The intravenous injection technique was also developed with the major aim of overcoming the potential trauma associated with the intracarotid artery injection technique. A bolus injection of 370 MBq (10 mCi) of xenon-133, dissolved in saline, is administered using a vein in the arm. As with the inhalation method, the observed clearance curves from the head are modified by arterial recirculation of tracer and extracranial contamination, and as a result Eq. (5), the general solution to the Fick equation, is again applicable. The clearance curves are analyzed using methods similar to those used for the inhalation technique to provide values of gray matter flow and mean flow.

The major advantage of the intravenous technique is that it utilizes a simple method of tracer administration. On the other hand, it suffers from the disadvantage that since so much of the injected tracer is excreted in the lungs prior to entering the cerebral circulation, the count levels detected at the head are significantly reduced. The simple tracer administration procedure makes the intravenous technique useful for investigating cerebral blood flow during neurosurgical procedures in which administration of gaseous xenon by way of the anesthetic equipment can cause problems from anesthetic gas dilution. The technique has been used with some success in monitoring cerebral blood flow during induced hypotension in aneurysm surgery and in the intensive care unit, in which difficulties with patient cooperation, such as that experienced with some head-injured patients, may make the use of the inhalation method impossible. Because of the count rate limitations, the use of the intravenous injection techniques is often limited to total brain blood flow measurements or interhemispheric comparisons.

5. Sources of Error and Limitations of the Xenon-133 Clearance Methods Used for the Determination of Cerebral Blood Flow

Both the xenon-133 inhalation and intravenous injection methods were developed and subsequently modified with the prime objective of achieving as far as possible the specification for the ideal cerebral blood flow measurement technique that was outlined previously. Both the inhalation and intravenous injection methods can be considered atraumatic. The two-component analysis method of Obrist (gray matter flow) and the Glasgow 2-min slope technique (mean flow) are attempts to reduce the investigation time as far as possible so that steady-state conditions can be considered to exist during the measurement period. Regional measurements are difficult to achieve with the intravenous technique, but they are certainly possible with the inhalation method. Repeated measurements in both the short and long terms are possible using both methods. However, the concept of precision is worth considering further.

In clinical practice the observed change in cerebral blood flow in response to a defined physiological stimulus can be more informative than the measured flow value at rest. However, the actual significance of these induced flow changes depends on prior knowledge of the random variation in flow measurement that exists when no deliberate attempt is made to change the level of blood flow. The precision of the applied method must be determined for the category of patients under investigation and can be estimated by taking two successive blood flow measurements in each patient under resting, control conditions after a defined interval and calculating the standard deviation of the mean difference in successive flow measurements for the subject group. The mean difference value gives an indication of any systematic error in the measurement, whereas twice the associated standard deviation gives a value for the precision within 95% confidence limits. This precision value depends not only on the consistency of the method in measuring cerebral blood flow, but also on the natural variation in flow that may occur in the category of patient under investigation. A typical precision value for normal subjects even with the carotid artery injection method is of the order of 20%, and if measured blood flow changes do not change by greater than that amount, after the application of a stimulus, one cannot claim that the blood flow has changed significantly from normal. The precision figure is much lower (repeatability greater) in geriatric patients, and can be even lower when these patients exhibit symptoms of dementia (McAlpine et al., 1981). In other categories of patient it may be so high that useful measurements cannot be carried out. Although it is impossible to achieve the fundamental accuracy of the carotid artery injection technique using the inhalation and intravenous injection methods, the latter techniques can provide the same level of precision, largely because of the basic physiological instability of the patient. This, together with the excellent statistical correlations obtained when these atraumatic measures are compared with the intracarotid artery injection technique, has led to the acceptance of both the inhalation and intravenous injection techniques in clinical practice.

However, there is one further disadvantage associated with both the Obrist method of analysis and the Glasgow 2-min flow analysis, and that is that the early part of the clearance curve is neglected in order to avoid the air passage artifact that results from facial cavities and passages during inhalation or intravenous studies. High flows in small cortical volumes can be measured by an initial slope flow index computed 10–50 s after injection, but not 30–90 s afterwards. The significance of this fact depends very much on the category of patients under investigation. However, even if these very high flows are not of interest, utilization of multiple extracranial detectors inevitably means that some of them will be "looking at" regions that include airways. With small detectors placed over frontal and temporal regions, a significant part of the recorded activity may result from tracer cleared via the lungs and airways. Nilsson et al. (1982) designed an algorithm for the detection and correction of airway contamination of clearance curves in order to obtain a more accurate determination of blood flow in regions subject to significant artifact. The basic assumption of this correction is that the xenon-133 tracer concentrations in the respiratory air determines the shape of a template artifact curve. When the artifact observed by the detectors situated around the head has a different shape from that recorded from the airways at the mouth, the algorithm will not give a fully accurate correction. This can happen, for example, when xenon gas leaks around the face mask. Furthermore, if the tracer concentration in the upper airways is inhomogeneous, because, for example, of slow tracer convection in the nasal and sinus cavities, only a partial correction for the airway can be determined by this method.

Risberg's algorithm for airway artifact compensation (Risberg and Prohovnik, 1981) utilizes the clearance of activity in the head detectors at the end of the inhalation period as an index of contamination. The average xenon-133 concentration in the respiratory air is used as an airway artifact reference. However, this method tends to be only suitable for xenon-133 administration by inhalation.

Difficulties experienced by some workers with exponential stripping of clearance curves in practice have led them to believe that they might be able to set up a sensitivity threshold at the boundary of cortex and white matter. The erroneous use of point source assessment of multidetector systems has given some credence to this concept. Based on the inverse square law, such an assessment may suggest that all the detected counts derive from gamma radiation emanating from localized regions of the brain situated close to the detector and therefore from the surface of the brain. However, it has to be remembered that in this measurement situation the source of radiation is not a point, and consequently point source assessment is grossly misleading. Proper consideration of the effect of thin, infinite, uniform sources arranged perpendicular to the axis of the detector shows that in these blood flow measurements the inverse square law is not applicable, and that the decrease in sensitivity with depth is caused largely by attenuation within the tissue (Gillespie, 1968). As a result, the reduction in detector response is not nearly as severe as that predicted from point source assessment (Fig. 4). When xenon-133 is used as a tracer, a radiation detector placed over the head will always detect gamma radiation from both gray and white matter. Furthermore, with the inhalation and intravenous injection techniques the tracer will be distributed throughout both cerebral hemispheres. Although this permits measurement of cerebral blood flow from both hemispheres simultaneously, it also introduces the problem of "cross-talk" between hemispheres, i.e., detectors ostensibly looking at one hemisphere detect activity originating from the other hemisphere. Using the geometrical symmetry of the head, cross-talk corrections can be deduced (Wyper and Cooke, 1977). Short-range beta radiation such as that from krypton-85 may be used in laboratory animals to monitor cortical flow only.

The attempt to measure flow in small regions of the brain is

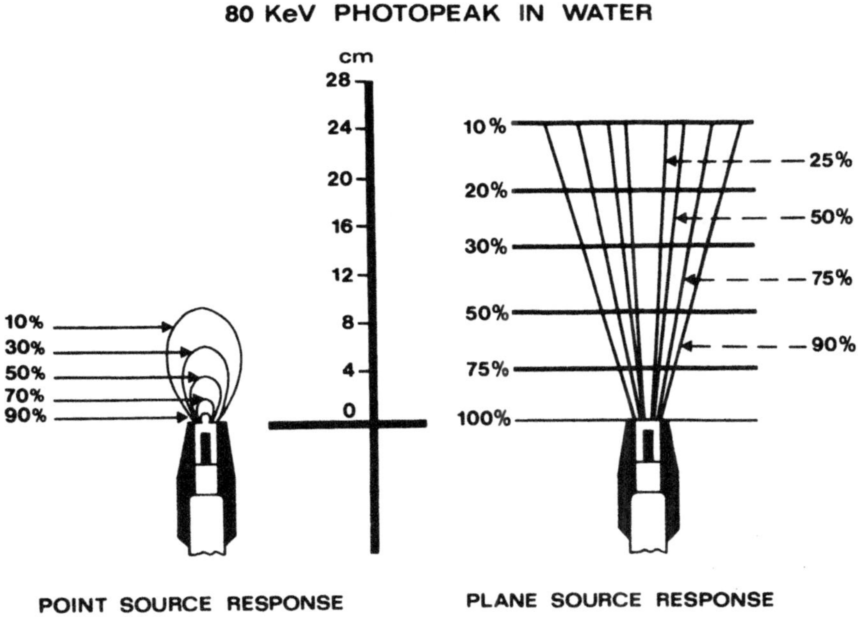

Fig. 4. Point source and plane source responses for the same scintillation detector monitoring 80 keV gamma radiation from various depths in water (from Gillespie, 1968).

further undermined by the fact that the 81 keV gamma radiation from xenon-133 undergoes significant Compton scatter in tissue. Limitations on the regionality of the measurement can therefore result from the use of wide pulse analyzer windows since a significant proportion of the detected photons will have come from outside the defined field of view. If all first-order scatter greater than 60° is to be avoided, the analyzer discriminator threshold must be set to a value corresponding to 75 keV. Inevitably this will reduce the counting sensitivity, and consequently the minimum size of the region from which useful flow measurements can be made will be compromized. The use of the 203 keV gamma radiation of xenon-127 alleviates this problem to some extent.

The smaller the region of inhomogeneous tissue, the greater the accuracy of the xenon-133 clearance methods depend on the uniform distribution of tracer partial pressure. The partial pressure will decrease more rapidly where flow is high and, as a result, partial pressure gradients will tend to be set up at tissue boundaries. There will then be a tendency for tracer to diffuse from high partial pressure regions to low partial pressure regions, modifying the mean rate of tracer clearance. This puts a further constraint on the minimum volume of tissue from which worthwhile cerebral blood flow measurements can be obtained.

The tissue–blood coefficient (λ_i) may vary in different categories of disease since its value depends on tissue lipid content and hematocrit. Correction for large vessel hematocrit can be made (Andersen and Ladefoged, 1965). However, the ratio of local cerebral level hematocrit to large vessel hematocrit is, in general, unknown. The magnitude of the changes in tissue–blood partition coefficient must be put into perspective. If the water content of the brain tissue increases by a factor of the order of 5%, as it may do in the initial phases of ischemia, this results in only a 2% decrease in partition coefficient (Tomita and Gotoh, 1981). The alteration in partition coefficient is small unless the tissue is infarcted, causing a considerable change in tissue components. In these situations the tissue can no longer be regarded as homogeneous. Furthermore, it should be noted that if a proportion of the tissue volume under investigation is not perfused with blood, it will not receive any tracer and will not be taken into account in the calculation of clearance half-times. This is referred to as the "look through" phenomenon.

Strictly

$$k_i = m_i \frac{f_i}{\lambda_i} \tag{16}$$

where m_i is the diffusion limitation factor.

In the case of gases, m_i approximates to unity (Ohta and Farhi, 1979), and provided the tissue is uniformly perfused with tracer, only a small error arises from the assumption that tracer equilibrium between tissue and blood occurs rapidly. However, when blood flow is severely reduced, uniform mixing of tracer and tissue may not occur. Thus cerebral blood flow estimates using these techniques have only limited accuracy at ischemic levels. At very high flow levels, m_i may drop below unity (Eklof et al., 1974), again affecting the accuracy of the methods.

The compromizes made in the data analysis associated with the application of the xenon-133 clearance techniques have on occasion come in for some criticism. However, anyone who insists on a totally rigorous mathematical interpretation of clearance curves on every occasion completely overlooks the variable nature of the data which is being acquired. Indeed, this attitude can give a technique a false air of respectability. For example, a completely monoexponential clearance curve in a patient does not necessarily mean that there is no gray matter flow. Furthermore, physiological instability is an ever-present problem in the clinical situation and a realistic compromize often has to be made. A number of workers have demonstrated that flow indices based on initial slopes of clearance curves correlate well with estimates obtained using a more rigorous type of analysis, thus permitting flow to be monitored in situations in which it would be otherwise impracticeable.

6. Tomographic Techniques

The widespread acceptance of high-resolution X-ray computerized tomographic scanning into clinical practice provided an impetus for the development of analogous techniques for the measurement of cerebral blood flow. In devising these techniques it was hoped that some of the major disadvantages of counting over the scalp with externally placed probes could be overcome. The potential advantages of tomography are that contamination of regional cerebral blood flow values by scalp flow can be avoided, errors caused by Compton scatter and tissue absorption can be decreased, and identification of low-flow or zero-flow areas deep in the brain is possible. Furthermore, by selecting regions of interest from the brain as a whole, it is possible to measure highly localized blood flow and with some techniques, partition coefficients in cortex, white matter, brain stem, thalamus, and cerebellum.

6.1. Xenon-133 Single Photon Tomography

Atraumatic techniques for cerebral blood flow measurement involving administration of xenon-133 (or xenon-127) by inhalation or intravenous injection have been modified further by the introduction of moving detector systems. Well-defined cortical and subcortical focal cerebral blood flow abnormalities have been demonstrated utilizing techniques that were introduced with the basic aims of investigating substantial focal reductions in blood flow and overcoming the fundamental problems posed by the "look through" phenomenon.

The instrumentation involved employs a rotating detector (or detectors), and in order to cope with the continuous changes in tracer concentration, the rotation must be rapid enough so that the basic assumption, that every projection has been recorded at the same time, is valid. Data from a number of rotations are accumulated to obtain a sufficient number of counts for reconstruction of an image. Generally, recordings are taken from two to three slices of the brain simultaneously. Great care has to be taken in the design of the tomograph to ensure that the counting sensitivity is sufficient to enable satisfactory tomographic reconstruction.

In the system described by Lassen et al. in 1981, the detector head consists of four rotating gamma cameras, and a series of four 1-min pictures is taken during and after inhalation of 370 MBq (10 mCi) of xenon-133 per L of air. Thus each picture obtained represents a map of the distribution of xenon-133 concentration averaged over time. The count rate was quoted as 600,000 cpm per slice in the second 1-min period when the count rate is at its maximum. The associated spatial resolution was stated as 1.7 cm full-width half-maximum. Three slices, 2 cm thick, are recorded simultaneously with the midslice plane separated by 4 cm. The tomographic images are constructed using a filtered back projection algorithm.

With this particular technique, the xenon-133 input curve is monitored using a single stationary detector over the right lung. The detector is positioned at a distance of 30 cm to minimize the effect of movement of the chest wall.

In 1979, Kanno and Lassen proposed two different methods of calculating regional cerebral blood flow from these tomographic images. The first approach, which the authors entitled the "sequence of pictures" method, utilizes a series of four consecutive 1-min integrated images. The exponential decay constant

$$\left(k_i = \frac{f_i}{\lambda_i}\right)$$

is calculated for each pixel assuming a monoexponential clearance. The four corresponding tracer concentration values recorded in any given pixel are normalized, by dividing by their sum, thereby disregarding the absolute value of the count rate. The second approach, termed by the authors the "initial picture" method, assumes that the count rate distribution reflects the distribution of cerebral blood flow. Thus a low count rate in a given pixel in a picture produced from the first and second 1-min tomograms is taken as being indicative of a low flow.

The "sequence of pictures" method has been found to be unreliable in detecting low levels of blood flow such as that found in ischemic areas in stroke patients. This is thought to be caused by the effect of Compton scatter and the resultant limitation it places on the focal nature of the measurement. However, low count rate areas and, hence, low flow rate areas were readily observed during the "initial picture" method.

Celsis and coworkers (1981) have subsequently devised an algorithm combining the two calculation methods. In actual fact, this combined approach involves only a minor modification of the "early picture" method (Celsis et al., 1981).

To investigate focal areas of low cerebral blood flow, tomographic systems must possess high sensitivity and provide good spatial resolution. In many ways these are opposing requirements, and in most situations spatial resolution suffers in the attempt to obtain the required count rate sensitivity. Improved spatial resolution can be obtained by using xenon-127 rather than xenon-133, but the resultant increases in cost can be substantial, and economic considerations may preclude its use.

Accurate measurement of low levels of cerebral blood flow in focal areas of ischemia remains a challenging target for radionuclide techniques and it is doubtful, even with the help of tomography, if the xenon methods will ever achieve the desired level of spatial resolution. However, the basic advantage of the single-photon tomography method is that it uses gamma-emitting radionuclide tracers that are readily available for use on a 24-h basis with minimum preparation.

6.2. *Short-Lived Radionuclides and Dual Photon Tomography*

Short-lived positron-emitting radionuclides, for example oxygen-15 (half-life, 2.1 min), nitrogen-13 (half-life, 10 min), carbon-11 (half-life, 20.3 min), and fluorine-18 (half-life, 1.83 h) are being used more and more to investigate regional cerebral blood flow and metabolism in conjunction with dual photon photography systems that rely on the detection of the pairs of gamma ray photons that result from positron annihilation. Because of their very short half-lives, if they are to be of any real advantage in clinical practice these radionuclide agents must be produced on site using a cyclotron installed locally.

Dual photon tomography systems (or positron tomographs) employ circular arrays or multiple banks of detectors. Positron annihilation coincidence events are identified by detectors situated directly opposite in the array. This provides "electronic" collimation such that the origin of the detected gamma radiation is observed to be within well-defined regions between the opposing detectors. Relatively uniform resolution and sensitivity across the slice of the tissue under investigation is obtained. The activity recorded by each detector is collected at a number of linear and angular positions around the head in as short a time as possible with corrections being applied for photon attenuation. The tangential scan-rotation or rotation-precession motions of the detectors are controlled by computers. After processing the regional count rate and angular information, the distribution of tracer within a number of defined slices of brain tissue can be reconstructed.

Methods have been devised to study regional cerebral perfusion and oxygen utilization. Continuous inhalation of oxygen-15 can be used to provide estimates of regional metabolic uptake of oxygen in cerebral tissues, whereas regional cerebral blood flow measurements can be obtained following the administration of oxygen-15-labeled carbon dioxide.

These methods provide better three-dimensional resolution than the single-photon technique, and furthermore, positron-emitting radionuclides can be tagged to a range of clinically useful compounds. The major disadvantage of the methodology is the need for ready access to a local cyclotron, thus demanding a considerable financial investment.

The methods currently in use are described in detail in the chapter by Herscovitch in the volume.

6.3. Stable Xenon and X-Ray CT Scanning

Since xenon absorbs X-rays, it can be used as an enhancement agent in CT scanning procedures. Two basic techniques, which attempt to utilize the good spatial resolution of CT scanners, have been described. One method provides estimates of local cerebral blood flow levels derived from a series of time-dependent enhanced images obtained during or after inhalation of xenon, and the other is fundamentally an in vivo autoradiographic technique (Meyer et al., 1981).

After denitrogenation by breathing 100% oxygen for 20 min, the patient inhales a subanesthetic mixture of 35% xenon and 65% oxygen for 7–8 min. Preliminary denitrogenation is employed to increase the rate and efficiency of tissue saturation with xenon and to avoid build-up of nitrogen in the rebreathing system. After xenon administration, the recorded changes in X-ray absorption coefficients of the various cerebral tissues are assumed to be dependent on tissue perfusion and xenon solubility. These changes, along with the measurements of xenon concentration in end-expired air, are used to provide estimates of cerebral blood flow. The end-expired air concentrations can be determined by utilizing thermal conductivity meters or mass spectrometer systems. Partition coefficient values can be calculated from tissue and blood X-ray absorption values estimated at saturation.

With the serial image method, either the uptake phase or the clearance phase can be used with a sequence of scans covering periods extending from 5 to 25 min in order to obtain xenon uptake and clearance profiles. Some investigators prefer to use the in vivo autoradiographic technique because the computer analysis involved is much simpler. $C_{i(T)}$ is determined from data from a single enhanced image and iterative methods are used to drive k_i in Eq. (5). With the autoradiographic technique, the optimum scanning time for high blood flow levels is between 1.5 and 2.5 min after administration of xenon. For accurate determination of slow flow values, a second delayed scan is desirable.

The stable xenon tomographic method has the potential to provide three-dimensional cerebral blood flow estimates using in vivo measurements of local partition coefficient values for abnormal tissue. In common with other tomographic methods, when compared with stationary detector methods, focal areas of very low flow are more likely to be detected and, again, rapid scan times are necessary for the investigation of gray matter flow levels.

Although in this case there is no radiation dose to the patient resulting from tracer administration, repeated scans at the same level can result in relatively high X-radiation exposure. Depending on the CT scanner used, the irradiation to the center of the slice of brain under consideration can be of the order of 1 rad per min. Furthermore, a significant number of subjects under investigation may experience subanesthetic effects, such as numbness of extremities, light-headedness, and restlessness. Other disadvantages include the effect of patient movement artifact, reduction of which requires patient cooperation. This can seriously degrade the quality of CT xenon enhancement data.

With the autoradiographic technique, tissue heterogeneity can be a major problem. At high flow rates, tissue heterogeneity may lead to measurement errors in excess of 20%. CT noise may further complicate the issue and this, together with patient movement, can lead to unpredictable uncertainties in the measurements (Rottenberg et al., 1982). Rottenberg and colleagues claim that a multiple-scan wash-in–wash-out analysis, based on a two-compartment convolution integral model that takes into account the variable gray matter–white matter composition of brain tissue, is less affected by tissue heterogeneity and unfavorable signal/noise ratio, and as a consequence provides results that are much more satisfactory than either the autoradiographic or multiple-scan wash-in protocols.

However, perhaps the greatest criticism of the stable xenon/CT scanning method results from the report given by Obrist at the Twelfth International Symposium on Cerebral Blood Flow and Metabolism held in Lund in 1985. In the abstract of their paper, Obrist and colleagues described how, when using the xenon-133 intravenous injection method, they observed substantial increases in cerebral blood flow when eight adult male subjects inhaled 30–35% stable xenon over a period of 6 min. These increases were nonfocal and varied from 8 to 39% (mean 24%) when initial slope flow indices were employed, and from 4 to 27% (mean 15%) when a 15-min clearance of radioactivity was analyzed (CBF_{15}). The bolus of xenon-133 was given after 1 min of stable xenon inhalation, and the initial slope flow index was therefore considered to be more sensitive to the clearance of xenon-133 during the period of stable xenon inhalation. The magnitude of the increases in cerebral blood flow was dependent on changes in p_aCO_2 levels, with subjects with the greatest reductions in p_aCO_2 displaying the smallest increases in cerebral blood flow. When a p_aCO_2 correction (3% per mmHg) was applied to the calculated flow values, the mean increase in initial slope indices was 41% and in CBF_{15}, 29%.

Clearly this vasoactive reaction together with the subanaesthetic effects of xenon, when used in the quantities required for CT scan determination of cerebral blood flow, must be given very serious consideration.

7. Hydrogen Clearance Methods

Measurement of blood flow by monitoring the clearance of hydrogen was first described by Aukland and colleagues in 1964. The various methods that have been devised rely on the measurement of the partial pressure of hydrogen in tissue by polarized platinum electrodes. The electrode currents, generated by the oxidation of hydrogen, have values proportional to hydrogen partial pressure levels and can be measured after amplification by sensitive DC amplifiers. This permits the tissue desaturation process to be monitored after the administration of hydrogen has ceased, and clearance curves analogous to those obtained using the xenon-133 methods can be recorded. Hydrogen can be administered by inhalation (Bozzao et al., 1968; Haining et al., 1968; Pasztor et al., 1973; Griffiths et al., 1975;) or by injection, using hydrogen-saturated normal saline or hydrogen-saturated heparinized blood (Shinohara et al., 1969; Meyer et al., 1972; Halsey et al., 1977;).

Hydrogen is an ideal tracer for blood flow measurements in many ways. It is metabolically inert and not normally present in body tissues. In addition, since it dissolves readily in lipids and diffuses rapidly in tissues it can penetrate nervous tissues effectively. The pulmonary circulation is able to remove hydrogen rapidly from arterial blood because of its low water:gas partition coefficient of 0.018 (Lawrence et al., 1946). Hydrogen is relatively easily detected polarographically, and because of the small diameter of the electrodes (0.05–0.1 mm) that are used, the methods have the major advantage of being able to measure cerebral blood flow in small discrete regions of the brain. Again, because of the size of the electrodes, little change to the surrounding brain tissue should ensue after electrode insertion. Nevertheless, general ethical considerations limit the use of the methods in most centers to the measurement of blood flow in laboratory animals.

7.1. Principles Underlying Hydrogen Polarography

The fundamental basis for hydrogen polarography is that electrons are generated by the oxidation reaction as follows:

$$H_2 \rightarrow 2H^+ + 2e^- \tag{17}$$

If a platinum electrode is present, it will receive electrons and, as a result, a current will flow. The receipt of these electrons by the electrode is optimal when it is held at a positive voltage. When such a positively polarized electrode is placed in a solution containing hydrogen molecules, those closest to the surface of the electrode oxidize, forming H^+ ions. A concentration gradient is set up between the bulk solution and the area immediately adjacent to the electrode. This latter location becomes hydrogen-depleted and, as a result, hydrogen migration takes place, the rate of which depends on the diffusion coefficient of hydrogen in tissue. With rapid hydrogen diffusion the electrode current becomes diffusion-limited.

The situation can be described mathematically using Fick's first law of diffusion.

$$f_H = D_H \left| \frac{dC_H}{dx} \right| \tag{18}$$

where f_H is the hydrogen flow in one dimension (expressed in mol/s), dC_H/dX is the concentration gradient, and D_H is the diffusion coefficient.

Using Faraday's law

$$f_H = \frac{i_e}{zF} \tag{19}$$

where i_e is current density per unit area of electrode surface, z is the number of charges transferred by each hydrogen molecule, and F is Faraday's number.

Thus

$$i_e = zF\, D_H \left| \frac{dC_H}{dx} \right| \tag{20}$$

Assuming that the concentration gradient is linear, this is simplified to

$$i_e = zF\, D_H \left(\frac{C_{H\infty} - C_{HO}}{L} \right) \tag{21}$$

where $C_{H\infty}$ and C_{HO} are, respectively, the hydrogen concentrations in the bulk solution and at the electrode surface, and L is the diffusion layer thickness. If the oxidation process is such that the hydrogen concentration at the electrode surface approaches zero, the equation can be written as

$$i_e = zF\, D_H \left(\frac{C_{H\infty}}{L} \right) \tag{22}$$

This equation demonstrates that the electode current is directly proportional to the concentration of hydrogen in the bulk solution. The underlying assumption is that there is complete correlation between the reaction at the electrode and the electrode current output. In practice, part of the current resulting from a particular reaction will be dissipated as heat and a further fraction may not be recorded because of current leakage. Furthermore, a poorly constructed reference electrode may absorb some of the current generated. Attempts must be made to keep these signal losses to a minimum.

7.2. Electrodes and Recording Circuit

The simplest electrodes for cerebral blood flow measurement can be made from Teflon-coated platinum wire, 0.1 mm in diameter, with the terminal 0.2–1 mm scraped bare. The tip may be platinized to increase the catalytic surface area. Halsey et al. (1977) used 0.075 mm diameter wires consisting of 90% platinum and 10% iridium and insulated in glass. In practice polarization voltages vary from +200 mV to +700 mV.

Ideally, the reference electrode should have no electrical resistance, be capable of passing current in either direction, and must not generate current intrinsically. Reference electrodes made from stainless steel needles or brass screws should be avoided since such electrodes produce junction potentials and current rectification. The reference electrode should be in good electrical contact with the animal preparation as a whole, but must be isolated from the chemical environment of the tissue under investigation. Clearly, since the reference electrode is held at a negative potential with respect to the platinum hydrogen electrode, it may reduce oxygen and be sensitive to changes in oxygen levels in the environment under investigation. In practice a silver/silver chloride reference electrode is often placed subcutaneously in the animal's back. Another approach is to place the reference electrode in a glass pipet filled with concentrated potassium chloride, using agar as a bridge between the solution and the animal preparation.

In order to prevent hydrogen diffusion into the surrounding atmosphere, the area immediately surrounding the brain electrodes is covered either with mineral oil or dental acrylic. The acrylic also helps to anchor the electrode rigidly to the skull. Where mineral oil is used, the electrodes must be secured in place using electrode holders attached to the head holder for the animal.

The small currents (10^{-9} to 10^{-12} amps) obtained from these electrodes can be amplified and monitored using a multichannel DC amplifier system and chart recorder. Great care must be taken to ensure that all the generated electrode current is measured. Any current leakage may result in significant loss of signal. The monitoring system used by Griffiths et al. (1975) is fairly typical and is shown in Figs. 5A and 5B. Every amplifier channel has an input balance control so that the output to the chart recorder can be set to zero to correspond with zero hydrogen concentration in tissue. The gain of each amplifier can be preset to give an output of 1 V for an input of 1 μamp. Each recorder channel has an independent variable gain control so that the hydrogen concentration equilibrium plateau level can be set to full-scale deflection without affecting the original zero concentration setting. The amplifier band width is restricted to 0–1 Hz to limit the effect of high-frequency noise.

7.3. Hydrogen Administration

7.3.1. By Inhalation

Hydrogen gas can be introduced into the anesthetic circuit before the respiratory pump. During hydrogen inhalation the nitrous oxide level may have to be reduced and the level of oxygen increased to prevent critical reduction in arterial oxygen levels. Variable concentrations of hydrogen have been given (up to 80%) over varying periods ranging from a single breath to 15 min. Depending on the type of tissue and the state of perfusion, concentration equilibrium can generally be achieved in 3–15 min. The hydrogen can then be turned off, nitrous oxide and oxygen returned to their former levels, and hydrogen clearance curves recorded.

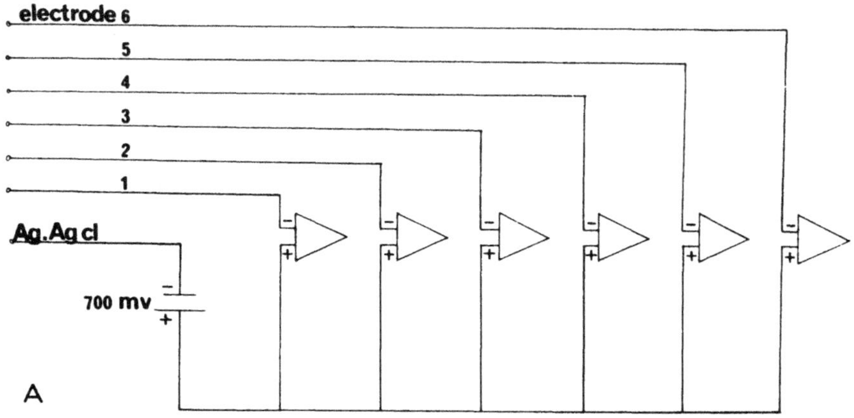

Fig. 5A. Schematic diagram of a six-channel hydrogen detection system.

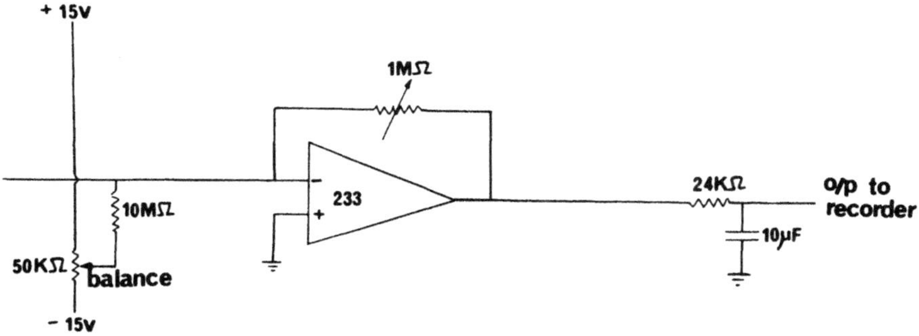

Fig. 5B. Circuit diagram of a hydrogen electrode current amplifier (from Griffiths et al., 1975).

If hydrogen concentrations in arterial blood and expired air are monitored by suitably designed electrodes (Griffiths et al., 1975; Halsey et al., 1977), it can be demonstrated that arterial and expired air clearance occurs rapidly after cessation of hydrogen administration, 90% arterial clearance occurring within 45 s, and expired air clearance within 30 s. Therefore, in general the clearance curves detected at the electrodes in the brain do not need to be corrected for arterial recirculation of tracer.

7.3.2. By Intracarotid Injection

Hydrogen can be administered via the intracarotid artery either by injection of hydrogen-saturated normal saline or hydrogen-saturated heparinized blood. Halsey et al. (1977) found that the latter method of administration provided approximately a tenfold greater delivery of hydrogen to the brain than that achieved using hydrogen-saturated saline. These investigators, however, have reported observing a significant number of distorted curves after intracarotid injection or ultrashort (less than 0.5 min) inhalation. These distortions were observed less frequently with longer inhalation periods and not at all after saturation equilibrium had been reached. The distortions were considered to be the result of partial pressure gradients within tissue causing hydrogen diffusion to occur through tissue independent of perfusion. This partial pressure gradient effect that was referred to in terms of limiting the regionality of measurements by the xenon-133 methods is more pronounced with hydrogen because of its rapid diffusion characteristics and the smaller volumes of tissue being investigated by this particular method. If saturation equilibrium is achieved in all tissue compartments, intercompartmental diffusion will not occur as readily, although there could still be a problem at boundaries between tissues with significantly different flow levels, as was found when Griffiths et al. (1975) attempted to measure blood flow in the gray matter of the spinal cord and found that they could only obtain an average gray/white matter flow value.

7.4. Clearance Curves

Since hydrogen clears rapidly from the arterial blood after cessation of administration, it is generally assumed that no correction for arterial recirculation of tracer is required and Eq. (7) applies. If the electrode is inserted into homogeneous tissue, then monoexponential clearance curves should be obtained provided the tissue layer is thick enough. At locations near tissue boundaries or when tissue is heterogeneous, then multiexponential clearance curves will be observed. The normal exponential stripping process is carried out to obtain the clearance half-times for each of the components, and since the partition coefficient is taken to be unity for both gray and white matter, Eq. (12) can be simplified to

$$\text{flow (mL/100 g/min)} = \frac{A}{t_{1/2}} \qquad (23)$$

where $A = \log_e^2 \times 100 \times 60$ and $t_{½}$ is measured in seconds.

When each compartment is saturated with tracer, the relative weights of each of the compartments are given by the equations

$$W_g = \frac{I_g}{I_g + I_w} \tag{24}$$

$$W_w = 1 - W_g \tag{25}$$

where W_g and W_w are the weights of gray and white matter, respectively, and I_g and I_w are the intercepts of the fast and slow compartments on the y axis. With a bolus injection

$$W_g = \frac{I_g/f_g}{I_g/f_g + I_w/f_w} \tag{26}$$

where f_g and f_w are the gray and white matter flow values. The weighted mean flow can then be calculated from the equation

$$f = W_g f_g + W_w f_w \tag{27}$$

Any analysis of local cerebral blood flow that depends on exponential stripping of tracer clearance curves has two major disadvantages:

1. When observed flow values vary over a wide range, the ability to resolve the different components also varies so that when flow is substantially reduced it may be possible to define only one component, and when flow is extremely fast it may be possible to extract more than two components.
2. As has already been stated, when two different tissue compartments lie adjacent to the electrode location or when the tissue under investigation is heterogeneous, diffusion of tracer between slow and fast compartments may be significant. This can also lead to inaccuracies in the resolution of flow components.

In these situations consideration has to be given to the use of an initial slow flow index. Rowan et al. in 1975 and Rowan in 1977 demonstrated highly statistically significant correlations between component flow indices, initial slow flow indices, weighted mean flow indices, and height/area indices calculated from hydrogen and xenon-133 clearance curves, whereas La Morgese et al. (1975) demonstrated excellent correspondence between cerebral blood flow values obtained using the hydrogen clearance technique and the radioactive microsphere injection method.

7.5. Advantages

The hydrogen clearance method can provide estimates of blood flow in any cerebral tissue in which a small platinum electrode can be inserted. [For example, Rowan and Teasdale (1977) have inserted electrodes into the brain stem of baboons for the purpose of measuring brainstem blood flow in intracranial hypertension.] Furthermore, because of the rapid clearance of hydrogen from body tissues, tracer recirculation and background problems are minimal and, as a result, multiple flow estimations can be carried out at regular intervals over long periods of time. It is important, however, that the electrode response remains constant over the period of each clearance measurement. The associated measuring equipment is simple and inexpensive, and the electrodes themselves are readily constructed. The tracer is easily obtained and requires no special preparation.

7.6. Limitations and Sources of Error

Great care has to be taken when inserting the electrodes to ensure that no significant damage is caused to the surrounding brain tissue. Clearly the smaller the diameter of the electrode used, the less chance there will be of significant tissue damage, and for this reason electrodes with diameters of 0.1 mm or less are generally employed.

In some pathological situations, clearance of hydrogen from the arterial system may be delayed, causing a recirculation artifact at the beginning of the clearance curve. Some workers ignore the first minute or so of the clearance curve in this situation. In monoexponential clearance situations, any recirculation artifact is clearly identified.

The intensity of the chemical reaction at the electrode is determined to some extent by the polarization voltage. With hydrogen oxidation, the more positive the polarizing voltage, the more likely electron transfer will occur. This situation exists until the supply of hydrogen to the electrode becomes diffusion-limited. In the non-diffusion-limited situation the electrode response theoretically increases linearly with hydrogen concentration. However, the main disadvantage here is that, although independent of diffusion factors, the observed current is very dependent on electrode characteristics, including capacitance, speed of response, polarization level, and linearity. In general, the polarization voltage is set so that the hydrogen concentration at the electrode approaches zero and the electrode reaction rate is controlled primarily by diffusion factors. It should be noted, however, that the greater the polarization voltage, the longer the electrode takes to reach equilibrium after it has been inserted into tissue.

A polarographic electrode will respond in general to any chemical that will exchange electrons with it. The transfer of electrons is limited by the potential energy barrier that binds the electrons to the reacting molecule or to the electrode. If the level of polarization voltage matches this potential energy barrier, there is a greater chance of electron transfer. Therefore, an electrode polarized to a level of, say, +200 mV will, in addition to hydrogen, oxidize such substances as ascorbic acid, catecholamines, and hemoglobin. Furthermore, the same electrode system can reduce substances such as oxygen. Thus when measuring hydrogen clearance in brain tissue, substances other than hydrogen may contribute to the overall electrode current. To reduce the sensitivity of the electrode to oxygen, and thus the significance of the oxygen contribution to the electrode response, many investigators use polarization levels of 650–700 mV. However, at these polarization voltage levels the sensitivity to ascorbic acid is optimized. Whether or not ascorbic acid in physiological concentrations provides a significant contamination problem in the detection of hydrogen clearance in cerebral tissue remains controversial (Young, 1980).

As with other inert gas clearance techniques, at very fast flow rates (i.e., greater than 100 mL/100 g min), the diffusion limitation factor m_i (Eq. 16) may drop below unity, affecting the accuracy of the measurement. Also, with very high flow levels and multiexponential clearance curves, the first minute of the clearance curve is important in the accurate resolution of the fast components. In these situations it is necessary to monitor hydrogen concentration in the arterial blood or in expired air in order to provide a correction for tracer recirculation.

The fact that hydrogen is readily diffusible in cerebral tissues tends to limit the localization of the blood flow measurement, and experimental results suggest that the hydrogen clearance method can only localize cerebral blood flow measurements to volumes of the order of 5 mm^3 (Young, 1980).

A fundamental assumption with the hydrogen technique is that the electrode signal returns to a constant level when there is no hydrogen present. As indicated earlier, this polarographic current level is used as a zero reference on the associated chart recorder. Clearly, if this level changes during the measurement period, errors are introduced. It is important therefore to ensure that steady-state conditions prevail before a measurement is taken and that baseline current levels before and after a measurement are carefully monitored. Variations in baseline current levels can result from changing ascorbic acid concentrations in brain tissues resulting from pathological stress or from tissue impedence changes from the formation of edema. Both situations may arise at ischemic levels of blood flow, but tissue impedence changes should only have a significant effect when large electrodes having impedences of the order of 10^4 Ω are being used. Careful checks on electrode insulation can help to ensure that the frustration of observing a sudden large increase in electrode current part way through the investigation period does not occur.

References

Andersen A. and Ladefoged J. (1965) Relationship between hematocrit and solubility of ^{133}Xe in blood. *J. Pharm. Sci.* **54,** 1684–1685.

Aukland K., Bower B. F., and Berliner R. W. (1964) Measurement of local blood flow with hydrogen gas. *Circ. Res.* **14,** 164–187.

Bozzao L., Fieschi C., Agnoli A., and Nardini N. (1968) Autoregulation of Cerebral Blood Flow Studied in the Brain of Cat, in *Blood Flow Through Organs and Tissues* (Bain W. H. and Harper A. M., eds.) Livingstone, Edinburgh.

Celsis P., Goldman T., Henriksen L., and Lassen N. A. (1981) A method for calculating regional cerebral blood flow from emission computed tomography of inert gas concentrations. *J. Comput. Assist. Tomogr.* **5,** 641–645.

Eklof B., Lassen N. A., Nillson K., Siesjo B. K., and Torlof P. (1974) Regional cerebral blood flow in the rat measured by the tissue sampling technique: A critical evaluation using four indicators C^{14}-antipyrine, C^{14}-ethanol, H^3-water and xenon-133. *Acta. Physiol. Scand.* **9,** 1–10.

Gillespie F. C. (1968) Some Factors Influencing the Interpretation of Regional Cerebral Blood Flow Measurements Using Inert Gas Clearance Techniques, in *Blood Flow Through Organs and Tissues* (Bain W. H. and Harper A. M., eds.) Livingstone, Edinburgh.

Glass H. I. and Harper A. M. (1963) Measurement of regional cerebral blood flow in cortex of man through intact skull. *Br. Med. J.* **1,** 593.

Griffiths I. R., Rowan J. O., and Crawford R. A. (1975) Spinal cord blood flow measured by a hydrogen clearance technique. *J. Neurol. Sci.* **26,** 529–544.

Haining J. L., Turner M. D., and Pantall R. M. (1968) Measurement of local cerebral blood flow in the unanaesthetised rat using a hydrogen clearance method. *Circ. Res.* **23,** 313–324.

Halsey J. H., Capra N. F., and McFarlane R. S. (1977) Use of hydrogen for measurement of regional cerebral blood flow. Problem of intercompartmental diffusion. *Stroke* **8,** 351–357.

Kanno I. and Lassen N. A. (1979) Two methods for calculating cerebral blood flow from emission computed tomography of inert gas concentrations. *J. Comput. Assist. Tomogr.* **3,** 71–76.

Kety S. S. and Schmidt C. F. (1945) The determination of cerebral blood flow in man by use of nitrous oxide in low concentrations. *Am. J. Physiol.* **143,** 53–66.

La Morgese J., Fein J. M., and Shulman K. (1975) Polarographic and Microsphere Analysis of Ultraregional Cerebral Blood Flow Rate in the Cat, in *Blood Flow and Metabolism in the Brain* (Harper A. M., Jennett B., Miller J. D., and Rowan J. O., eds.) Churchill-Livingstone, Edinburgh.

Lassen N. A., Hoedt-Rasmusson K., Sorensen S. C., Skinhoj E., Cronqvist S., Bodforss B., and Ingvar D. H. (1963) Regional cerebral blood flow in man determined by a radioactive inert gas (krypton 85). *Neurology* (Minneap.) **13,** 719–727.

Lassen N. A., Henriksen L., and Paulson O. (1981) Regional cerebral blood flow in stroke by ^{133}xenon inhalation and emission tomography. *Stroke* **12,** 284–288.

Lawrence J. H., Loomis W. F., Tobias C. A., and Turpin F. H. (1946) Preliminary observations on the narcotic effect of xenon with review of values for solubilities of gases in water and oils. *J. Physiol.* **105,** 197–204.

Mallet B. L. and Veall N. (1963) Investigation of cerebral blood flow in hypertension using radioactive xenon inhalation and extracranial recording. *Lancet* **ii,** 1081–1082.

McAlpine C. J., Rowan J. O., Matheson M. S., and Patterson J. (1981) Cerebral blood flow and intelligence rating in persons over 90 years old. *Age & Ageing* **10,** 247–253.

Meyer J. S., Fukuuchi Y., Kanda T., Shimazu K., and Hashi A. (1972) Regional cerebral blood flow measured by intracarotid injection of hydrogen—comparison of regional vasomotor capacitance from cerebral infarction versus compression. *Neurology* (Minneap.) **22,** 571–584.

Meyer J. S., Haymann L. A., Amano T., Nakajima S., Shaw T., Lauzon P., Derman S., Karacan I., and Harati Y. (1981) Mapping local blood flow of human brain by CT scanning during stable xenon inhalation. *Stroke* **12,** 426–436.

Nilsson B. G., Ryding E., and Ingvar D. H. (1982) Quantitative airway artefact compensation at regional cerebral blood flow measurements with radioactive gases. *J. Cer. Blood Flow Metabol.* **2,** 73–78.

Obrist W. D., Thompson H. K., King C. H., and Wang H. S. (1967) Determination of regional cerebral blood flow by inhalation of ^{133}Xenon. *Circ. Res.* **20,** 124–135.

Obrist W. D., Thompson H. K., Wang H. S., and Cronqvist S. (1970) A Simplified Procedure for Determining Fast Compartment rCBFs by ^{133}Xenon Inhalation, in *Brain and Blood Flow* (Ross Russell R. W., ed.) Pitman, London.

Ohta Y. and Farhi L. E. (1979) Cerebral gas exchange: Perfusion and diffusion limitations. *J. Appl. Physiol.* **46,** 1164–1168.

Olesen J., Paulson O. B., and Lassen N. A. (1971) Regional cerebral blood flow in man determined by the initial slope of the clearance curve of intra-arterially injected ^{133}Xe. *Stroke* **II,** 519–540.

Pasztor E., Symon L., Dorsch N. W. C., and Branston N. M. (1973) The hydrogen clearance method in assessment of blood flow in cortex, white matter and deep nuclei of baboons. *Stroke* **4,** 554–567.

Risberg J. and Prohovnik J. (1981) rCBF Measurements by ^{133}Xe Inhalation: Recent Methodological Advances, in *Progress in Nuclear Medicine,* vol. 7 *Neuronuclear Medicine* (Juge O. and Donath A., eds.) Karger, Basel.

Rottenberg D. A., Lu H. C., and Kearfot K. J. (1982) The in vivo autoradiographic measurement of regional cerebral blood flow using stable xenon and computerized tomography: The effect of tissue heterogeneity and computerized tomography noise. *J. Cer. Blood Flow Metabol.* **2,** 173–178.

Rowan J. O. (1981) *Physics and the Circulation.* Hilger, Bristol.

Rowan J. O. (1977) Mean Cerebral Blood Flow Indices Using the Hydrogen Clearance Method, in *Cerebral Function, Metabolism, and Circulation* (Ingvar D. H. and Lassen N. A., eds.) Munksgaard, Copenhagen.

Rowan J. O. and Teasdale G. M. (1977) Brain Stem Blood Flow During Raised Intracranial Pressure, in *Cerebral Function, Metabolism and Circulation* (Ingvar D. H. and Lassen N. A., eds.) Munksgaard, Copenhagen.

Rowan J. O., Miller J. D., Wyper D. J., Fitch W., Grossart K. W., Garibi J., and Pickard J. D. (1975a) Variability of Repeated Clinical Measurements of CBF Using the ^{133}Xe Intracarotid Injection Technique, in *Cerebral Circulation and Metabolism* (Langfitt T. W., McHenry L. C., Jr., Reivich M., and Wollman H., eds.) Springer-Verlag, New York.

Rowan J. O., Reilly P., Farrar J. K., and Teasdale G. (1975b) The Xenon-133 and Hydrogen Clearance Methods—A Comparitive Study, in *Blood Flow and Metabolism in the Brain* (Harper A. M., Jennett B., Miller J. D., and Rowan J. O., eds.) Churchill-Livingstone, Edinburgh.

Shinohara Y., Meyer J. S., Kitamura A., Toyoda M., and Ryn T. (1969) Measurement of cerebral hemisphere blood flow by intracarotid injection of hydrogen gas—validation of the method in the monkey. *Circ. Res.* **25,** 735–745.

Tomita M. and Gotoh F. (1981) Local cerebral blood flow values as estimated with diffusible tracers: Validity of assumptions in normal and ischemic tissue. *J. Cer. Blood Flow Metab.* **1,** 403–411.

Wyper D. J. and Rowan J. O. (1976) The construction and use of nomograms for cerebral blood flow calculations using a ^{133}Xe inhalation technique. *Phys. Med. Biol.* **21,** 406–413.

Wyper D. J. and Cooke M. B. D. (1977) Compensation for Hemisphere Cross-Talk When Measuring CBF, in *Cerebral Function, Metabolism and Circulation* (Ingvar D. H. and Lassen N. A., eds.) Munksgaard, Copenhagen.

Wyper D. J. Lennox G. A., and Rowan J. O. (1976) Two minute slope inhalation technique for cerebral blood flow measurement in man. I. Method. *J. Neurol. Neurosurg. Psychiat.* **39,** 141–146.

Young W. (1980) H_2 clearance measurement of blood flow: A review of technique and polarographic principles. *Stroke* **11,** 552–564.

Zierler K. L. (1965) Equations for measuring blood flow by external monitoring of radioisotopes. *Circ. Res.* **XVI,** 309–321.

Measurement of Pial Vessel Hemodynamics

Ludwig M. Auer

1. Introduction and Historical Survey

Two reasons might account for the fact that the investigation of brain blood flow is one of the youngest branches of physiological research of circulation. One is that the brain was for many years ethically inaccessible both for surgical treatment and basic research. The second equally important reason is certainly the brain's encasement within the bony skull, which makes its in vivo exposure and observation under physiological conditions difficult. As in other fields of scientific research, mere observation of phenomena formed the beginning. In the 19th century, Donders (1849) described reactions of superficial cerebral vessels *in situ*. Further systematic observations under in vivo conditions required a certain standard of microscopical technology, which was first applied to the brain by Florey (1925) and described by Forbes, Wolff, and Fog in a series of publications from 1928 onward (Wolff, 1929; Forbes and Wolff, 1928; Wolff and Lennox, 1930; Forbes et al., 1933, 1941; Fog, 1937, 1939a,b; Forbes, 1937a,b, 1938, 1954). Photographic documentation through the microscope provided the only possibility for collection of data. More sophisticated microscopic techniques, such as fluorescence angiography, had to await adequate optic systems and light sources. Heine's (1932) dark ground illuminator and Gottschewki's (1953) high-pressure mercury lamp were important steps in this direction. A high standard of intravital microscopy was subsequently developed by Peters (1955), Zweifach (1954, 1957), Illig (1961), Brånemark and coworkers (Brånemark, 1962, 1965a,b, 1966; Brånemark and Jonsson, 1963; Birch et al., 1968a,b), and others, including the development of kinematography and fluoresence angiography.

More recent technical developments for the analysis of caliber variations of superficial brain vessels (pial vessels) aimed at a more direct quantitation of reactions other than those done by measurement of calibers from photographic reproductions. These included the creation of optic systems such as the image splitting device

(Barer, 1960; Dyson, 1960; Baez, 1966), photometric techniques (Auer, 1978c), or electronic devices in combination with videosystems (Auer and Haydn, 1979; Gotoh et al., 1982).

The application of these techniques to in vivo observation of pial vessels was bound to the development of adequate surgical procedures for the exposure under physiologic conditions, the so-called cranial window technique. Since Florey's 1925 experiments, a variety of techniques for acute and chronic experiments have been developed; by Forbes, Wolff, and Fog, as mentioned above, and by Finesinger (1932, 1933), Pool (1934), Beickert (1953), Chorobski and Penfield (1932), Clark and Wenstler (1938), Villaret (1939), Shelden et al. (1944), Minard (1954), Heppner and Marx (1958), and Heppner (1959, 1961).

Today the exposure and direct in vivo investigation of pial vessels is still a valid and irreplaceable experimental procedure, especially when used in combination with other parameters such as blood pressure, intracranial pressure, and regional cerebral blood flow.

2. Recent Methods for Direct Investigation of Brain Vessels In Vivo

2.1. The Cranial Window Technique

Implantation of a cranial window is usually started with a parietal longitudinal skin incision. After exposure of either one or both parietal bones, depending on the species used, a trepanation is made with a hand-driven or liquid-cooled electric dental drill. Bone wax is applied to prevent bleeding from diploic vessels. The cat is especially suitable for cranial window studies because of the large head and brain in relation to body weight. This makes the microsurgical procedure safe and reduces the probability of disturbed cerebrovascular behavior caused by mechanical trauma. The last paper-thin layer of the internal lamina is removed with small forceps. Further surgical procedures are performed with an operating microscope under 16- to 40-fold magnification. We use glass microscalpels for opening the dura by hooking up the dura with one and incising with a second blade. Additional incisions for a star-shaped opening of the dura can be made with microsurgical scissors. The edges of the dura are then coagulated with bipolar microforceps, as used in microsurgery, under irrigation with mock cerebrospinal fluid (CSF) (Wahl et al., 1972). When the dura is reflected, small adhesions between dura mater and arachnoid membrane usually cause escape of CSF. This can be used to allow the trepanation to fill with autologous CSF. From this point, the procedure as used by different groups varies. The window may be left open and filled with mineral oil after forming a crater with wax or dental cement or by hooking up the incised skin (Schmidt, 1956; Rosenblum, 1969; Wahl et al., 1970; Adler, 1974; Kuschinsky and Wahl, 1974; Mackenzie et al., 1976). This procedure is used either for its simplicity or because it allows for a larger trepanation and exposure of more pial vessels or microtopical injection studies. Alternatively, a glass window can be inserted to cover the bone defect (Rosenblum, 1969; Wahl et al., 1970; Raper et al., 1971; Mackenzie et al., 1976; Dora and Kovach, 1981; Lavasseur et al., 1975). The glass is either fixed in a metal frame that can be screwed into the trepanation hole, or a cover glass is fixed and sealed with acrylic adhesive or dental cement (Auer, 1978a) (Fig. 1). Closure of the bone defect also has the advantage, among others, of reducing to a minimum brain and vessel movements induced by respiration and pulse. When no elevation of intracranial pressure (ICP) is expected, the cover glass may be larger than the trepanation and positioned over the external lamina as soon as the bone window has filled with CSF. In the case of experiments performed under conditions of elevated ICP, a smaller glass window must be placed at the level of the internal lamina by leaving a thin bony rim at the trepanation's periphery as support. The water-tight fixation of the glass is somewhat more complicated, but is possible if the acrylic adhesive is applied in small portions. The acrylic must not contact the CSF under the glass because it immediately covers all wet surfaces and abolishes vascular smooth muscle contractility.

The sensitivity of vessels to changes in CSF, pH, Ca^{2+}, and K^{+} requires an adequate composition of artificial CSF whenever used; a detailed description has been given by Kuschinsky and Wahl (1978).

Experiments involving superfusion of the brain surface have been performed with either open trepanations or glass windows that contain an inlet and an outlet needle for infusion and maintenance of a normal ICP. Discrepancies between superfusion studies of various groups and between superfusion and microtopical application studies are likely to be caused by a poorly recognized fact: When the dura mater is removed under great care to preserve the arachnoid membrane, only very small communications will

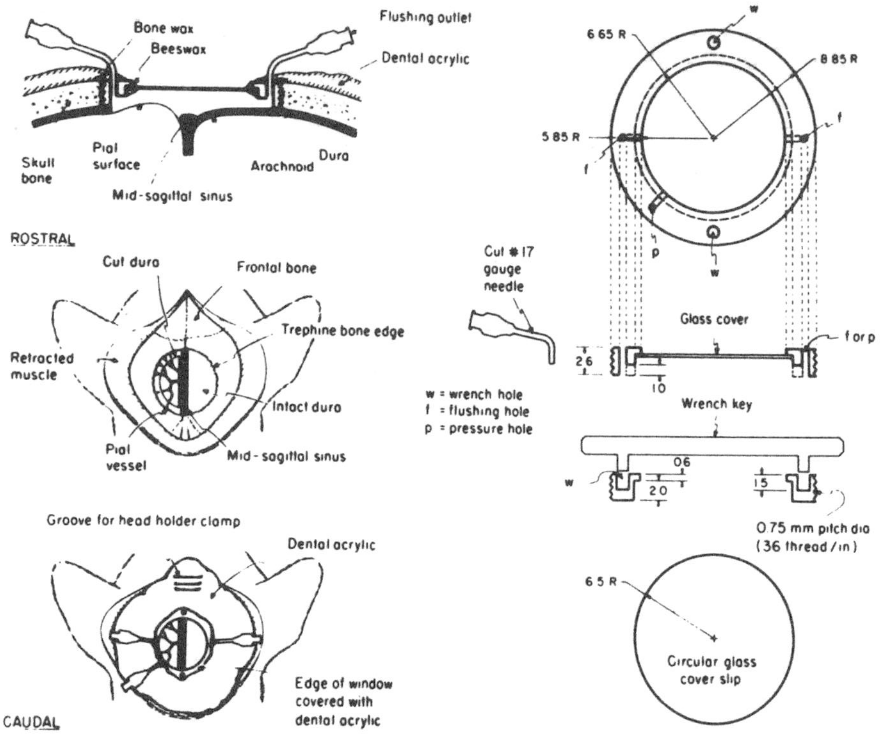

Fig. 1A. (Right) Schematic drawing of the cranial window for acute preparations. The dimensions are given in millimeters. (Left) Cross-section of implanted window (above) and top view of preparation before (middle) and after (below) placement of window (with permission, from Levasseur et al., 1975).

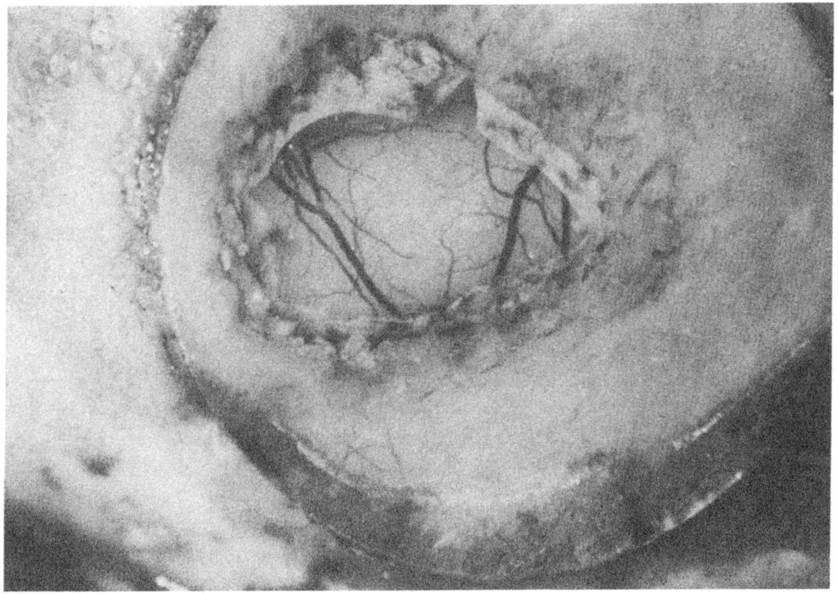

Fig. 1B. Parietal glass window in a cat, sealed with acrylic, reveals a single gyrus (with permission, from Auer, 1978a).

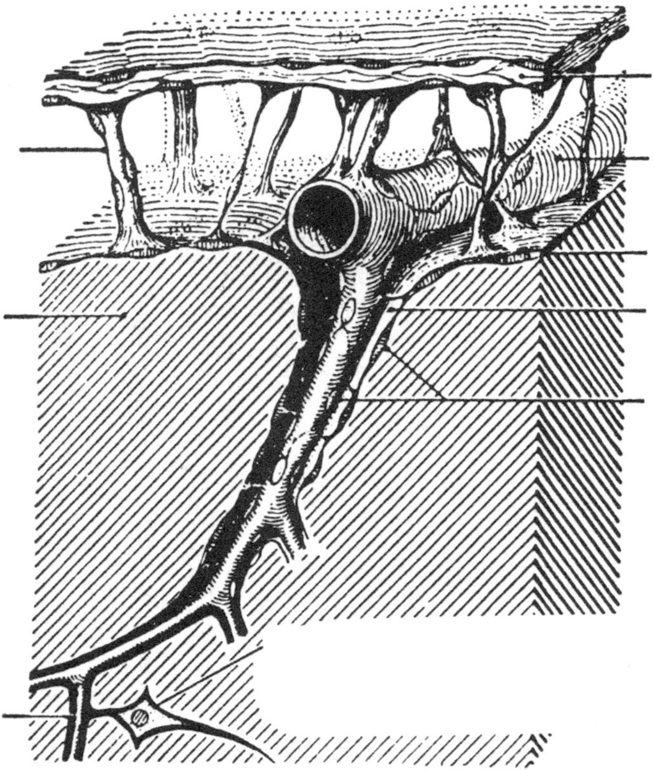

Fig. 2. Schematic drawing of pia-arachnoid architecture (with permission, from Ranson and Clark, 1959).

exist between the subdural space, where the superfusion takes place, and the subarachnoid space, which represents the perivascular space for the pial vessels. Any effect of a superfusate will therefore be widely underestimated with a substantially intact arachnoid membrane compared to a widely opened arachnoid or subarachnoid injection. The arachnoid membrane follows the architecture of the dura mater and therefore does not follow the brain surface into its sulci like the pia mater. It is over the sulci that the arachnoid can be opened without risks to the underlying tissue with a triangular cross-section (Fig. 2).

The extremely sensitive pial vessels suffer a disturbance of their autoregulation not only with the slightest mechanical trauma, but also with various kinds of stimuli and changes of their normal milieu. It is therefore necessary not only to preserve conditions of normal CSF composition, except for the factor under investigation, but also to control the temperature of the superfusate and the heat production of the light source.

There is no doubt, however, that one must consider that the cerebral vasculature will be reactive also to influences other than those under investigation. It is therefore necessary to compare observed phenomena not only with the vessels' own resting state, but above all with a control group; this is particularly important when only slight changes are seen in the experimental group.

An open cranial window introduces serious problems to the system whenever the experimental procedure includes important changes of intracranial volumes, i.e., cerebral blood volume, brain tissue, or CSF volume. An open window could also cause problems in the interpretation of data when pial venous reactions are of interest. Intracranial vessels within the skull behave like collapsible tubes guided through a closed cavity with a certain pressure, P_E. Flow through these tubes follows the rules of a Starling resistor (Starling, 1918; Holt, 1941; Noordergraaf, 1978). On the one hand, perfusion pressure, P_P, does not equal intraluminal pressure, P_i, in this system, but equals the difference between P_i and P_E; the latter is also called transmural pressure and is responsible for the vessel wall tension. On the other hand, vessels do not collapse in such a closed system unless transmural pressure becomes markedly negative. Even with a transmural pressure of zero, i.e., $P_E = P_i$, the vessels are not collapsed because pressure is transferred from the perivascular space to the intraluminal compartment, which leads, for example, to a parallel rise of cerebral venous pressure when pressure in the CSF space is elevated (for references, *see* Auer and Mackenzie, 1984). On the arterial side, rising perivascular pressure would cause a decreasing transmural pressure and a consequent smooth muscle cell relaxation, leading to vasodilatation. In an open system, i.e., with an open cranial window, any experimental circumstance that would induce a change of intracranial pressure, therefore, would lead to erroneous data because of the absence of perivascular pressure as a factor influencing vascular reactivity. Under such circumstances a closed cranial window technique should be given priority over an open technique.

2.2. Microscopes and Microscopical Techniques

2.2.1. Instruments

Depending on the technical and financial requirements for the method of data sampling, various types of instruments have been used for surface illumination microscopy. We and others (Suzuki,

1975; Suzuki et al., 1975; Auer, 1978a) have been working with a Leitz intravital microscope, many details of which have been worked out by Brånemark (Brånemark, 1966; Brånemark and Jonsson, 1963; Brandt et al., 1983). This instrument is mounted on a vibration-damped table and harbors two housings for light sources for rapid alternation of use (e.g., for flash-light photography). We use a 60-W low-voltage lamp, a Xenon lamp, a 200-W Hg high-pressure lamp, or a flash-light system, depending on the purpose. A mounting stand for documentation facilities such as photo, movie, or TV camera moves simultaneously with the objective during focusing. The microscope table allows for placement of even large animals (specially-made table necessary) and still remains moveable by macro- and microtransmission (Fig. 3). We also use the mobile microscope table as an operating table and have a stereotaxic head fixation and holders for pressure transducers mounted on that table (Fig. 4). The surgical procedures are performed after moving the microscope/operating table to where the operating microscope, drills, and coagulator are located (Fig. 5). Another instrument used widely is the Bausch and Lomb stereo zoom microscope (Wahl et al., 1972; Levasseur, 1979; MacKenzie et al., 1976; Dora and Kovach, 1981).

2.2.2. Objectives

We use the "Ultropak" system (Leitz) for normal microscopy. This is a dark-field system with a pancratic ring condenser. With special immersion caps, these objectives can be used in wet milieu, either for open-window studies or irrigation of the window's glass surface for control of temperature. For fluorescence angiography, we apply the "ploem-opak" system (Leitz) with the appropriate combination of filters.

An important problem is surface temperature and heat production by the light sources in use. Since different light sources are required for ordinary observation (photography, TV, use of a 16mm movie, and so on), different combinations of heat filters are required. These have been specified elsewhere in detail (Auer, 1978a). In addition, we have built an irrigation system with thermostatic control. Irrigation fluid is rinsed between the cranial glass window and the immersion cap of the objective, at the tip of which temperature is measured with a small thermosensor (DISA, Yellow Springs Instruments Co., Inc, Yellow Springs, Ohio, Model 46 TUC) (*see* Fig. 3B).

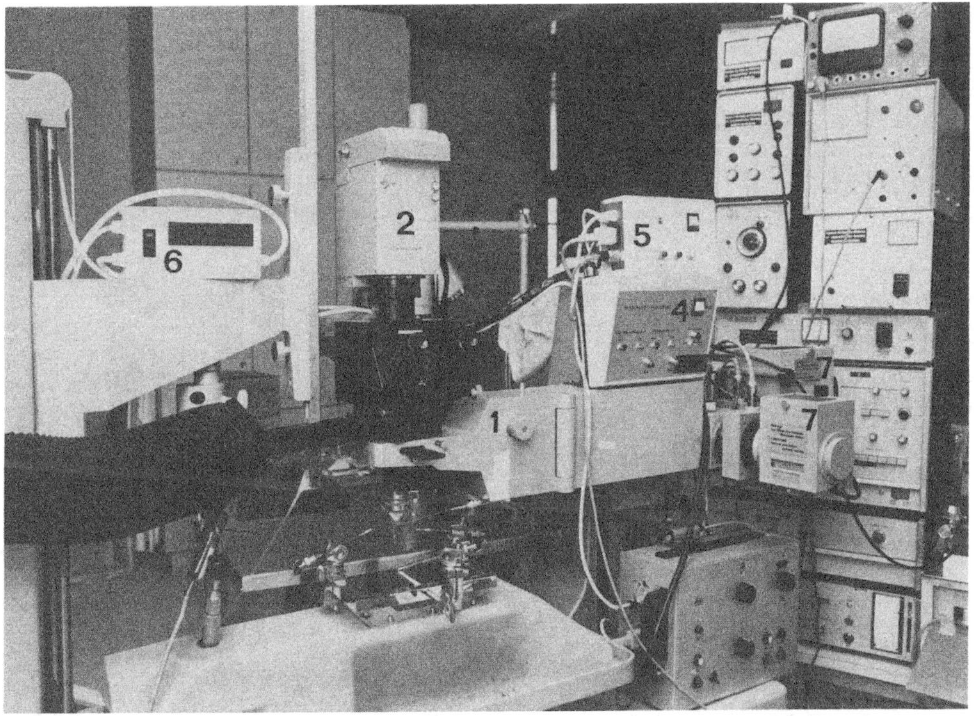

Fig. 3A. Leitz "Intravitalmikroskop" and physiological control unit. 1, three-way tube for observation, photography, and photometry; 2, still camera; 3, Ultropak objective; 4, control unit for electronic slips; 5, electronic timer for measurement of experiment duration; 6, digital display of mean arterial blood pressure; 7, light sources.

Fig. 3B. Ultropak objective with temperature sensor and irrigation tube (with permission, from Auer, 1978a).

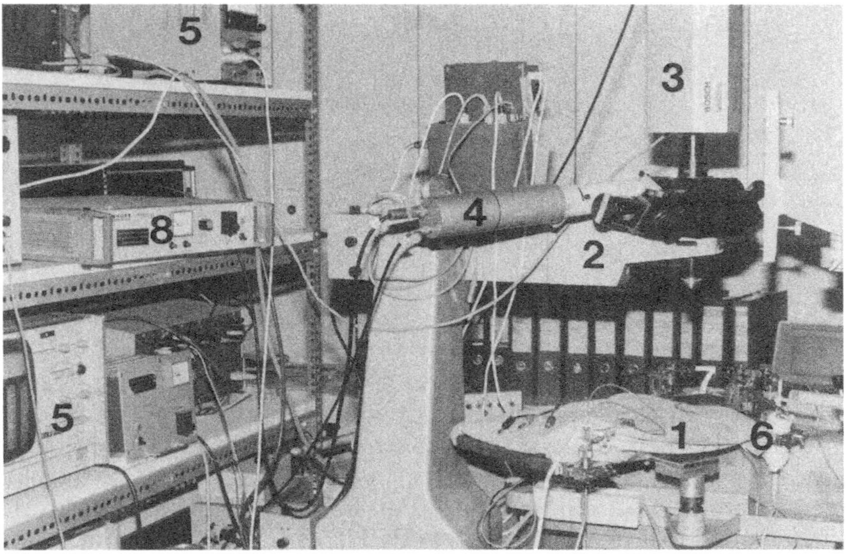

Fig. 4. 1, Mobile operating table, which also serves as the objective table for the intravital microscope; 2, intravital microscope; 3, TV camera; 4, photometer; 5, monitors for physiological parameters; 6, pressure transducer; 7, stereotaxic holders; 8, photometer power supply.

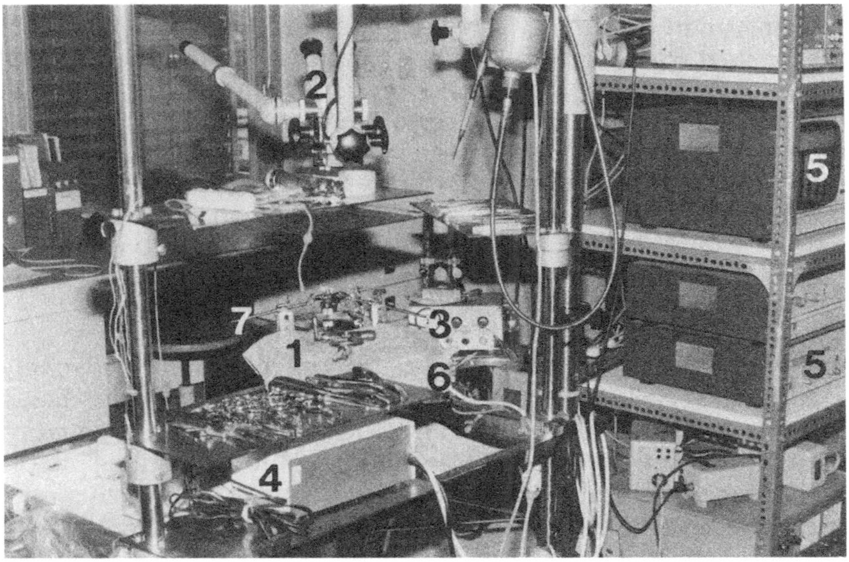

Fig. 5. 1, Mobile operating table, which also serves as the objective table for the intravital microscope; 2, operating microscope for surgical preparations; 3, respirator; 4, bipolar coagulator; 5, monitors for physiological parameters; 6, pressure transducer; 7, stereotaxic holders.

2.3. Imaging of Pial Vessels

2.3.1. Photography

Measurement from photographs taken at intervals is the technically easiest, but most time-consuming and imprecise, method of quantifying pial vessel caliber variations. Its major limitation is the lack of resolution both in time and space, which makes analysis of rapid reactions and observation of small vessels ($\leq$ 100 μm) difficult. The problem of light sources strong enough to allow for short exposure times can be solved in several ways: one is the use of adequate film material like Ilford Pan F, HP4, or FP4 for black-and-white and Kodachrome high-speed EHB135 or Agfachrome 50L for color slides. A second possibility is the synchronization of the photo camera's exposure time with a flash-light or a strong lamp such as Xenon light. We have built for the Leitz intravital microscope an electronic two-slip mirror beam reflector synchronized with a Leitz "Orthomat" camera, which closes the housing of the low-voltage lamp (used for ordinary observation) and opens the housing of the Xenon light source for the time of film exposure. For these short exposures, heat filters that reduce the light intensity considerably are not necessary. For serial photography, e.g., for angiography with fluorescent or other dyes, we have used a variotimer device from Bolex-Wild (Timer MBF-C). The same can be done with more modern motor-driven photo cameras. To display information such as time and blood pressure on the single photographs, we formerly used digital watches attached to the respective sources and mirrored into the photographic picture (Fig. 6).

2.3.2. 16mm Movie

We have used a Bolex 16mm camera for exposure of Ektachrome 7242 or Kodak Eastman 4-X7224 film. In our hands, this procedure has even more limitations for scientific evaluations than photography because of the poor resolution of single pictures and an endless workload, were these pictures to be evaluated. A sampling rate of caliber measurements of more than 1/s is unnecessary for most problems investigated with the technique, and this rate can be achieved with a photo camera, as mentioned above.

This method is, above all, adequate for the documentation of qualitative phenomena such as the number and distribution, the gross comparative size, and the shape of vessels, and qualitative observations on flow phenomena within them. Thus the rather conical shape of arteries and arterioles can be recognized by their

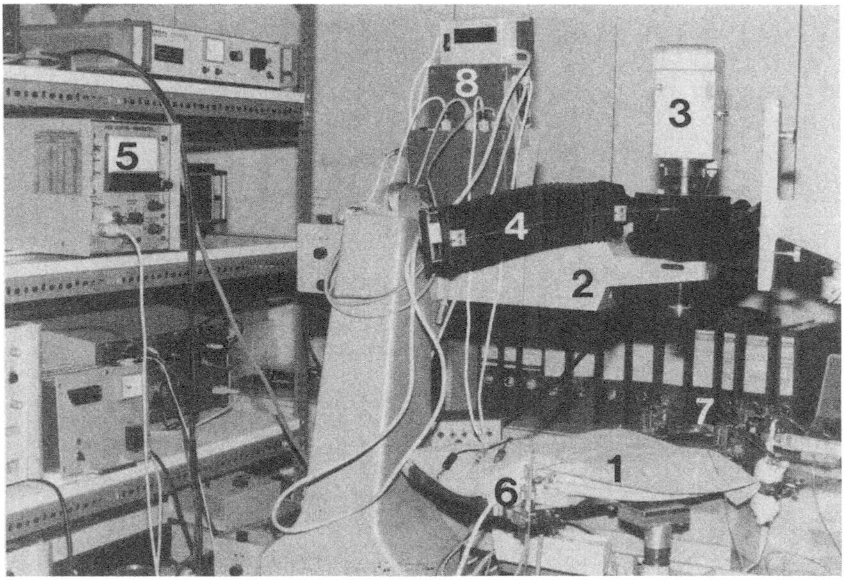

Fig. 6. 1, Mobile operating table, which also serves as the objective table for the intravital microscope; 2, intravital microscope; 3, photocamera; 4, phototube for electronic display and photographic imaging of pressure values; 5, monitor for physiological parameters; 6, pressure transducer; 7, stereotaxic holders; 8, electronic display.

curved course and dichotomic branching, in contrast to the veins that are more cylindrical, generally larger, and often branching rectangularly (Fig. 7). Interarterial connections are rare, whereas intervenous connections are rather frequent; transient stasis of the blood column within them is not a pathological event (Figs. 8–10).

Dye microangiography can be used for several purposes:

1. The differentiation of arteries and veins.
2. The measurement of transit times and local blood velocity.
3. The observation and measurement of dye extravasation as a test of blood–brain barrier integrity.

Injection of dyes via a cannulated lingual artery retrogradely into the carotid artery requires exact observation of dye movement and distribution. We have used 1 mL/kg 2% Evans blue for serial microangiograms to qualitatively observe its transition through the vascular tree under pathological circumstances (Auer, 1978a).

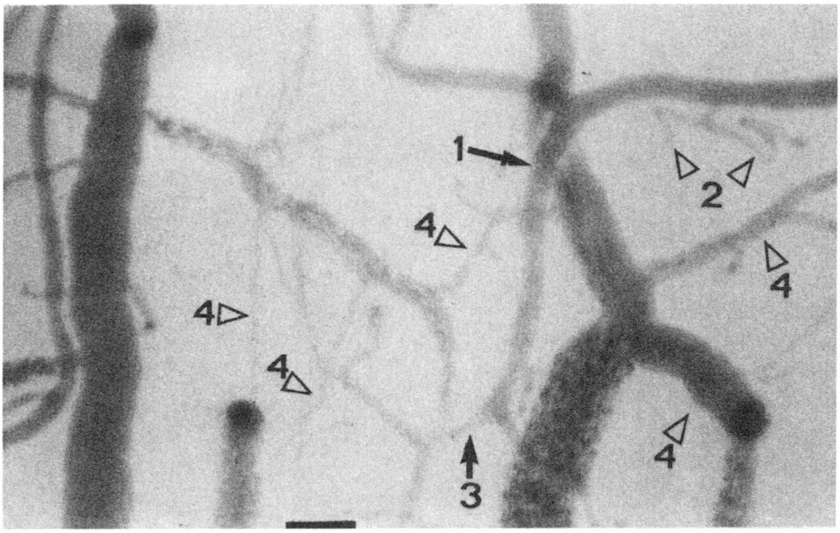

Fig. 7. Normal pial vessels of a cat. Calibration bar, 120 μm. 1, Typical conical shape of arteriole and diffuse staining of the latter caused by rapid flow; 2, small arteriolar branches forming a loop before entering the cortex; 3, one of the rate observations of an interarteriolar connection with slow flow, visible as granular structure of slowly moving blood cells; 4, intervenous connections. Venules are seen to be more numerous and larger, with different flow speeds in single branches. The branching angles differ widely and are often rectangular. The venous vessels are cylindrical rather than conical and sometimes irregular in diameter. Narrowing at the site of branching is sometimes visible (with permission, from Auer, 1978a).

Microangiography with sodium fluorescein has been widely used (Rosenblum, 1969, 1970; Sato et al., 1983; Wahl et al., 1983). By aid of a high-pressure Hg lamp as a light source and a Schott Bg 12-2 mm excitation filter, it has been used by us for the observation of flow characteristics in veno–venous connections (Auer, 1977, 1978b), laminar flow in pial veins (Fig. 11), and extravasation under pathological conditions such as acute arterial hypertension, in which disturbance of the blood–brain barrier was seen to commence at the postcapillary venular level (Auer, 1978a; Auer and Heppner, 1978) (Fig. 12).

2.3.3. Video Technique

Attaching a video camera instead of photo or movie cameras has made both the imaging and the image analysis easier and opened a new world of possibilities for using electronics for automatic measurement of pial vessel calibers. We are using a Bosch T6XK camera and a Sony video recorder (U-matic PAL system), which enables one to store all experimental observations from the pial surface on videotape and analyze all vessel portions visible in the cranial window by multiple replay of the tape. This technology makes it much easier to display data together with the image, using a video-RAM system connected to the sources of information, such as manometers for pressure recording (Fig. 13).

2.4. Analysis of Pial Vessel Caliber

2.4.1. Measurement from Serial Images (Photography)

Measurement from serial images can be done by magnification-projection of color slides or by projection of black-and-white negatives using a scale with 100-μm units. Calibration of measured values in micrometers can be done by photographing and projecting through the same optical systems. This procedure enables one to collect data from series for calculation of mean values, provided the time intervals between the photographs are the same in each experiment. Many vessel portions can be measured at each time interval from the start of an experiment; the time consumption is considerable, however.

2.4.2. Image Splitting

The use of an image-splitting eyepiece and a TV microscope as initially described by Baez (1966) (Fig. 14) has been successful in combination with the cranial window technique for several groups over the years (Busija et al., 1981; Kontos and Wei, 1982; MacKenzie et al., 1976). Using visual control, the vascular image on the video screen is sheared by an image-splitting eyepiece. The result is indicated by a micrometer with attached potentiometer as the difference between zero shear and the situation in which two images of the investigated vessel are lying exactly side by side. The image shear from zero to shear at one diameter is produced by the investigator turning the micrometer screw. The diameter information can be readily transferred from the potentiometer to a penwriter. The resolution of this method is satisfactory both in time

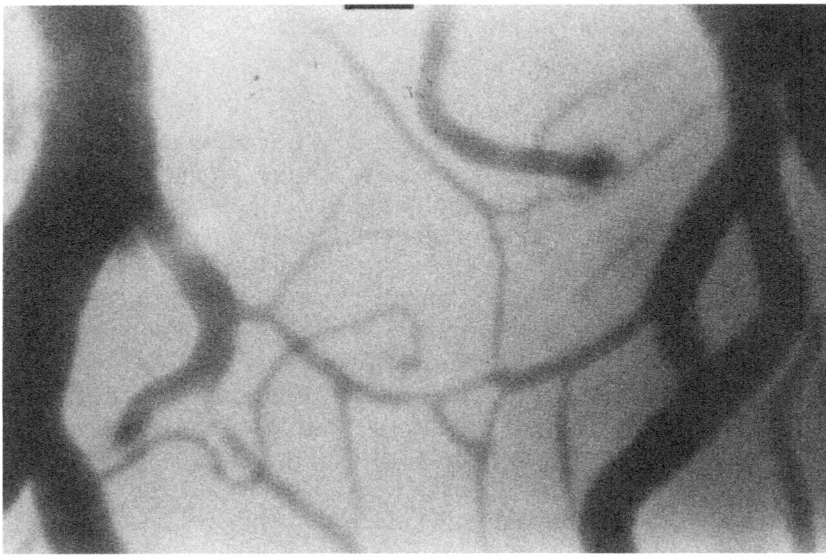

Fig. 8A.

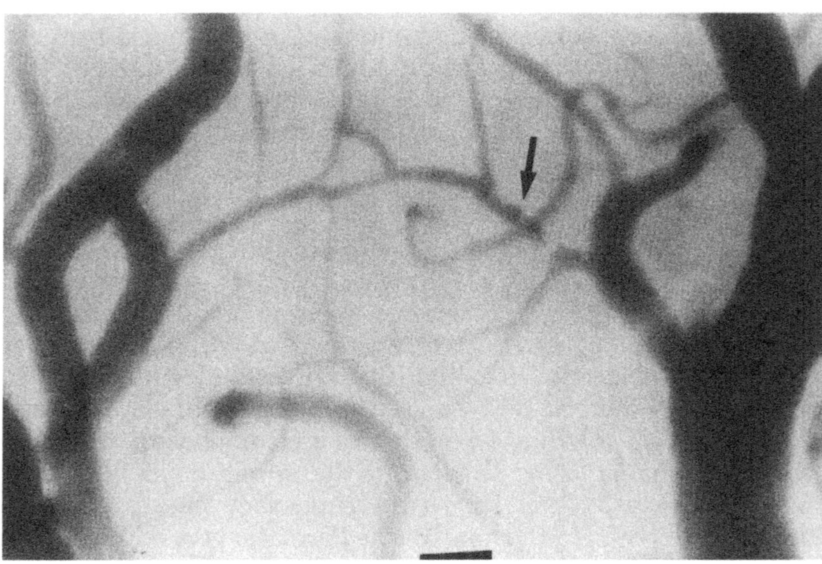

Fig. 8B.

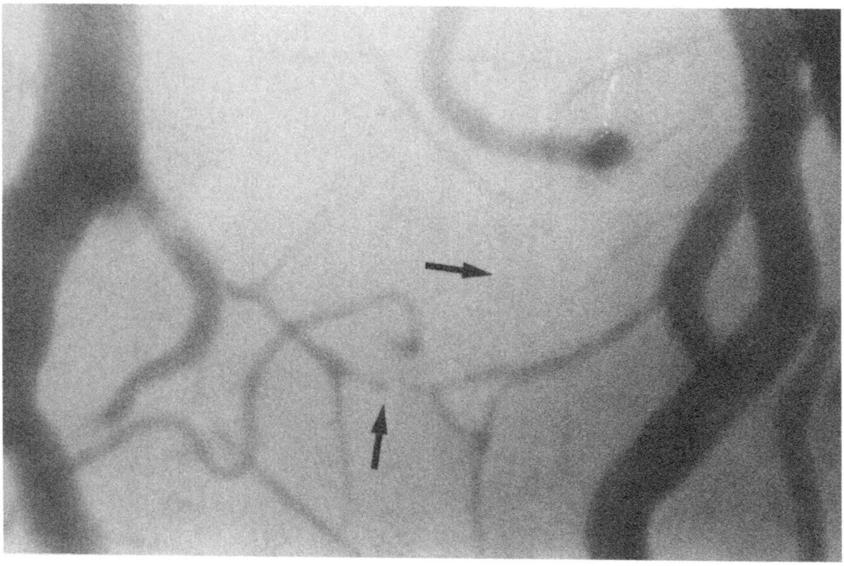

Fig. 8C.

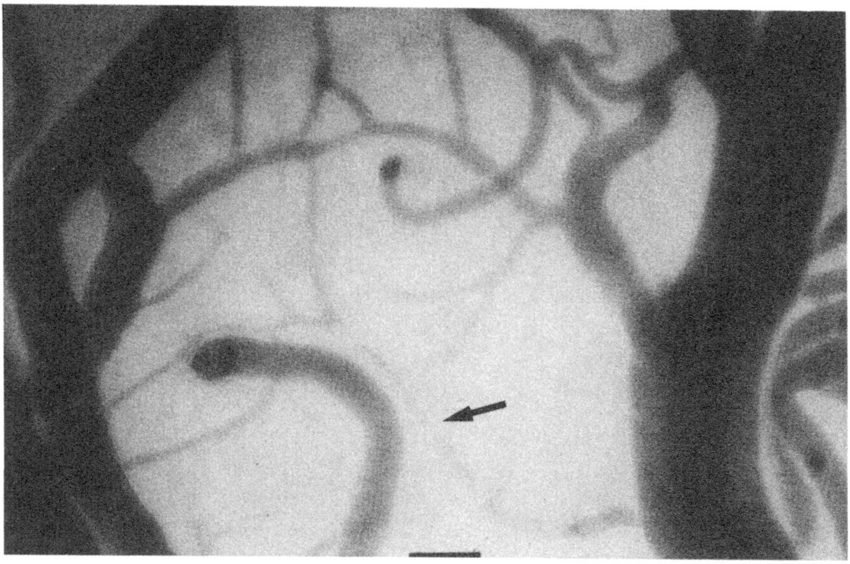

Fig. 8A–D. Normal pial vessels in the cat. Calibration bar, 66 μm. Different flow patterns in a network of intervenous connections. Depending on flow direction and quantity within individual branches, prestasis or stagnant flow occurs in intervenous connections. Flow is restored after varying periods of time without any activation of the coagulatory system or clumping of red blood cells (with permission, from Auer, 1978a).

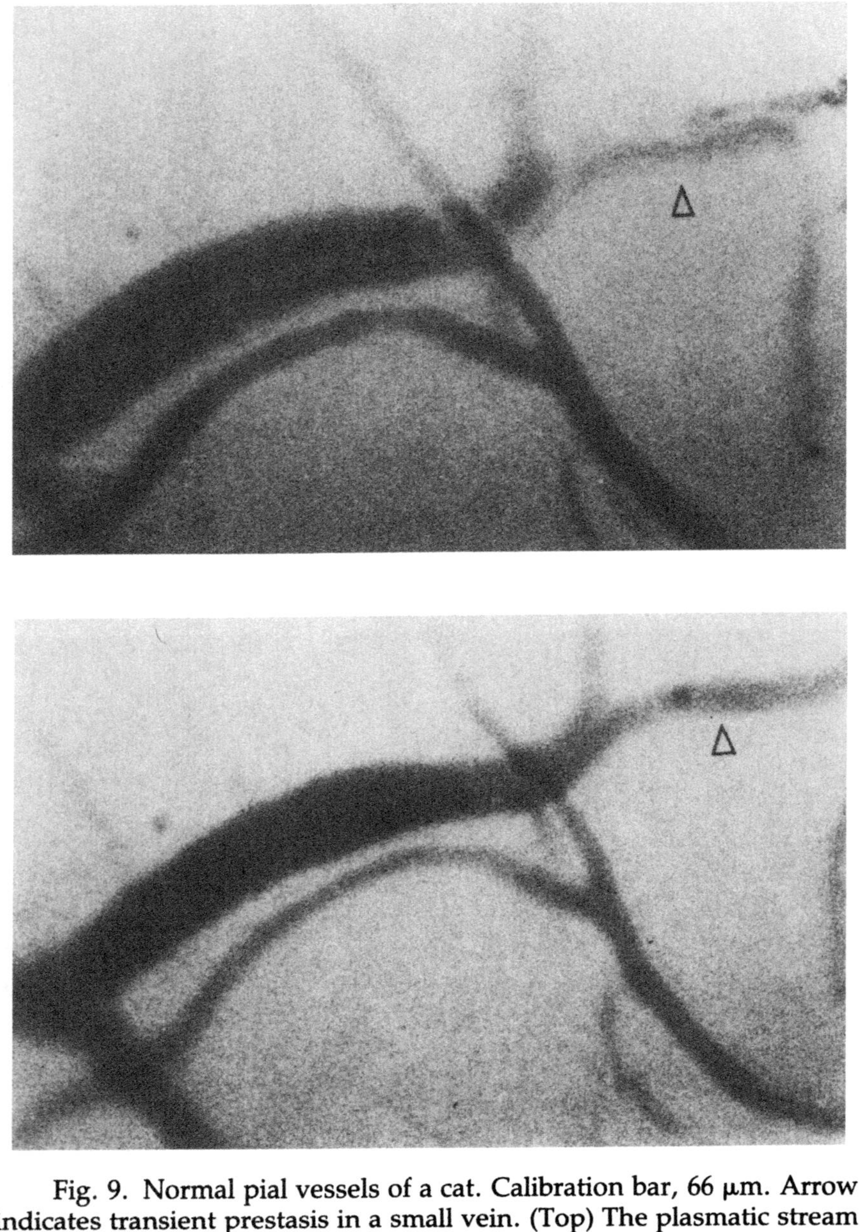

Fig. 9. Normal pial vessels of a cat. Calibration bar, 66 μm. Arrow indicates transient prestasis in a small vein. (Top) The plasmatic stream can be followed over a long distance after it enters a longer vein. (Bottom) A few minutes later, flow is fully restored. A small vein leaving the cortex can be seen left of the arrow, distributing its blood in both directions. The plasmatic stream within the larger vein has changed its position (with permission, from Auer, 1978b).

and space; it has the disadvantage, however, that one person is required full-time during measurements. Moreover, recording is only possible from one vessel portion per experiment or by moving from one vessel to the other and readjusting the system for each vessel with considerable intervals of time between the single measurements.

2.4.3. Photometry

A photomultiplier attached to the microscope can be used for the qualitative analysis of pial blood volume by measuring the intensity of light reflected from the pial surface in the cranial window. If one focuses with an iris lens to a single vessel and its immediate surrounding, dilatation of the dark vessel on the bright background of the brain surface will reduce the overall light intensity arriving in the photomultiplier tube. Conversely, vasoconstriction will allow more of the brain surface behind the vessel to show up and reflect light. Light absorption or reflected intensity can thus be used as a semiquantitative measure of caliber changes. For this procedure, we have used a Leitz photomultiplier supplied by a Knott-type NSHN,BN600 high-stability power supply (Auer, 1979a,c) (Figs. 4 and 15). The light intensity changes as a direct measure for diameter can be recorded on a penwriter as a continuous analog signal. Calibration of the penwriter curve can be accomplished by repeated measurements with other techniques such as the image-splitting eyepiece. The interesting aspect of this method is, however, semiquantitative continuous observation of vessel calibers at very high resolution, far beyond resolution of all other available techniques, which then reveals spontaneous oscillations of pial arteries (Auer and Gallhofer, 1981), as has also been reported from other vascular beds (Figs. 16 and 17).

2.4.4. Video Angiometry

Video analyzing systems as basically developed by Intaglietta (Intaglietta and Tompkins, 1973) and applied to physiology by Assmann and Henrich (1978) and others have also been developed to analyze reactions of the pial vasculature: we have used our own system since 1978 (Auer and Haydn, 1979) in about 350 experiments. A very similar system has been developed by Okayasu and coworkers (1979). "Multichannel video angiometry" is based on the evaluation of the signal of a single TV line crossing the image of a pial vessel on the video screen. Schmidt Triggers are used to detect the abrupt gray value changes as the TV line moves from the

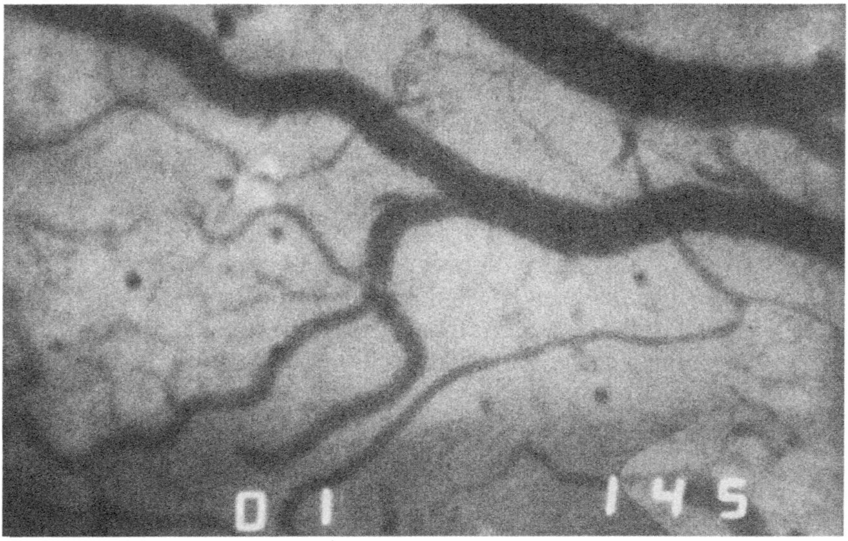

Fig. 10A.

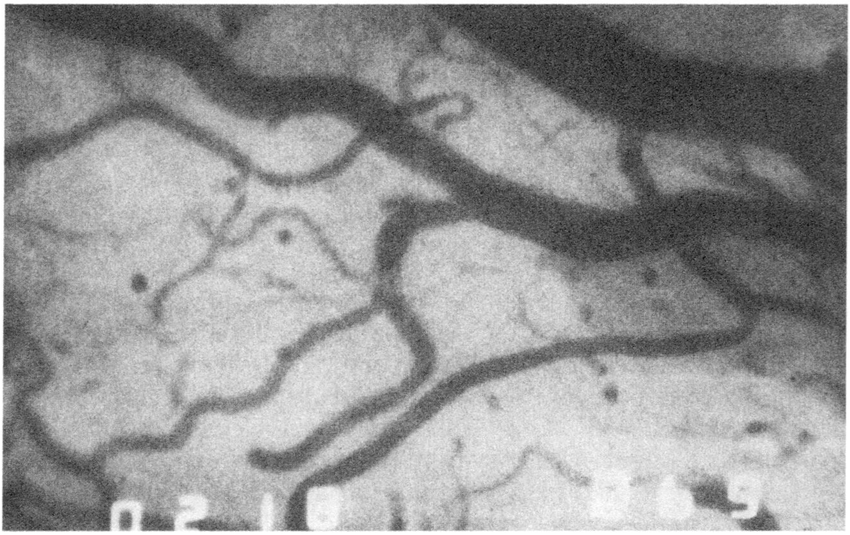

Fig. 10B.

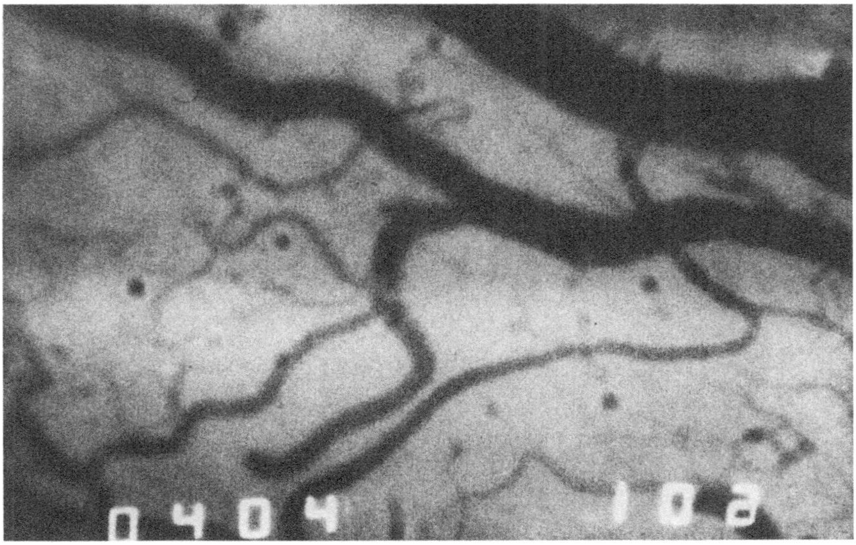

Fig. 10A–C. Normal pial vessels of a cat at various levels of mean arterial pressure: (A) 145 mm Hg (B) 69 mm Hg (C) 102 mm Hg (with permission, from Auer and Heppner, 1978).

white cerebral surface to the dark vessel and, after crossing the vessel, back to the white brain surface. These signals are used to start and stop an in-built oscillation counter. The number of oscillations from start to stop, i.e., from one end to the other of the TV line as it crosses the vessel, is used as a measure of diameter, which can be calibrated in micrometers. Three such analyzers built together in one instrument allow for the simultaneous and continuous analysis of three different vessels either during the experiment or during replay of the experiment from videotape. For each replay, the video measuring lines can be freely moved on the TV screen to new vessel segments. These measuring lines are made visible on the TV screen, the same as the vessel segments' diameters indicated in micrometers (Fig. 13). The analog signal from each channel can be recorded on a penwriter together with other parameters (Fig. 18). By multiple video replay, reactions of many different vascular segments from the same experiment can be brought together on one paper chart (Figs. 19 and 20). At high penwriter paperspeed, a high temporal resolution can be achieved for the comparison of

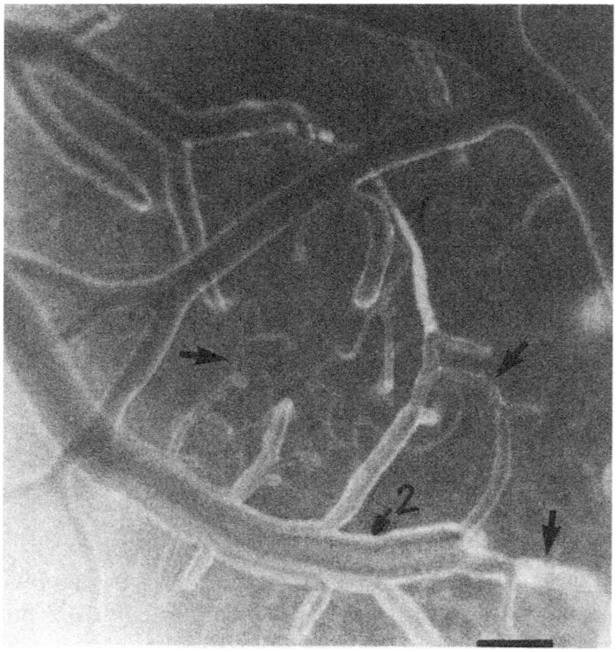

Fig. 11. Fluorescence angiogram of normal pial vessels. Calibration bar, 183 μm. 1, Stagnant plasma in intervenous connection; the arrow indicates other intervenous connections; 2, typical pattern of laminar flow in larger venule. The blood of an entering branch is seen to force the blood column of the larger vein toward the center; the entering blood remains isolated over a long distance without mixing with the other blood. It is thus easy to determine the direction of flow (with permission, from Auer, 1978a).

reactions of small and large vessels (Auer et al., 1985a) (Fig. 21). Above all, this technique allows for the analog graphical demonstration of classical physiological or pathophysiological reactions of pial vessels to experimental stimuli (Fig. 22).

Fully digitalized video systems will further improve the possibilities for image analysis, enabling the calculation of blood volume and volume flow in single vessels.

2.4.5. Observation of Brain Capillaries

Especially in small species such as the rat, vessels a few micrometers in diameter can be visualized with the cranial window

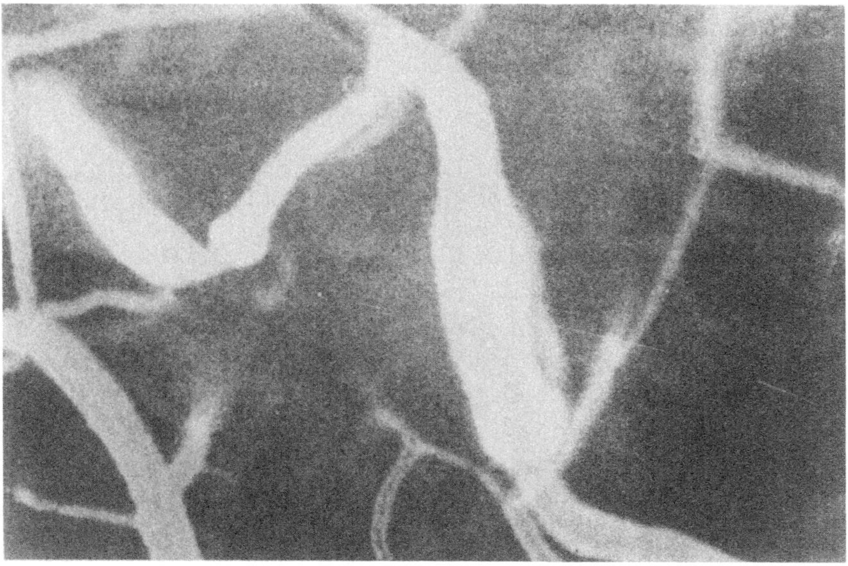

Fig. 12A.

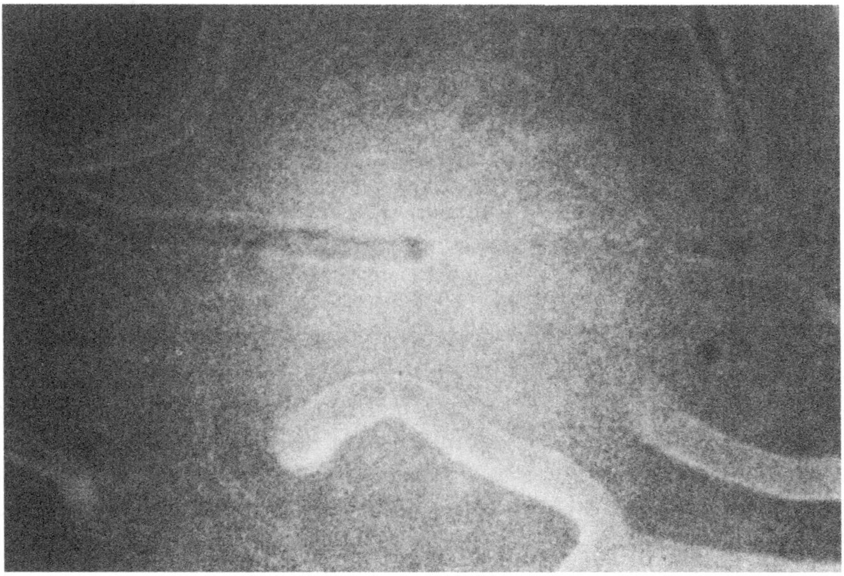

Fig. 12A and B. Fluorescence angiogram of cat pial vessels. (A) Arterial phase: sausage-string shaped pial artery during acute arterial hypertension. Note the lack of evident extravazation. (B) A small vein emerges from the cortex, surrounded by sodium-fluorescent extravazation during acute arterial hypertension (with permission, from Auer and Heppner, 1978).

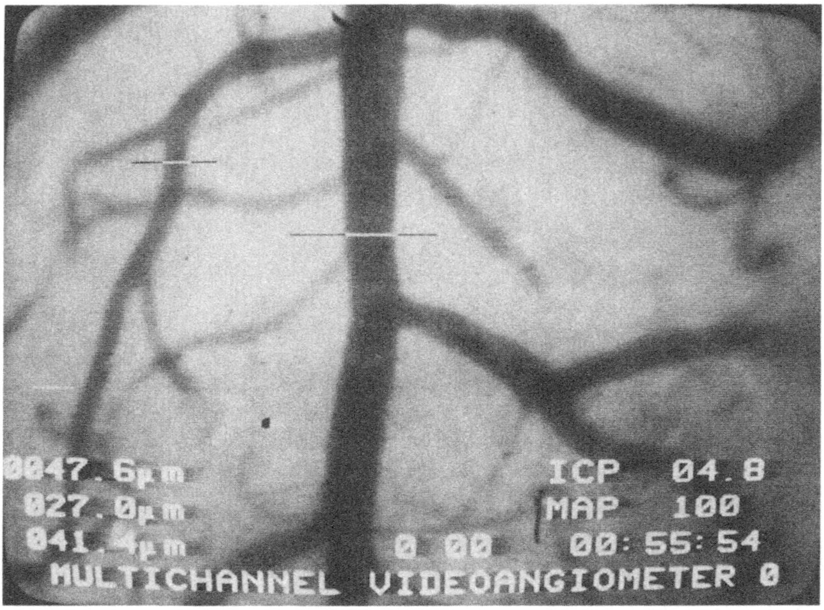

Fig. 13. TV-image of pial vessels with measuring bars of the angiometer. Sizes of measured vessels are indicated on the lower left. On the lower right, intracranial pressure (ICP), mean arterial pressure (MAP), and time are displayed (with permission, from Auer, 1984).

technique. For continuous recordings, however, the technical problems are considerable. With lower microscope magnifications, the resolution of any recording technique for caliber changes is insufficient. At high magnifications, brain movements make it difficult to record the same vessel over long periods of time. Observation of brain capillaries has been shown to be possible, however, especially with fluorescence techniques that allow one to follow single vessels a short distance into the cortex (Auer, 1978a; Rosenblum and Zweifach, 1963).

2.5. Measurement of Flow Velocity

Measurement of the velocity of the blood stream within single vessels provides important information in addition to the vessel's caliber, primarily because it allows for the calculation of volume flow within that vessel. The technical problem of registering velo-

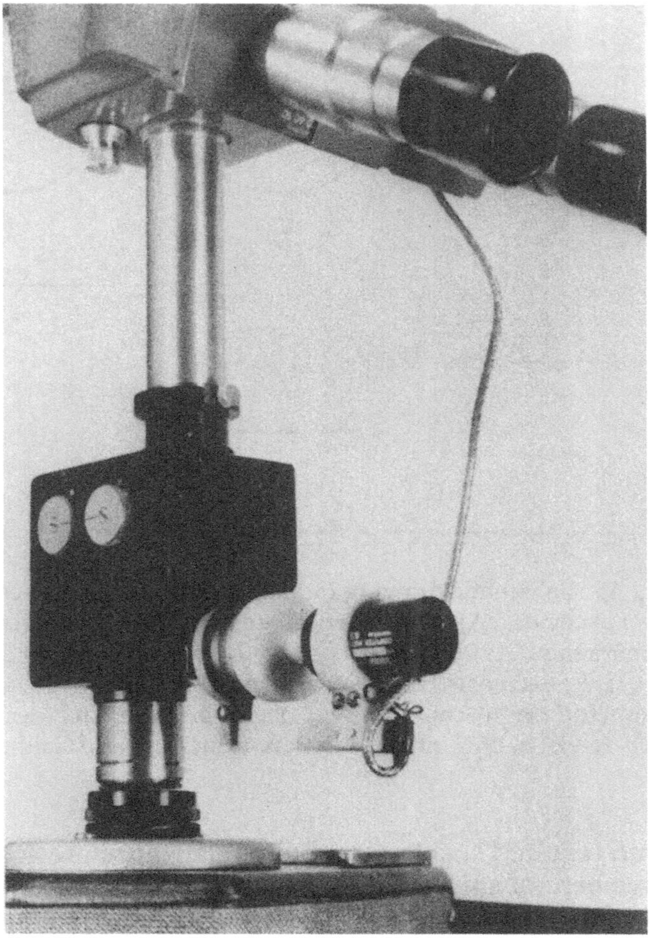

Fig. 14. Physical layout of assembled image-splitter eyepiece, TV camera, and microscope equipment. The potentiometer is coupled to the splitter micrometer head at left (with permission, from Baez, 1966).

city is different between small and large vessels and between vessels with rapid and slow flow. All methods developed so far are based on detection of the movement of erythrocytes. Earlier developments like Branemark's flying spot (1965a) or Munro's rotating prism (1966) were very much dependent on individual judg-

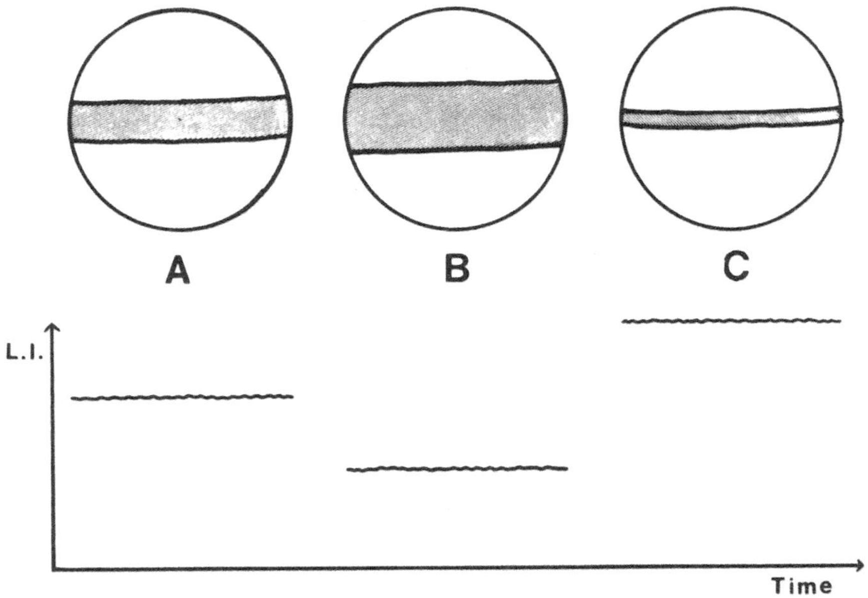

Fig. 15. Schematic drawing of pial arterial portion as seen through the microscope. (A) The normal vessels (B) Vasodilataion (C) Vasoconstriction. As the curves below the vessel portions show, light intensity (L.I.) decreases with dilatation and increases with vasoconstriction, since the bright compartment of the brain surface decreases and increases, respectively (with permission, from Auer and Gallhofer, 1981).

ment and required small vessels with relatively slow flow because the movement of single distinguishable erythrocytes is observed. Wiederhelm's photometric slit noise technique (1966) was already independent of observer judgment, but was only semiquantitative. The photometric techniques of Asano et al. (1964) and Wayland and Johnson (1967) were all restricted to small vessels with a maximal diameter of around 20 μm. Larger vessels could be investigated with Intaglietta's (1971) photometric cross-correlation technique and the photometric grating technique of Rockemann and Plesse (1973). None of these techniques has been used, however, for the investigation of the cerebral circulation. Koyama et al. (1982) used a laser Doppler microscope to measure blood velocity in single pial vessels (Fig. 23); this method provides continuous, but only semiquantitative, data without the possibility for calibra-

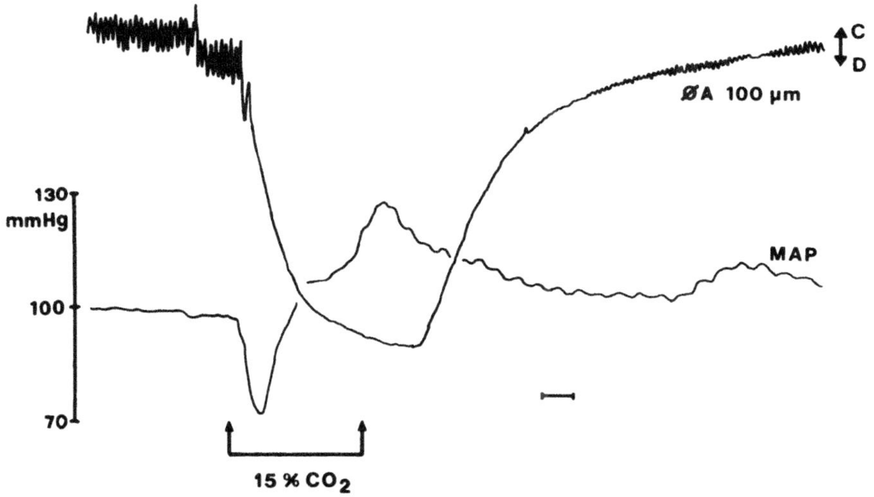

Fig. 16. 1-1/2-min MAP oscillation unrelated to rhythmic oscillation of a 100-μm arteriole with a 6-min rhythm. The vessel's rhythm is abolished during hypercapnic dilatation. At the end of CO_2 ventilation, the vessel constricts, and the rhythm reappears. Photometric recording: C, constriction; D, dilatation. Calibration bar, 1 min (with permission, from Auer and Gallhofer, 1981).

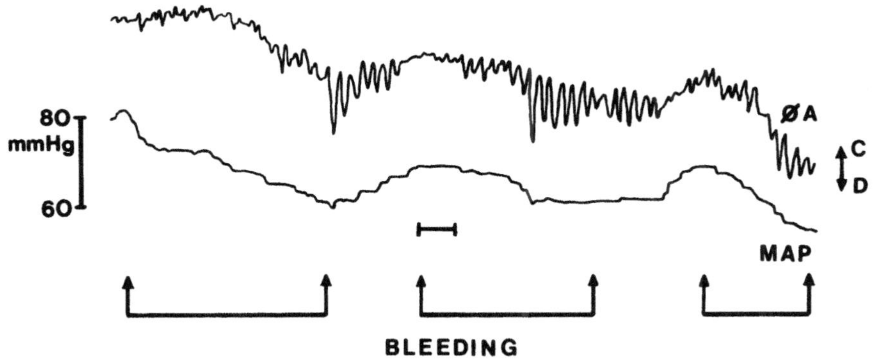

Fig. 17. A 3–4-min rhythm on a 70-μm arteriole increases in amplitude with falling pressure and vice versa. At the amplitude maximum, frequency tends to slow down. Photometric recording: C, constriction; D, dilatation. Calibration bar, 1 min (with permission, from Auer and Gallhofer, 1981).

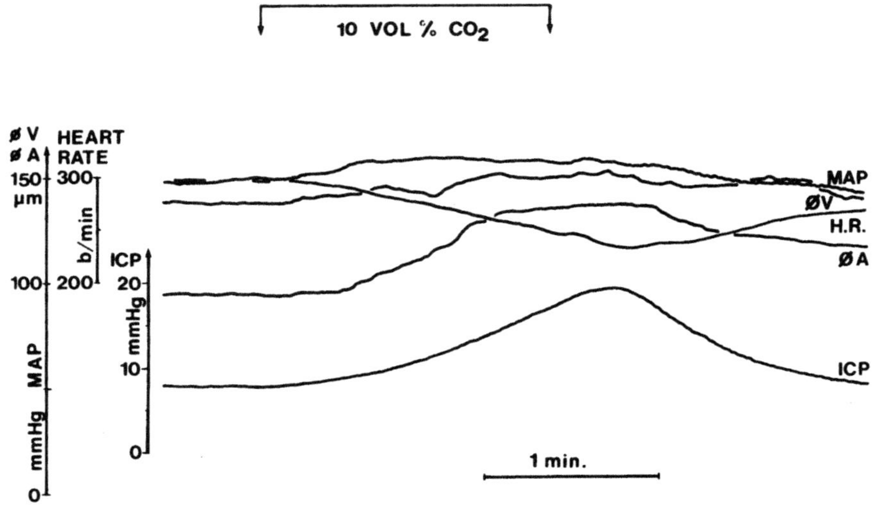

Fig. 18. Recording of diameter changes of a 95-µm pial artery (ØA) and a 140-µm pial vein (ØV) during an episode of hypercapnia (with permission, from Auer and Haydn, 1985).

tion in vivo. Other recent technical approaches on brain vessels included the use of a pulsed Doppler technique, video-based densitometry, and cross-correlation of video signals. Busija et al. (1981, 1982) placed a small-pulsed Doppler crystal under a large pial

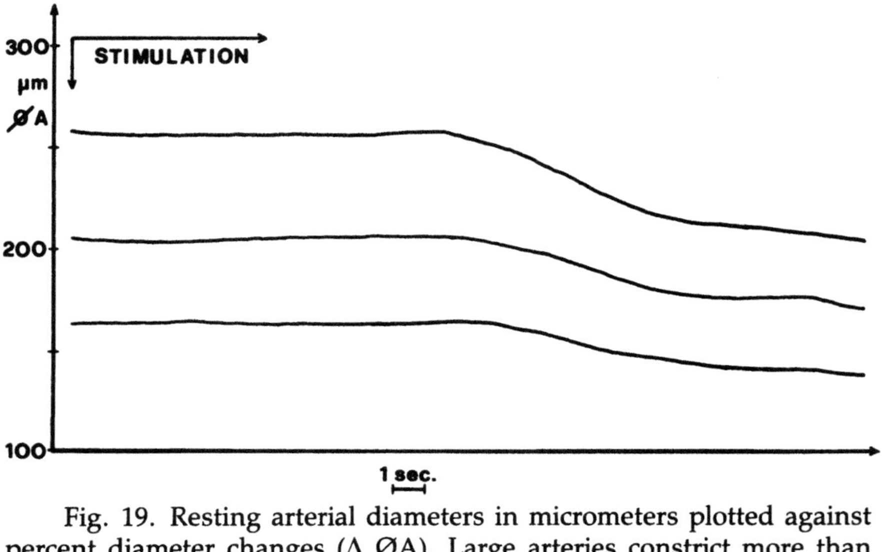

Fig. 19. Resting arterial diameters in micrometers plotted against percent diameter changes (Δ ØA). Large arteries constrict more than small arteries (with permission, from Auer et al., 1981).

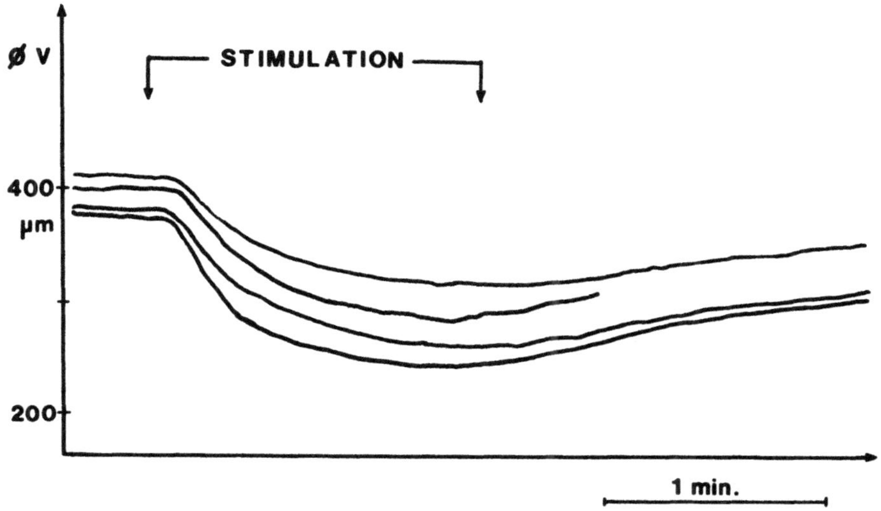

Fig. 20. Example of venous reactions to stimulation in a single experiment: Maximal constriction is reached at the end of stimulation. The return to the initial diameters is much slower than the change from normal to maximal constriction. The time interval from resting diameter to maximal constriction is markedly shorter than the period between the end of stimulation and return to initial diameters (with permission, from Auer et al., 1981).

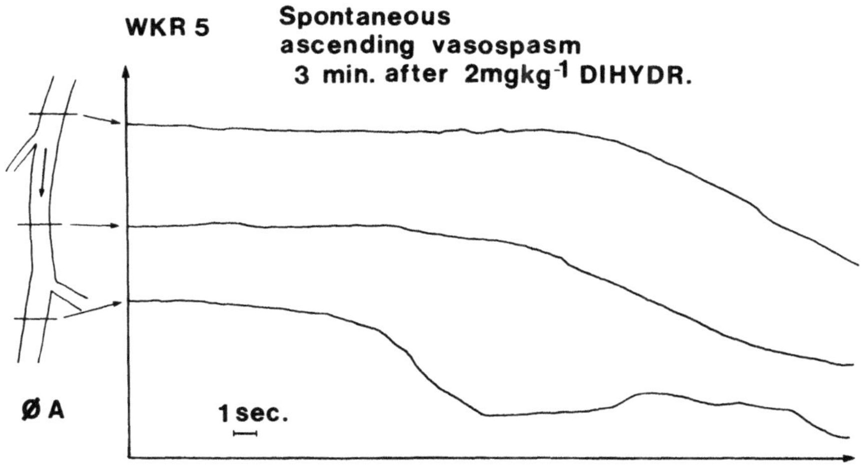

Fig. 21. Angiometrical curves of changes on three points of measurements of a pial artery (high-velocity recording). Vasoconstriction starts 10 s earlier at the peripheral limb of the vessel after giving off several branches (lowest curve) than on the larger upstream situated point of measurement (upper curve) (with permission, from Auer et al., 1985a).

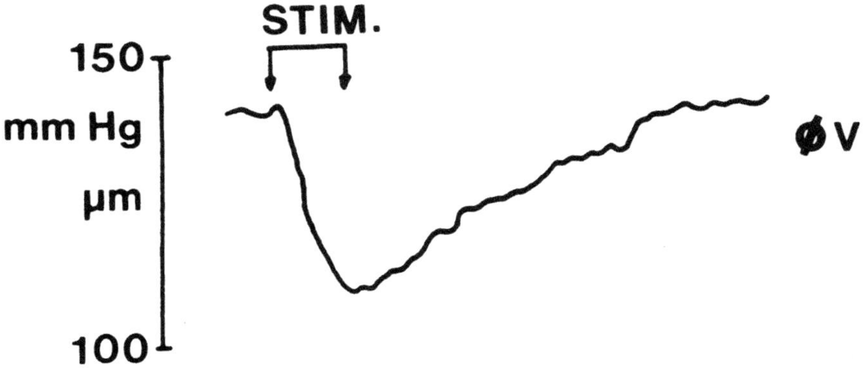

Fig. 22. Constriction of a pial vein (ØV) during cervical sympathetic stimulation (STIM) ipsilaterally to the cranial window.

artery. This method provides continuous, but only relative, data with no possibility for calibration. Moreover, only one vessel segment can be measured. This technique has been used for quasi-continuous calculation of volume flow together with a TV image splitter that allowed for measurement of caliber every 2–4 s. As a

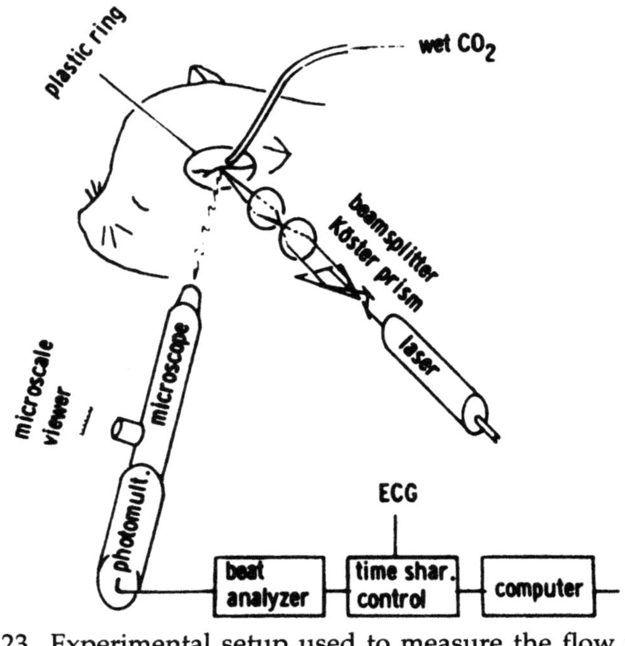

Fig. 23. Experimental setup used to measure the flow velocity in vivo for adult cats (with permission, from Koyama et al., 1982).

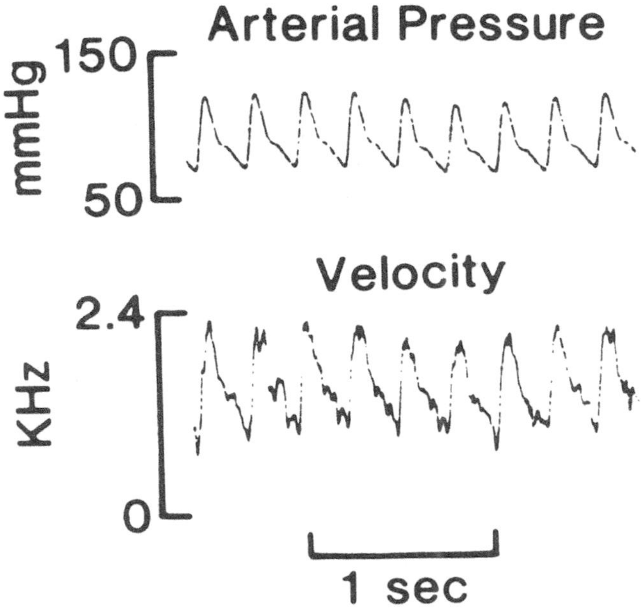

Fig. 24. Arterial blood pressure and blood viscosity in cat pial artery (diameter, 490 μm) (with permission, from Busija et al., 1981).

consequence of qualitative velocity data, volume flow is also qualitative or semiquantitative, and measurement is restricted to one exceptionally large artery per animal (Fig. 24). Moreover, it is not very likely that a Doppler crystal can routinely be placed under such arteries without producing functional impairment.

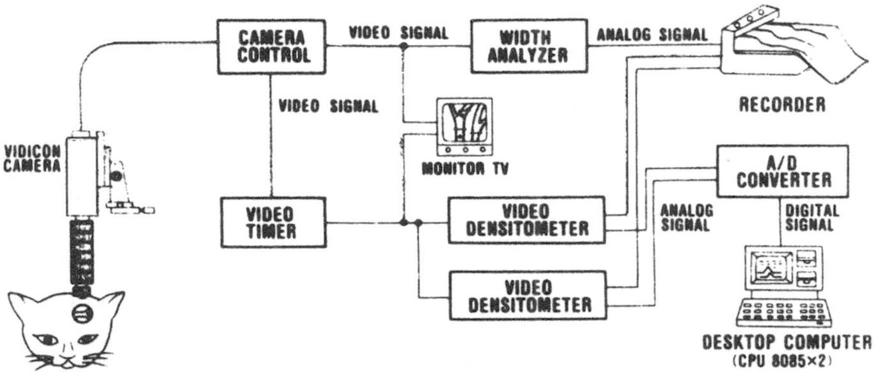

Fig. 25. Block diagram of the video camera system for simultaneous measurement of flow velocity and pial vessel diameter (with permission, from Gotoh et al., 1982).

Video densitometry has been attempted by Gotoh et al. (1982) (Figs. 25 and 26). The travel of a 0.1-mL bolus of saline injected retrogradely into the lingual artery was followed densitometrically between two defined points of a pial artery. Velocity was calculated from the measured travel time of the bolus between the two check points and the distance between the latter. This method provides absolute data of flow velocity in millimeters per second, but is not continuous. At intervals, when combined with video angiometrical registration of caliber, volume flow can be calculated in single vessels.

Cross-correlation of video signals obtained from a defined image-window of the video picture placed over a vessel under observation also analyzes intravascular contents. Here, the movement of a pattern of optical densities produced by dark red blood cells and bright plasma gaps is observed between two check points. The characteristic wave form of optical density as sampled from the two check points is calculated off-line by special cross-correlation

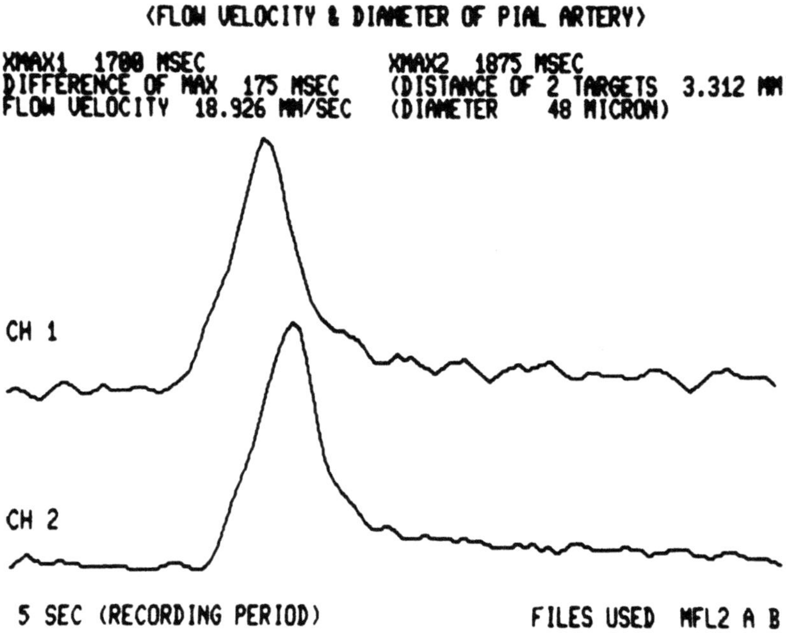

Fig. 26. Example of the CRT display of the computer showing the output of the densitometers and results of subsequent calculation. CH 1 and 2 are the recordings at the upstream and downstream points of the pial artery, respectively (with permission, from Gotoh et al., 1982).

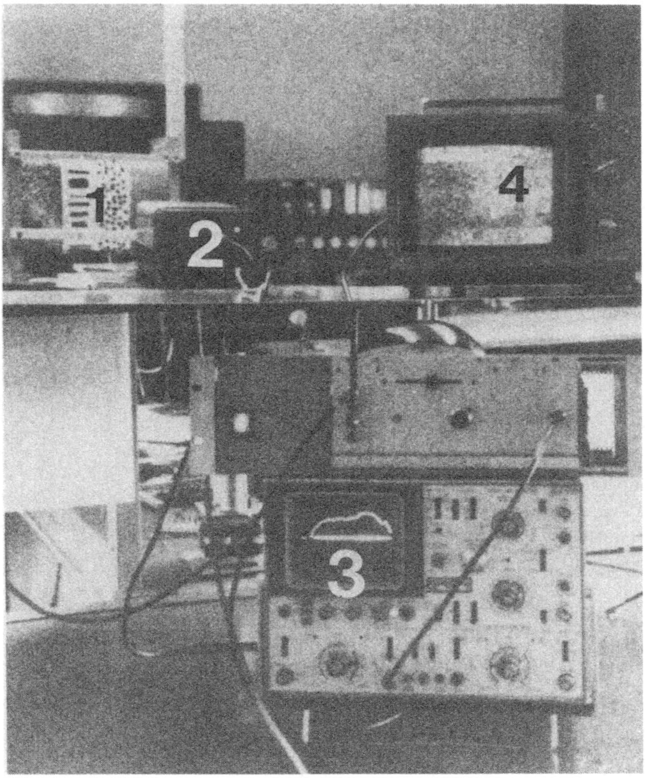

Fig. 27A. Experimental setup for continuous estimation of flow velocity by aid of TV techniques. 1, Rotating samples for simulation of granular flow; 2, TV camera; 3, oscilloscopic picture; 4, TV picture of granular flow with video window.

(Auer and Haydn, 1985) (Fig. 27). This computer-assisted version of a frame-to-frame method uses the spatial and time resolution of TV imaging, the former by the known distance between the TV scanlines, the latter by the frequency of TV frames per unit of time, thus allowing for description of the movement of an object with the following formula:

$$V[\text{mm/s}] = \frac{xd}{nt}\,[\text{mm/s}] \qquad (1)$$

where V = velocity of the object; x = number of TV scanlines; d =

Fig. 27B. Magnified TV picture with video window and display of flow velocity in mm/s (with permission, from Auer and Haydn, 1985).

distance between TV scanlines; n = number of frames; and t = time between frames.

Since the quality of measurements depends on the contrast between corpuscular elements and plasma between them, this method is also restricted to small vessels with slow flow, such as venules.

Cross-correlation of a photometric signal with the additional use of a grating system such as that built in an instrument by Leitz has been used by us to measure flow velocity in larger pial arteries (Fig. 28). This method provides absolute and continuous data on flow velocity within a defined portion of a pial vessel. But this technique has a limitation; the light intensity required for long-term observation allows for adequate recording in only few instances. The curve shown on Fig. 28B is a smoothed and averaged curve from the original recording. The method does not lend itself to simultaneous measurement of caliber because the required light intensity does not allow for beam splitting for the use of a second technique such as video angiometry.

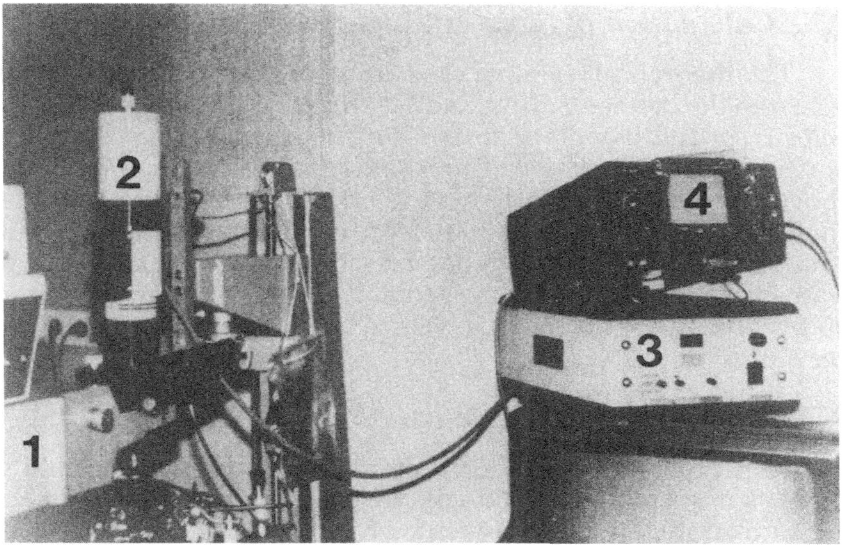

Fig. 28A. Experimental setup of photometrical correlation technique for measurement of flow velocity. 1, Microscope; 2, photometer; 3, correlator with digital display of velocity; 4, oscillometer.

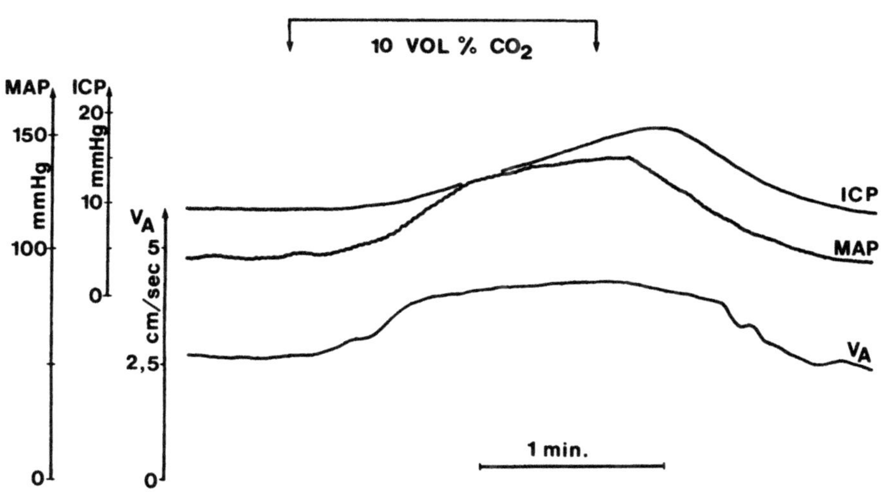

Fig. 28B. Penwriter recording of blood velocity in a pial artery (VA) (drawing) in cm/s (with permission, from Auer and Haydn, 1985).

2.6. Calculation of Flow in Single Pial Vessels

The theoretically simple calculation of flow in single pial vessels from the diameter of a cylindrical vessel segment and averaged velocity of the blood stream within this segment has so far, in practice, not been achieved in a satisfactory way. As mentioned above, Gotoh et al. (1982) have tried the combination of video angiometry and video densitometry to obtain intermittent absolute data from a pial vessel segment as seen in a cranial window, whereas Busija et al. (1981, 1982) obtained qualitative or semiquantitative, but quasi-continuous, values from one single segment of a very large pial artery.

2.7. Measurement or Calculation of Pial Blood Volume

Semiquantitative though continuous information about changes of whole pial blood volume changes in a defined area of the cortical surface can be obtained by the photometric technique described above (Auer, 1978c). The technical setup is shown in Fig. 4. The photodetector measures the light intensity reflected by a cortical area, which is higher during vasoconstriction and lower during vasodilatation. A differentiation between arterial and venous volume is not possible to a reliable degree for all experimental circumstances because differentiation would be based on the degree of oxygen saturation, which varies considerably in luxury perfusion states.

Calculation of blood volume is possible by the aid of TV techniques when calculating the volume of a cylinder between two points where the vessel's caliber is measured. Alternatively, "television reflectometry" can be used to calculate blood volume from all vessels visible in a cranial window (Eke, 1982, 1983). This method allows for a differentiation between arterial and venous volume when a bolus is injected retrogradely into the carotid artery and images are analyzed during the arterial and venous phase of bolus transition (Fig. 29).

3. Pial Vessels and Cerebral Blood Flow—Analysis of Relevance of Techniques

Although the discussion of factors that regulate, modulate, or influence cerebral blood flow in some way started with the

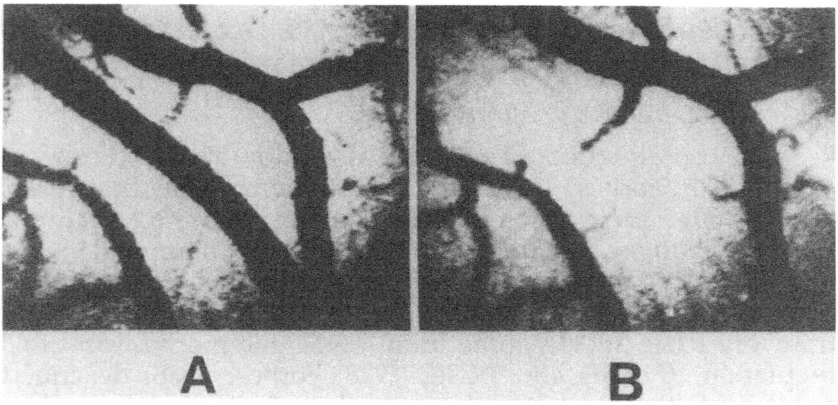

Fig. 29. Television images of 6 mm^2 of the brain cortex of an anesthetized cat before (A) and at the peak of the arterial transit of a hemodiluted bolus (B) induced by an intracarotid injection of isotonic dextran-saline solution. Note that on image B the blood is completely eliminated by the bolus from the artery and that the parenchymal and venous blood content was not yet affected at all (with permission, from Eke, 1983).

observation of pial vessels under various experimental circumstances (Forbes and Wolff, 1928; Wolff, 1929; Wolff and Lennox, 1930; Fog, 1937, 1939a,b; Forbes, 1937a,b, 1938, 1954; Forbes et al., 1933, 1941), the relevance of hypotheses and data derived from observations of pial vessels through cranial windows has been questioned many times (Sokoloff, 1959; Edvinson and MacKenzie, 1976). Some believe that the real resistance vessels of the brain are located intracortically, and pial vessels are all too large to act as resistance vessels. A wide range of vessel calibers is found, however, on the surface of the mammalian brain as related to body weight. Another widely accepted opinion has been that regulation of flow must depend on metabolic demand and therefore takes place within the brain parenchyma; it is thus not visible on the surface. Finally, the relevance of pial vascular reactions for the regulation of cerebral blood flow has been questioned with the argument that a role of segmental resistance and compensatory response of intraparenchymal vessels, in contrast to reactions of pial vessels, has not been ruled out. The following section thus analyzes the relevance of observations from the pial surface by comparison with results obtained with other techniques, such as measurement of cerebral blood with tracer or clearance techniques.

3.1. Relevance of Direct Pial Vessel Observation for the Definition of Normal CBF Autoregulation

3.1.1. Myogenic Regulation

The consequence of Fog's cranial window studies (Forbes and Wolff, 1928; Forbes, 1938) has been the hypothesis that cerebral arteries alter their vasomotor tone in response to intraluminal pressure changes, i.e., dilatation with falling intraluminal pressure and increasing vasomotor tone resulting in vasoconstriction as a response to rising intraluminal pressure (Fig. 30). In addition to this very basic and hypothetically myogenic mechanism of CBF regulation, (Forbes and Wolff, 1928; Forbes, 1938) described a number of other important factors influencing the pial vessels to a varying degree. This was by no means self evident in the year 1928, when the view of brain vessels as only passive tubes was widely accepted.

It was first shown by measurements with the Kety-Schmidt technique that cerebral blood flow remains constant within an arterial pressure range of 60–160 mm Hg (Kety, 1950). In order to

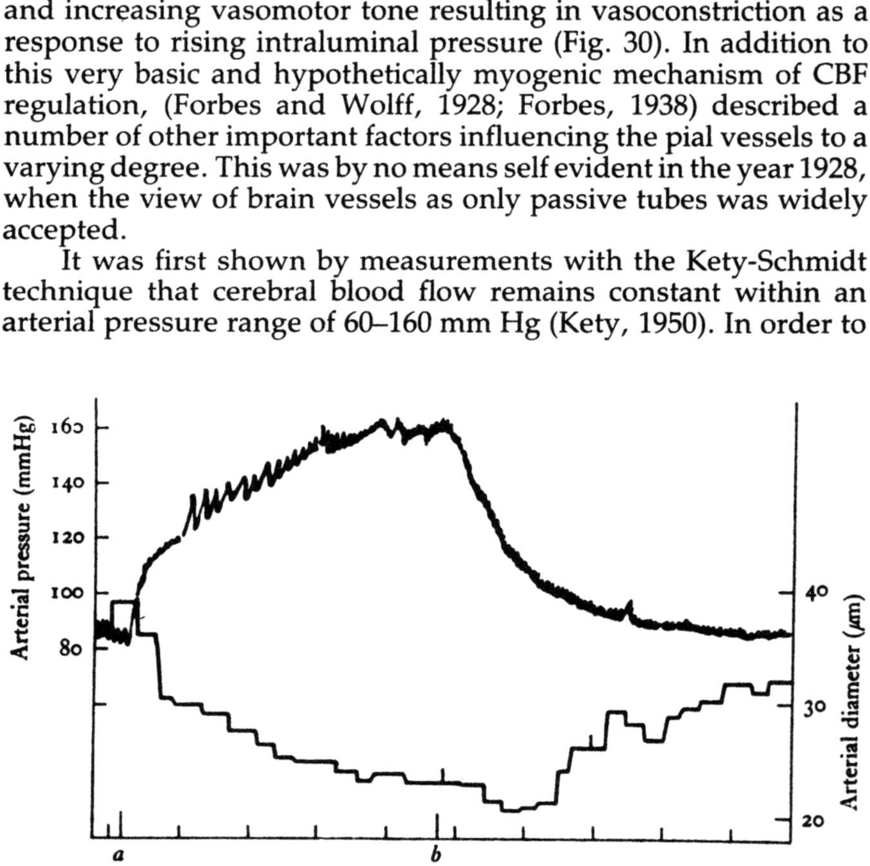

Fig. 30. The changes in caliber of a pial artery (below) that accompanied an increase in arterial pressure (above) when adrenaline was injected intravenously between *a* and *b*. Time in minutes (with permission, from Fog, 1939b).

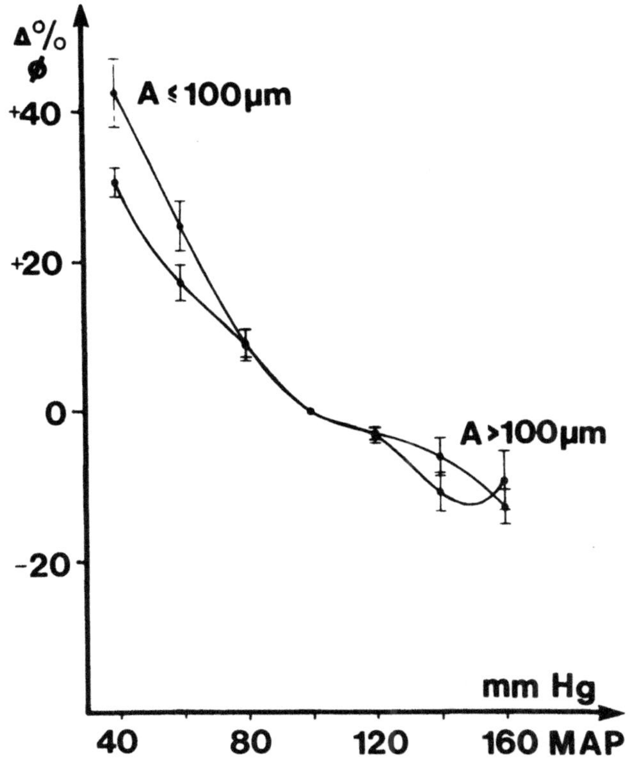

Fig. 31. Diameter variations of small (ØA ≤ 100 μm) and large (ØA > 100 μm) pial arteries during changes of mean arterial pressure (abcissa).

maintain constant cerebral blood flow within this range, vessels must dilate with falling and constrict with rising intraluminal pressure. This reaction has in fact been observed with the cranial window technique (Figs. 10, 30, and 31).

The relationship between intraluminal pressure and vessel caliber is not linear; although vessels smaller than 100 μm diameter and those between 100 and 200 μm react in a similar way, the smaller vessels tend to react nearer the lower and upper limit of the autoregulatory range. Pial veins show very little change (Auer, 1984; Auer et al., 1987a). This accordance of observations and measurements does not imply, though, that the basic regulation mechanism is of a myogenic nature, because dilatation could also be caused by the accumulation of metabolites with falling blood pressure.

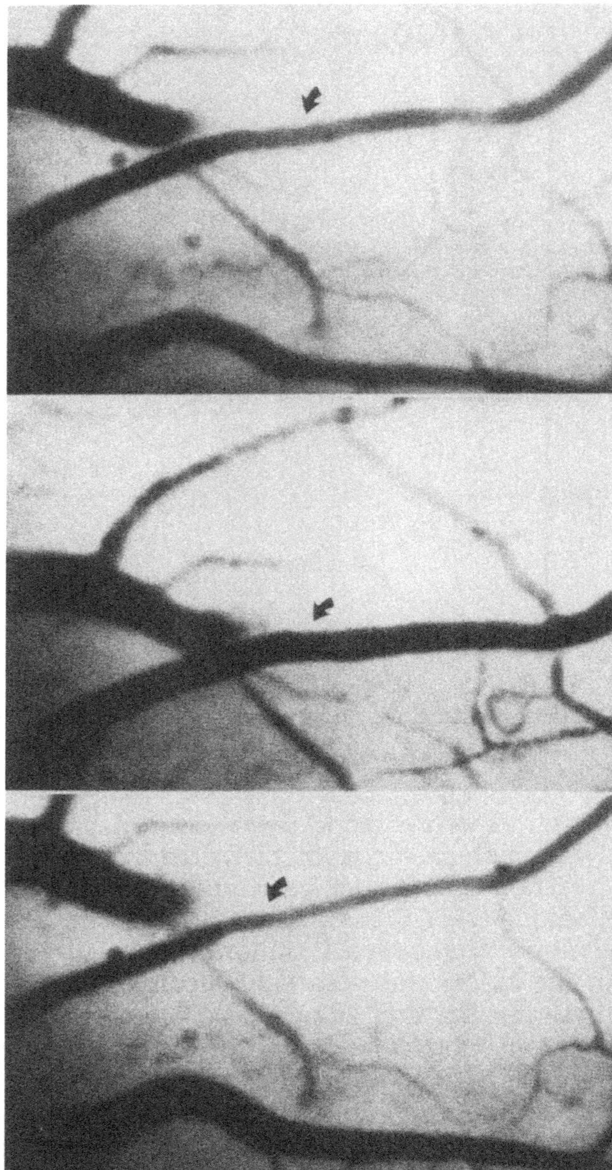

3.1.2. Metabolic Regulation

The most potent metabolic factor influencing both pial vessel caliber and cerebral blood flow—although not a regulator of CBF in the actual sense—is the partial pressure of carbon dioxide in the blood. Pial arteries dilate during hypercapnia and constrict during hypocapnia, small vessels more than large ones (Fig. 32), and veins react less than arteries (Fig. 33) (Auer, 1978d). Since blood pressure remains more or less stable during variation of $PaCO_2$, one would expect from observations of pial arteries a rise of CBF during hypercapnia and a fall of CBF during hypocapnia. This has, in fact, been shown in several studies, and there is a linear correlation between a wide range of $PaCO_2$ and the level of cerebral blood flow (Fig. 34). One of the main metabolic factors regulating regional cerebral blood flow response to metabolic demand is extracellular pH: Lowering of perivascular pH to 4.8 results in pial arterial dilatation, whereas an increase of pH beyond neutral leads to arterial constriction (Kuschinsky et al., 1972; McCulloch et al., 1982). Parallel with these reactions of pial vessels, cerebral blood flow rises and falls (Lassen, 1968). Adenosine has been described as a likely candidate for metabolic regulation of cerebral blood flow, and again, observations made with the cranial window technique correlate well with measurements of cerebral blood flow: Pial arteries dilate with rising concentrations of adenosine (Morii et al., 1985) and so does cerebral blood flow (Winn et al., 1981) (Fig. 35).

3.1.3. Sympathetic Regulation

Long before nerve fibers were demonstrated in the walls of cerebral vessels (Owman et al., 1974; Edvinsson et al., 1974; Nielsen and Owman, 1967), Forbes and Wolff showed pial arterial constriction during cervical sympathetic stimulation in the cat (Forbes and Wolff, 1928). Only very recently has it been shown that pial veins constrict even more than arteries (Auer and Johannsson, 1980, 1983) (Fig. 36). Both pial arteries and veins also constrict during topical microapplication of noradrenaline into the perivascular space (Wahl et al., 1972; Ulrich et al., 1982) (Fig. 37). Measurements of pial vessel reactions and cerebral blood flow using the

←

Fig. 32. Pial vascular pictures at different levels of arterial carbon dioxide tension and constant MABP. Black line, 113 μm, arteriole. (Top panel) $PaCO_2$ 38 mm Hg; (center panel) $PaCO_2$ 67 mm Hg; (lower panel) $PaCO_2$ 16 mm Hg (with permission, from Auer, 1978d).

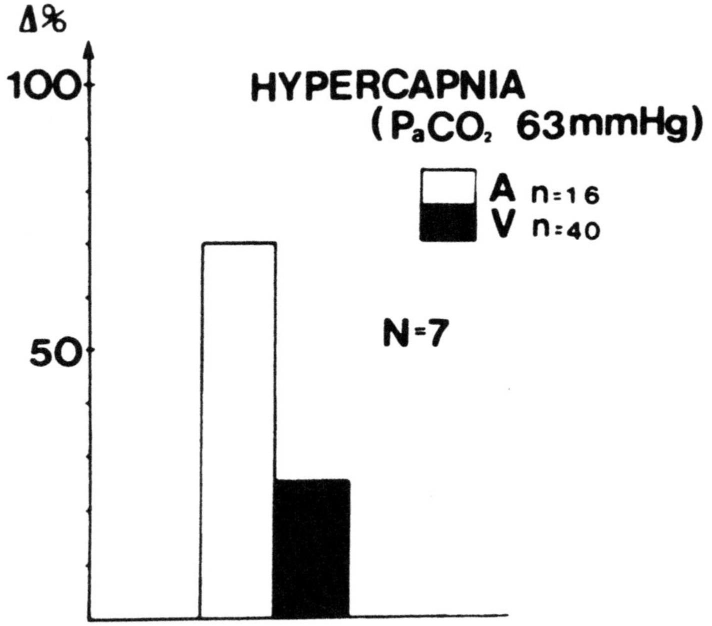

Fig. 33. Dilatation of cat pial arteries (A) and veins (V), presented as percent change (Δ%) from resting calibers, induced by hypercapnia. N, number of animals; n, number of vessels (with permission, from Auer and MacKenzie, 1984).

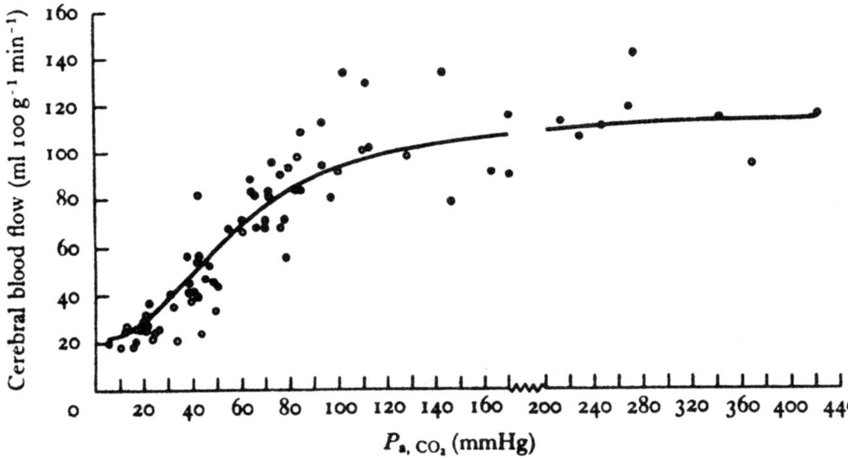

Fig. 34. The relation between cerebral blood flow (mL per 100 g per min) and $PaCO_2$ (mm Hg) obtained from a series of rhesus monkeys anesthetized with pentobarbitone and paralyzed with gallamine. The regression equation, $y = 20.9 + 92.8/1 + 10570e^{-5.251 \log x}$ describes the line of best fit (with permission, from Reivich, 1964).

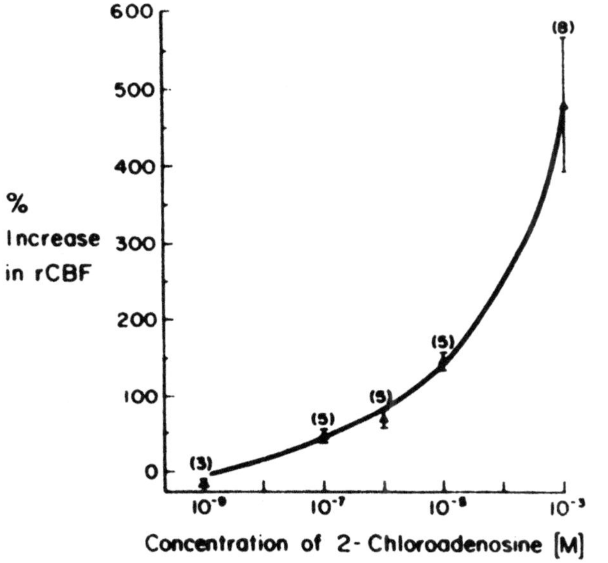

Fig. 35. With $10^{-9}M$ CHL-ADO, no increases in flow were obtained, but a 50% increase in rCBF occurred during perfusion with $10^{-7}M$ CHL-ADO. With $10^{-3}M$ CHL-ADO, a greater than 400% increase in rCBF was observed (with permission, from Winn et al., 1981).

hydrogen clearance technique show the decrease of CBF in parallel to the vasoconstriction (Auer, 1986). The stronger response of veins is in support of the present view that the sympathoadrenergic system is a regulator of cerebral blood volume rather than of cerebral blood flow (Auer and MacKenzie, 1984). On the other hand, ionic metabolic influences act only on arteries and remain ineffective in veins (*see* Figs. 38 and 39), thus reflecting mainly an influence on cerebral blood flow and only a minor effect on cerebral blood volume. The sympathoadrenergic effect is still, or even more, effective during hypercapnia (Kontos and Wei, 1982; Auer and MacKenzie, 1984; Auer, 1986) and acute arterial hypertension (Kontos and Wei, 1982) (Figs. 40–42).

Sympathetic stimulation contralateral to a cranial window fails to show both consistent vasoconstriction and a reduction of CBF (Auer, 1986) (Fig. 43). Bilateral sympathetic stimulation over 20 min revealed again a parallel behavior of small pial vessels and cerebral blood flow; namely, fading of the sympathoadrenergic effect, a so-called escape phenomenon, after several minutes (Sercombe et al., 1979; Auer, 1986) (Fig. 43).

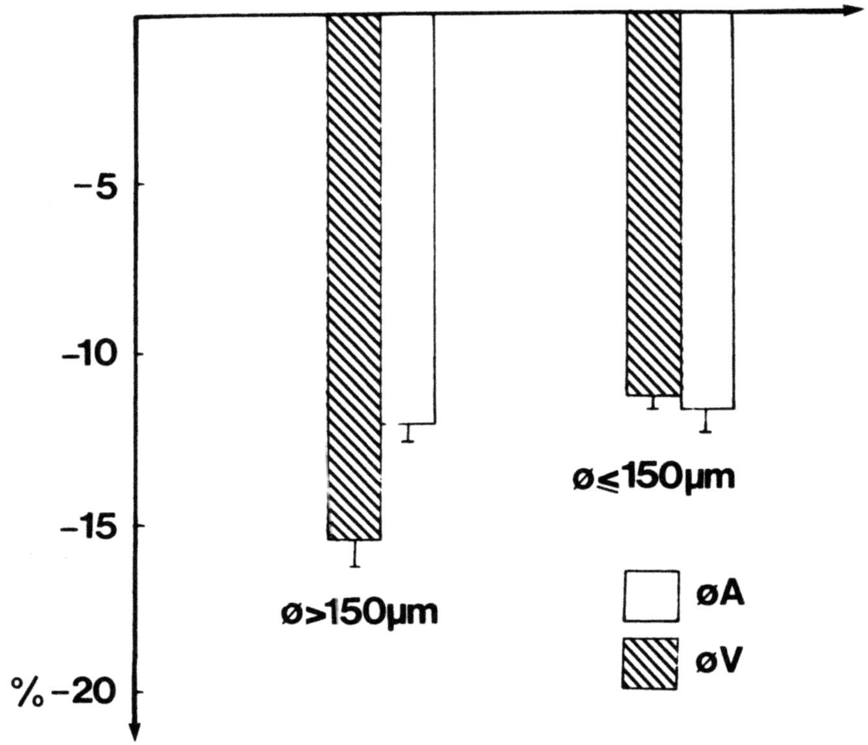

Fig. 36. Pial venous (hatched bars) and arterial (open bars) constriction during cervical sympathetic stimulation in percent of resting diameter. >150 μm, large vessels; ≤150 μm, small vessels. Bars indicate mean values ±SEM (with permission, from Auer and Johansson, 1983).

3.2. Observations of Pathological Reactions

Congruent hypotheses have also been put forward based on pial vessel and CBF measurements under pathological experimental circumstances. Beyond the upper limit of autoregulation (Lassen and Agnoli, 1973); the mechanical force of suddenly elevated blood pressure distends pial arteries either like "sausage strings" (Fig. 44) or in a diffuse manner (Fig. 45), depending on the percentage and acuteness of arterial pressure increase (Auer, 1978a).

3.2.1. Arterial Hypertension

Acute arterial hypertension results in an increase of cerebral blood flow (Skinjoe and Strandgaard, 1973; Strandgaard et al.,

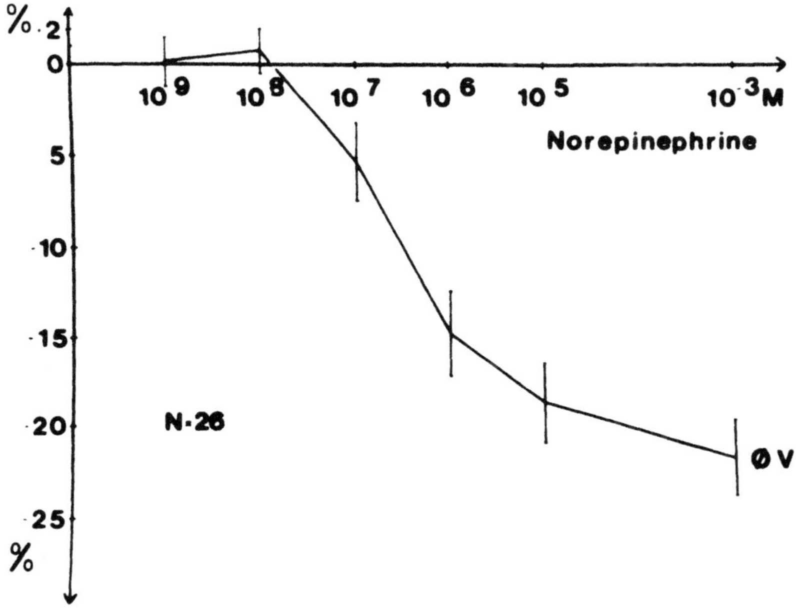

Fig. 37. Average concentration–response curve of cat pial venous constriction to perivascular microapplication of norepinephrine (noradrenaline): Each value represents the mean of three measurements taken at 20, 40, and 60 s after application. Ordinate, percentage pial venoconstriction; ØV, pial venous diameter; N, number of portions of vessels investigated at each step of norepinephrine concentration (with permission, from Ulrich et al., 1982).

1973, 1974; MacKenzie et al., 1976; Collman et al., 1978), which can reach up to 10-fold of normal in areas with the greatest pathological cerebrovascular response and blood–brain barrier disturbance (Hossmann et al., 1977). Pial vessel studies (Auer, 1978a) and measurements of cerebral venous pressure and intracranial pressure (Auer et al., 1980) showed that the blood–brain barrier disturbance in acute arterial hypertension probably starts in the postcapillary venules because of congested venous outflow and an increase of venous transmural pressure.

3.2.2. Arterial Hypotension

Below the lower limit of autoregulation, cerebral blood flow falls because cerebral vessels have reached a state of maximal dilatation and smooth muscle relaxation (*see* Figs. 31 and 46). Thus

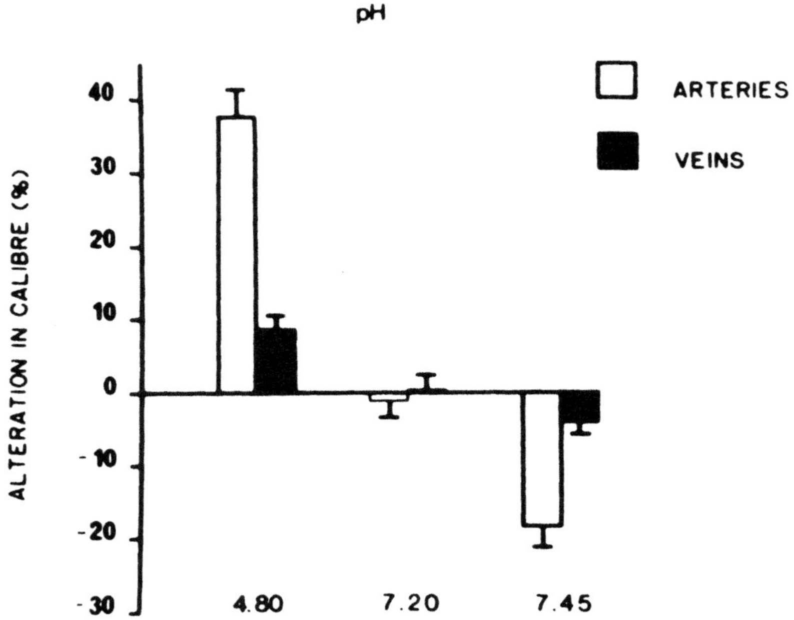

Fig. 38. Alterations in the caliber of pial arteries and pial veins to perivascular microinjections of CSF varying pH. The potassium concentration of each solution was 3 m*M*. Data are presented as means ±SE (with permission, from McCulloch et al., 1982).

further reduction in intraluminal pressure can no longer result in compensatory arterial dilatation.

3.2.3. Elevated Intracranial Pressure

When intracranial pressure is diffusely elevated, e.g., by experimental cisternal infusion of artificial CSF, pial arteries dilate markedly, whereas pial veins show minor changes (Auer et al., 1987a). Since arterial pressure remains constant, cerebral perfusion pressure, defined as the difference between arterial pressure and intracranial pressure, is reduced. From these observations one would expect the cerebral arterial dilation to act as compensation for the reduction of perfusion pressure to maintain constant CBF. Flow in this situation has in fact been shown to remain constant or to slightly increase, until autoregulatory vasodilatation has reached the state of complete smooth muscle cell relaxation. This occurs at an ICP level of about 50 mm Hg, beyond which no further

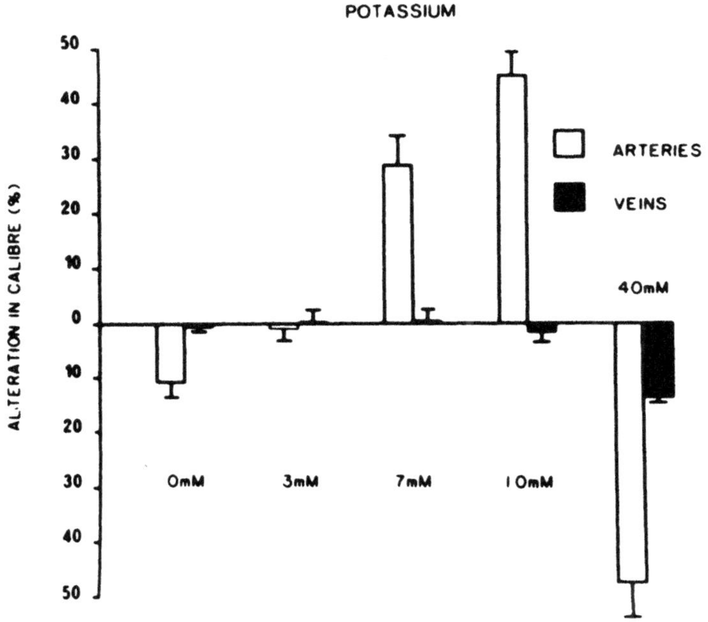

Fig. 39. Alterations in caliber of pial arteries and veins to perivascular microinjections of CSF containing varying concentrations of potassium. The pH of each solution was 7.20. Data are presented as means ±SE (with permission, from McCulloch et al., 1982).

dilatation occurs, and CBF is reduced (Langfitt et al., 1965a,b; Johnston et al., 1972; Grubb et al., 1975). Because of the Starling resistor phenomenon mentioned in the introduction, neither pial arteries nor veins collapse when CSF pressure reaches the level of arterial pressure (Fig. 47). Veins remain open because perivascular pressure, i.e., intracranial pressure, raises cerebral venous pressure, thus leaving cerebral venous transmural pressure unchanged over a wide range of intracranial pressures (Fig. 48). It has been suggested, from pressure measurements, that the reduction of CBF during elevated ICP is a result of collapse of bridging veins and lateral lacunae (Wright, 1928; Hedges and Weinstein, 1964; Langfitt et al., 1966; Shulman and Verdun, 1967; Yada et al., 1973). A sudden drop of intraluminal pressure was found between lacunae laterales and the superior sagittal sinus. A recent cranial window study in rats revealed, however, that this so-called cuffing mechanism of bridging veins exists only to a very minor extent (Auer et

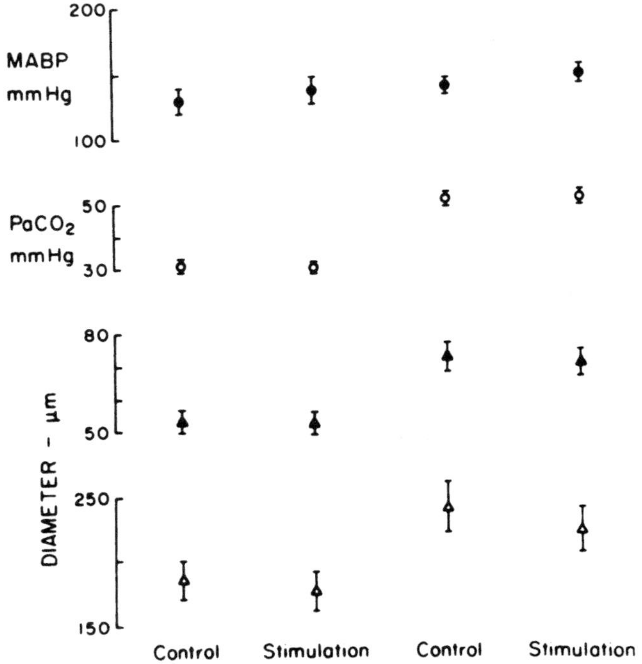

Fig. 40. Effects of stimulation of the ipsilateral superior cervical ganglion on pial arteriolar diameter during normocapnia and hypercapnia. All values are mean ±SE obtained from five cats. Shown from top to bottom are mean arterial blood pressure (MABP), $PaCO_2$, and diameter of small and of large pial arterioles (with permission from Kontos and Wei, 1982).

al. 1987b) (Fig. 49). Most likely, therefore, increased ICP above 50 mm Hg reduces CBF not because of vascular compression and collapse, but because of the fall in perfusion pressure.

4. Direct Pharmacological Cerebrovascular Effects

Calcium antagonists, also called calcium entry blockers or calcium channel blockers, are thought to interfere with the calcium channels in the membranes of vascular smooth vessels cells, thereby preventing the calcium-dependent actin–myosin coupling mechanism. The lipophilic dihydropyridine nimodipine has been

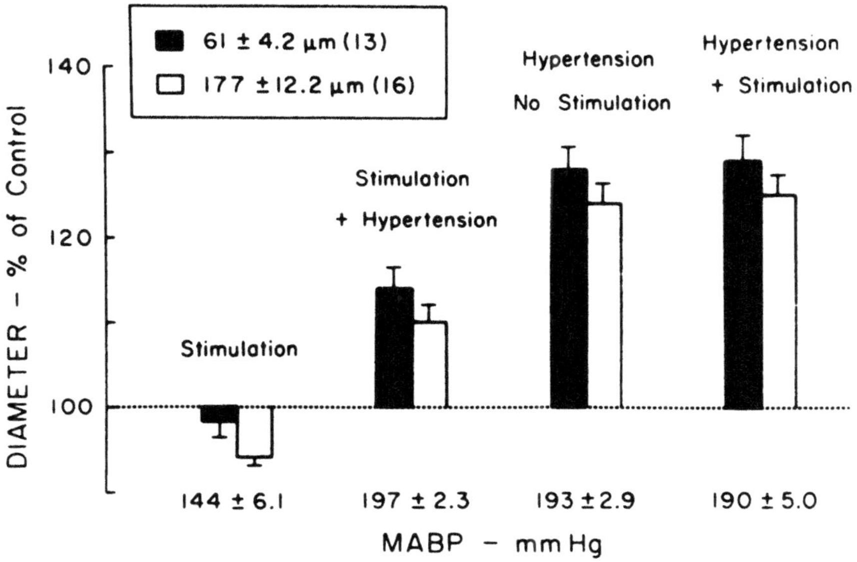

Fig. 41. Effects of stimulation of the ipsilateral superior cervical ganglion on pial arterioles with and without arterial hypertension. Means ±SE of vessel diameters are shown. Diameters are expressed as percent of the control value obtained under resting conditions when the arterial blood pressure was normal. These control diameter values are shown in the box in the left upper part of the figure. The mean arterial blood pressure (MABP) is shown in the lower part of the figure for each experimental condition. Shown from left to right are the effects of stimulation under conditions of normal arterial blood pressure, the effects of stimulation begun just before the onset of hypertension, the effects of hypertension in the absence of stimulation, and the effects of stimulation after hypertension had been established for several minutes (with permission, from Kontos and Wei, 1982).

found to be the most selectively cerebroarterial dilating substance upon both perivascular and intravascular (Auer et al., 1983; Auer and Mokry, 1986) administration. Intravenous doses of 1 mg/kg leave mean arterial pressure unchanged while producing a significant increase in CBF (Haws et al., 1983) (Fig. 50). A similar effect has been shown with the parent substance nifedipine, although this drug showed less cerebroarterial selectivity and induced marked dilatation of veins (Fig. 51). Many other so-called vasoactive drugs have been recommended for their dilatatory effect, but

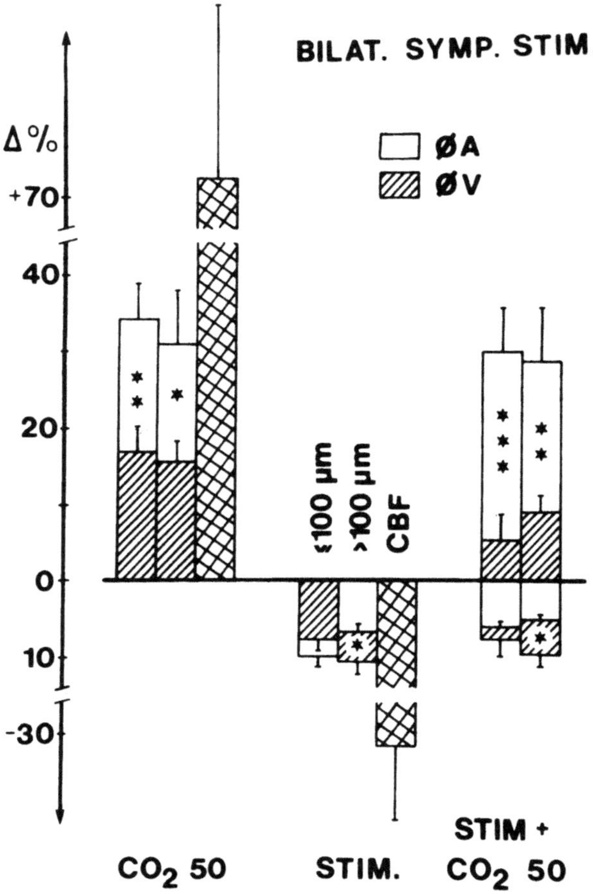

Fig. 42. Reaction of pial arteries (ØA), veins (ØV), CBF in percent of resting values during three experimental steps. (Left group of bars) Vasodilatation and rise of CBF during hypercapnia ($PaCO_2$ 50 mm Hg). (Middle group of bars) Vasoconstriction and reduction of CBF during 90 of cervical sympathetic stimulation (STIM). (Right lower group of bars) Vasoconstriction during the first 3 min of the second stimulation phase. (Right upper pair of bars) Induced hypercapnia during continuing stimulation of pial veins. $^{*}p < 0.025$, $^{**}p < 0.005$, $^{***}p < 0.005$ comparison between arteries and veins (with permission, from Auer and Ishiyama, 1986).

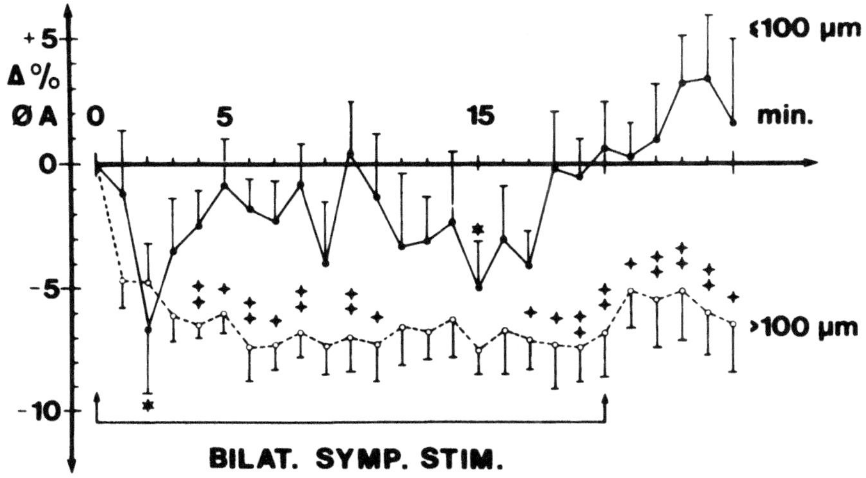

Fig. 43A.

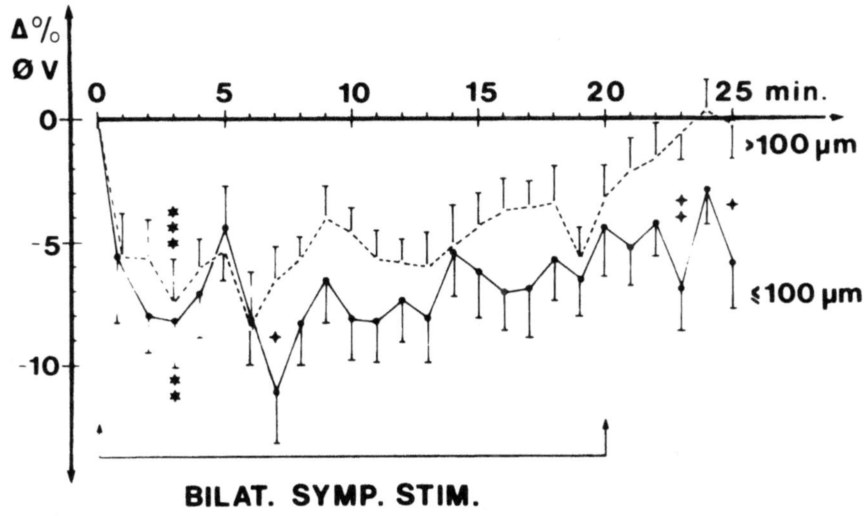

Fig. 43B.

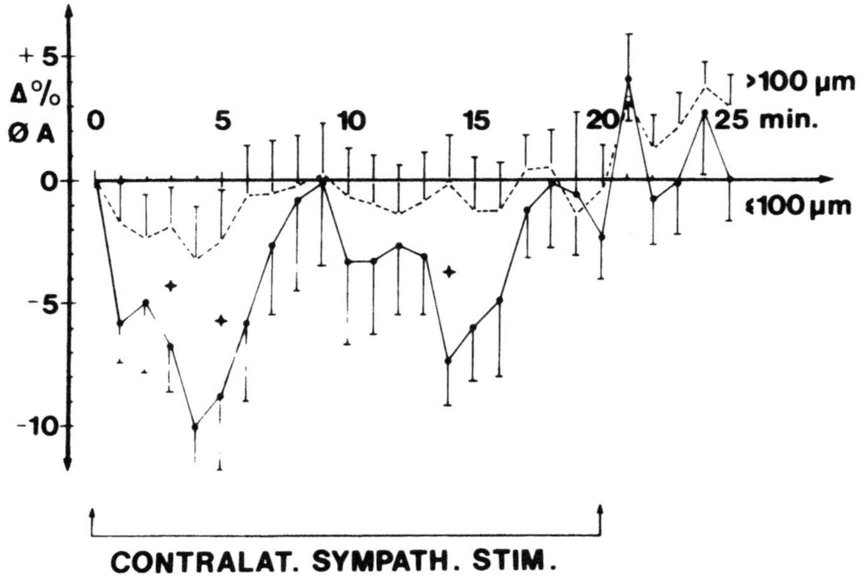

Fig. 43C.

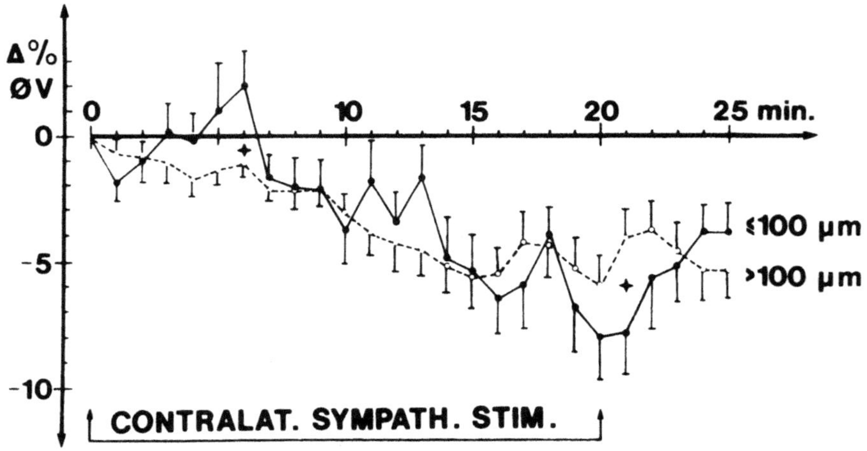

Fig. 43A–D. Behavior of pial arteries (A), (C) and veins (B), (D) during 20 min of bilateral [(A), (B)] and contralateral [(C), (D)] sympathetic stimulation. $^{*}p < 0.025$ comparison to resting; $^{+}p < 0.025$; $^{++}p < 0.005$ comparison between vessels up to 100 μm resting size (≤ 100 μm) and those larger than 100 μm (> 100 μm) (with permission, from Auer and Ishiyama, 1986).

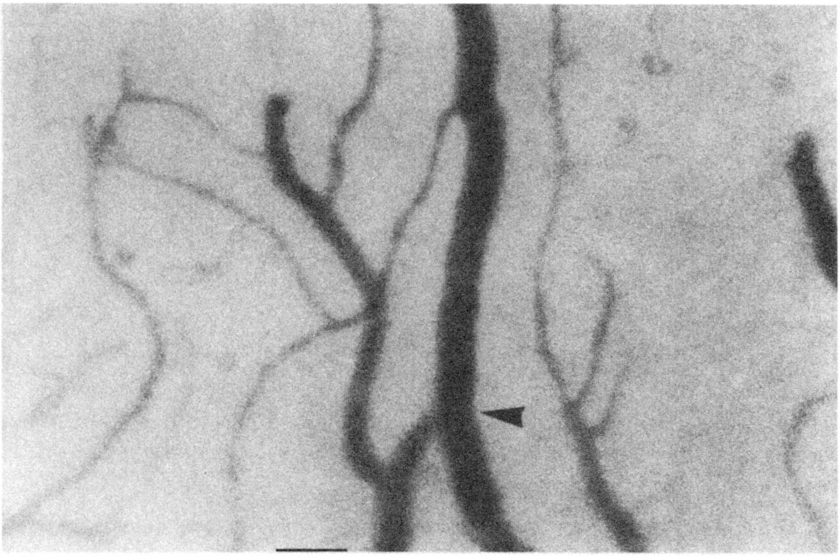

Fig. 44A. Normal situation of pial vessels at MABP of 120 mm Hg. Arrow indicates arteriole with branches. Calibration bar, 183 μm.

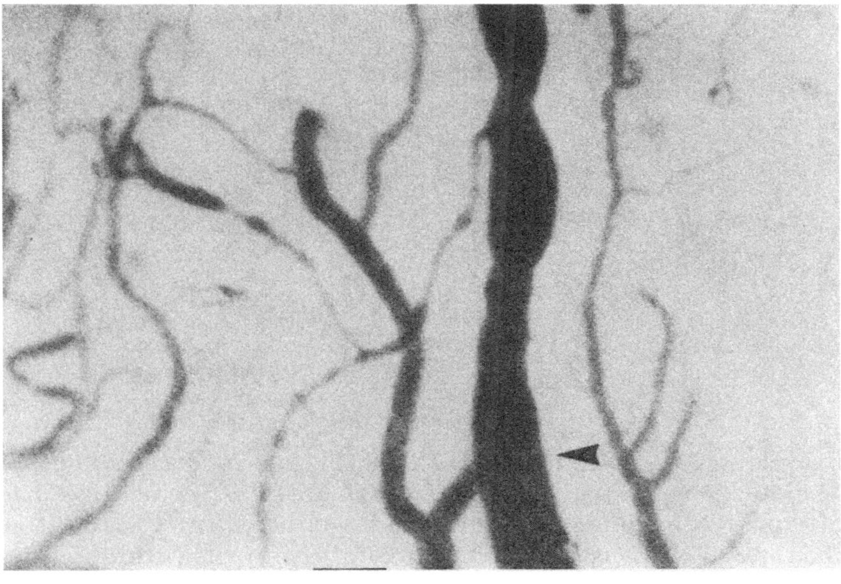

Fig. 44B. Multiple sausage-like dilatations are still present after 12 min at MABP of 120 mm Hg. Calibration bar, 183 μm (with permission, from Auer, 1978a).

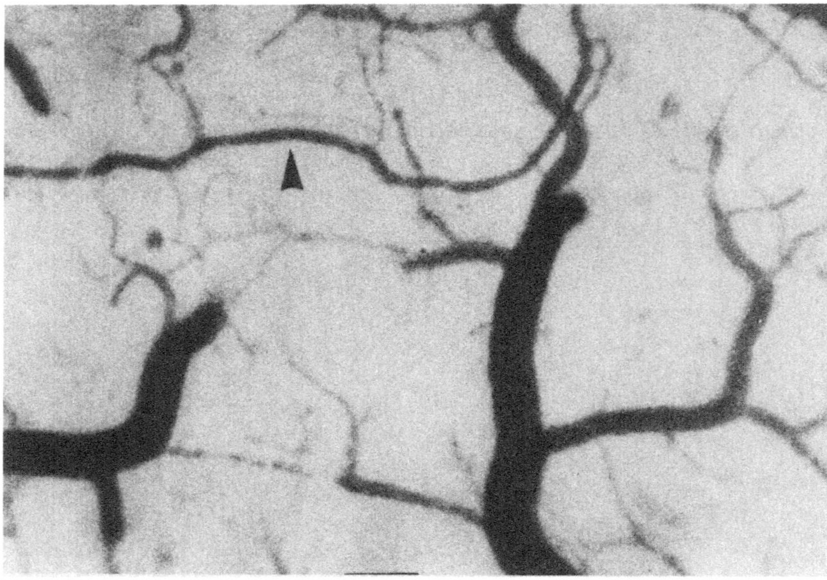

Fig. 45A. Normal situation of pial vessels before hypertension. MABP, 110 mm Hg. Arrow indicates arteriole with branches. Calibration bar, 183 μm.

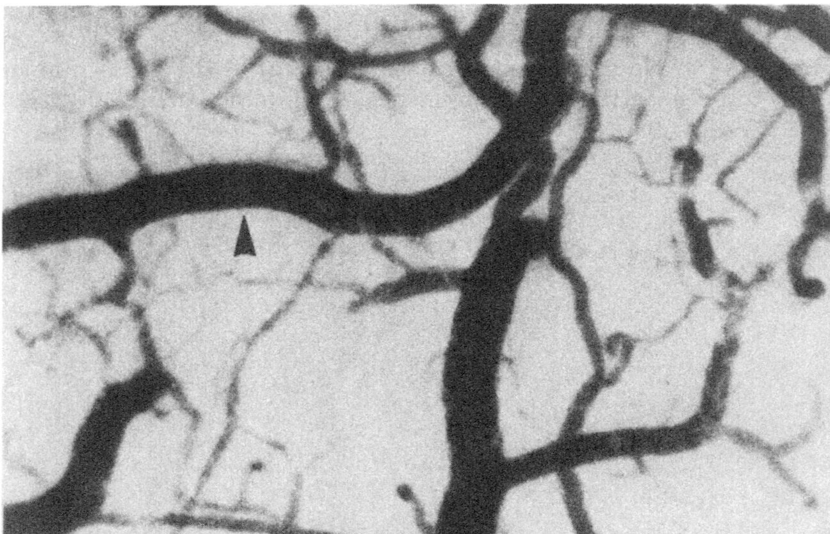

Fig. 45B. MABP increase to 230 mm Hg was produced within 24 s. The situation shown in the picture is at an MABP of 160 mm Hg 4.5 min after onset of hypertension. Venous dilatation is evident, yet less marked than arteriolar distension. Calibration bar, 183 μm (with permission, from Auer, 1978a).

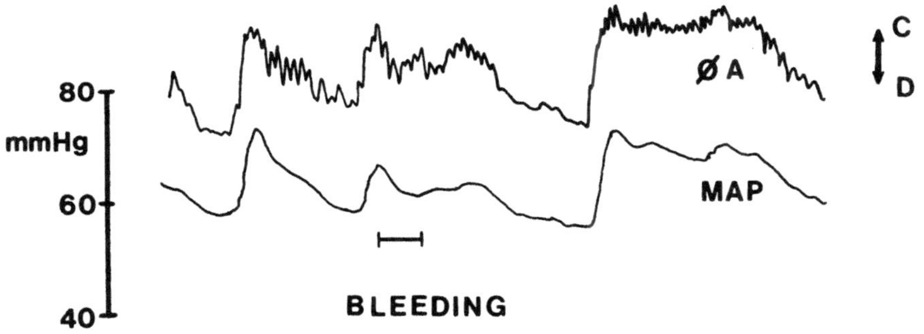

Fig. 46. 4–7-min rhythm in a 76-μm arteriole, the amplitude of which behaves clearly reciprocally to MAP. Photometric recording: C, constriction; D, dilatation; Calibration bar, 1 min (with permission, from Auer and Gallhofer, 1981).

most of them failed to show a net increase of CBF (Heiss and Podreka, 1978). Cranial window studies with administration of a compound in comparison to the vehicle for the compound can be very useful in describing the net effect of a drug by showing that the vehicle has no influence on the cerebral vessels (Auer and Mokry, 1986) (Fig. 52), or by showing an unexpected dominating effect of the vehicle (Auer et al., 1985b) (Fig. 53).

5. Conclusions and Summary

Pial vascular reactions as detected with the aid of the cranial window technique reflect cerebral hemodynamics in the vast majority of circumstances. Recent cerebral blood flow measurements give support to the hypothesis (based on the first systematic cranial window studies) forwarded by Forbes and Fog. If meticulously performed in the required manner, measurements of the physiologic and pathophysiologic behavior of cerebral vessels can be obtained from pial vessels. Cranial window studies are therefore still relevant in the era of noninvasive regional CBF measurements. Rapid cerebrovascular responses are still better seen by observation of pial vessels than by continuous measurement of CBF. Only cranial window studies were able to reveal the unexpected potential of cerebral veins to constrict actively and regulate their blood volume.

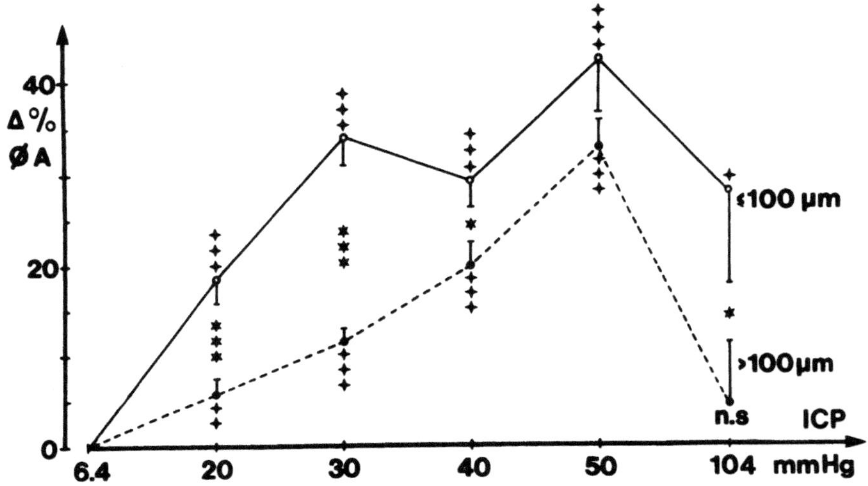

Fig. 47A.

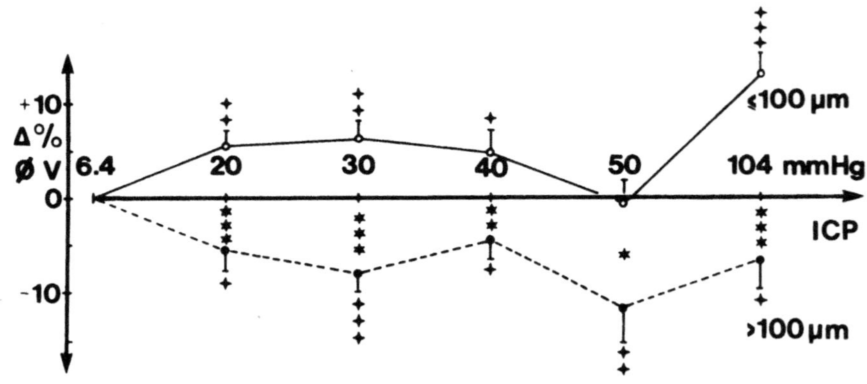

Fig. 47A and B. Changes in pial (A) arterial (ØA) and (B) venous (ØV) calibers (in percent of resting, %, mean ±SEM) during elevation of ICP by cisternal infusion of mock CSF. Dotted lines, vessels with resting calibers larger than 100 μm (>100 μm); solid lines, vessels up to 100 μm (≤100 μm). Comparison with resting: $^{+++}p < 0.0005$; $^{++}p < 0.005$; $^{+}p < 0.025$. Comparison between vessel groups: $^{***}p < 0.0005$; $^{**}p < 0.005$; $^{*}p < 0.025$ (with permission, from Auer et al., 1987).

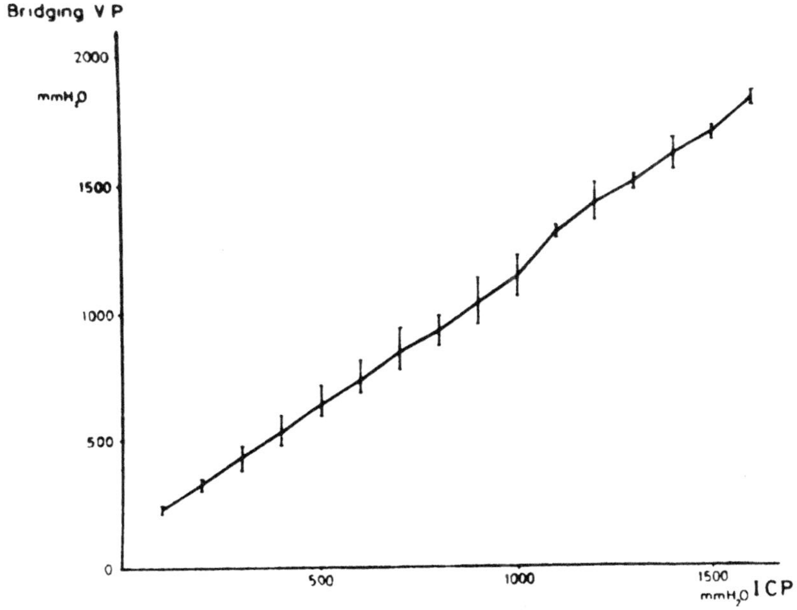

Fig. 48. Graph plotting the average pressures in the bridging veins compared to the intracranial pressure in six dogs (10 measurements). Vertical bars indicate upper and lower values at each level of ICP. Pressures in the bridging vein increase in almost linear relationship with the ICP as the ICP is elevated (with permission, from Nakagawa et al., 1974).

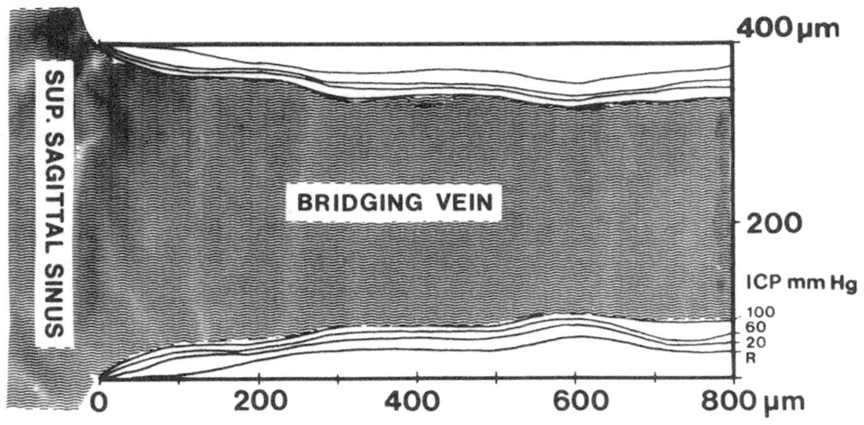

Fig. 49. Mean diameter of bridging veins near the entrance into the superior sagittal sinus in a series of rats during normal ICP (R) and various steps of ICP elevated by cisternal CSF infusion. Note that there is no significant narrowing ("cuffing").

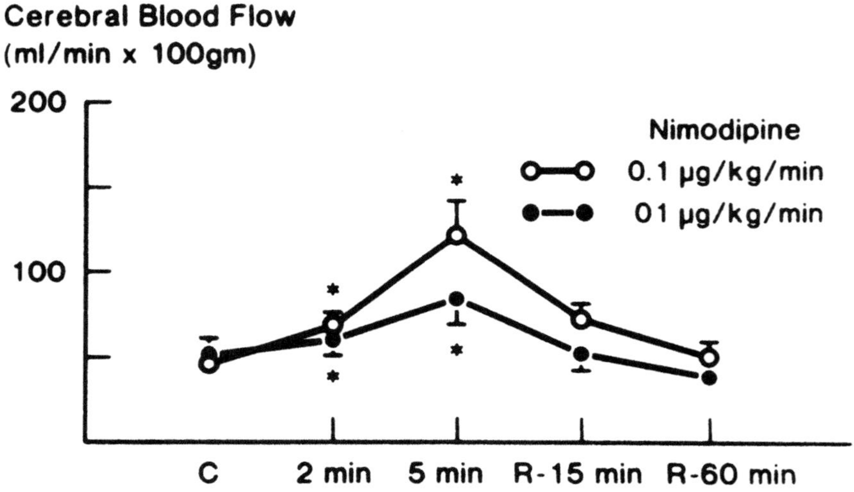

Fig. 50. Effect of intracarotid nimodipine on cerebral blood flow in rabbits. Values are means ±SE of five anesthetized rabbits for each dose. Cerebral blood flow was increased significantly from control with both high- and low-dose infusions. *$p < 05$ vs. control. C, control, R, recovery (with permission, from Haws et al., 1983).

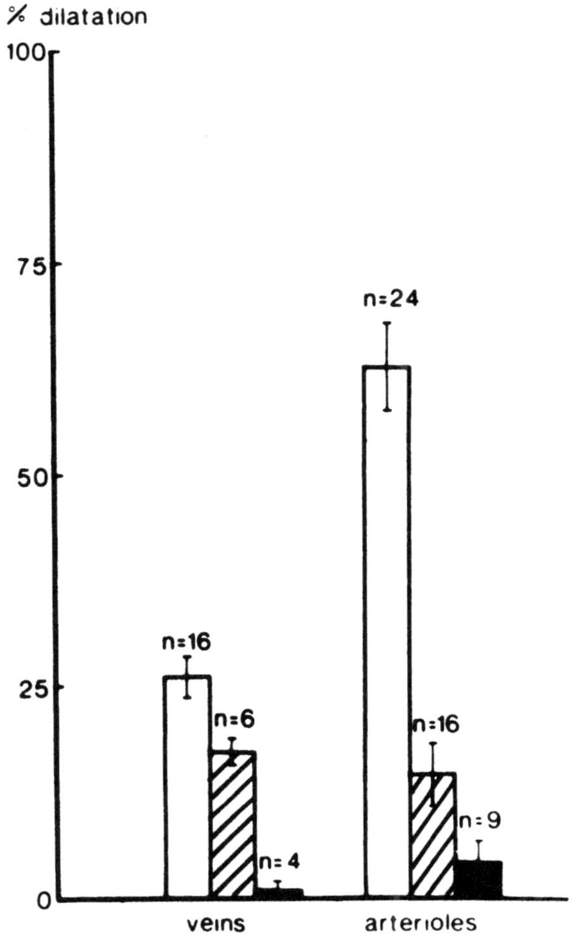

Fig. 51. Comparison of venular and arteriolar dilatatory responses to nifedipine (20 μM/mL, open columns), corresponding vehicle (hatched columns), and mock CSF (solid columns), respectively. Bars indicate mean values ±SEM (with permission, from Brandt et al., 1983).

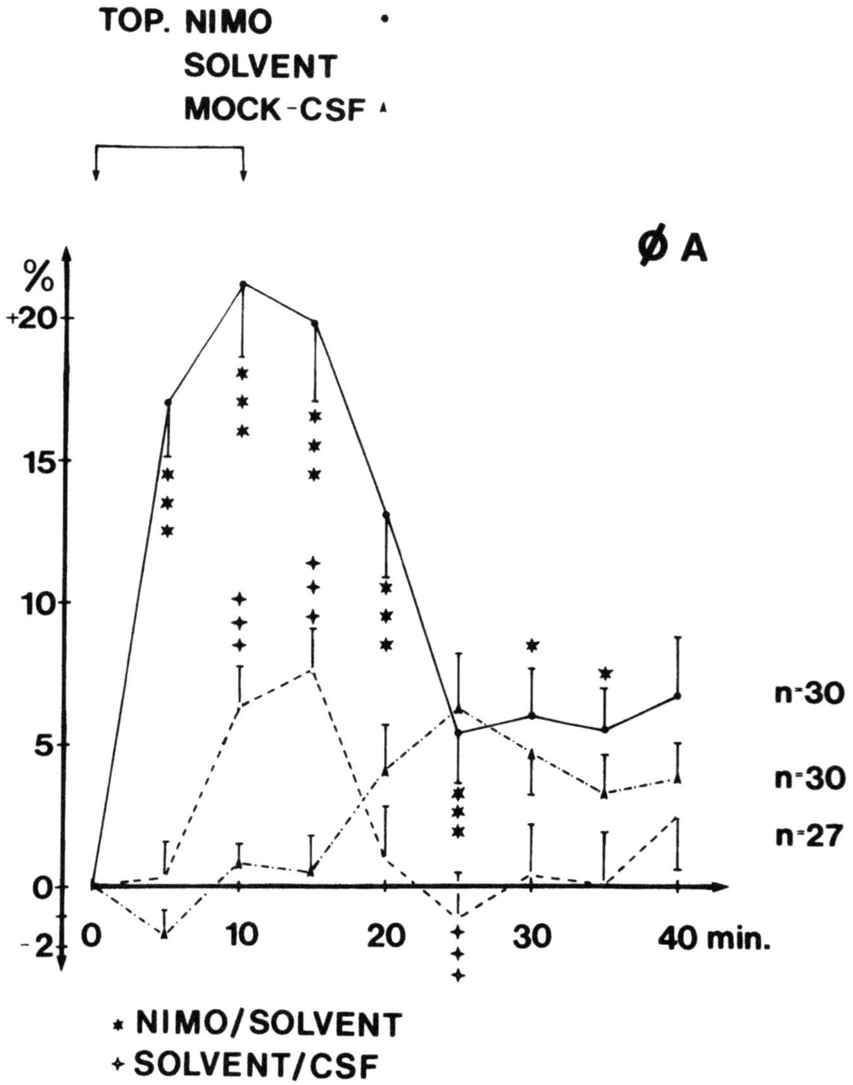

Fig. 52. Reaction of pial arteries ØA during and after superfusion with: ●, nimodipine + solvent + mock CSF; ○, solvent alone; or ▲, mock CSF alone; n = number of vessel segments studied; ***$p < 0.001$, *$p < 0.05$; for comparison between effects of nimodipine and solvent (*); for comparison between effects of solvent and mock CSF (☆) (with permission, Auer and Mokry, 1986).

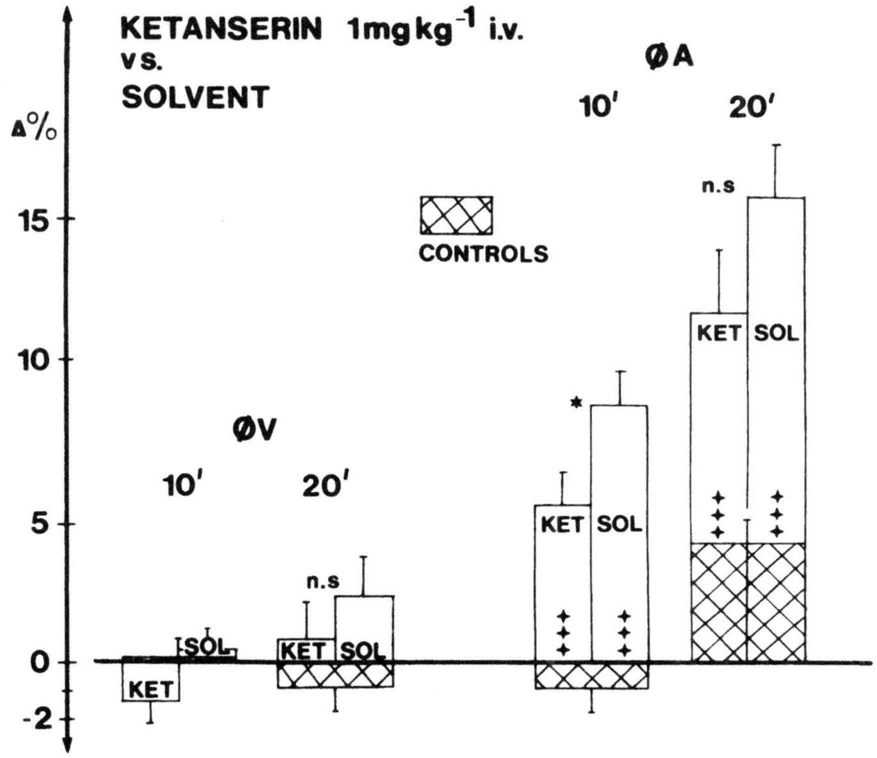

Fig. 53. Reaction of pial arteries (ØA) and veins (ØV) as percent of resting diameters (abscissa) 10 and 20 min after intravenous administration of Ketanserin (KET) or its solvent (SOL). $^{*}p < 0.025$, $^{+++}p < 0.0005$ (with permission, from Auer et al., 1985b).

References

Adler H. (1974) Untersuchungen zur Pathogenese des zerebralen Vasospasmus. *Neurochirurgia* **17,** 202–208.

Asano M., Yoshida K., and Tatai K. (1964) Blood flow rate in the microcirculation as measured by photoelectric microscopy. *Bull. Inst. Publ. Health* **13,** 201–204.

Assmann R. and Henrich H. (1978) A video-angiometer for simultaneous and continuous measurement of inner and outer vessel diameters—technical report. *Pflügers Arch.* **376,** 263–266.

Auer L. M. (1977) Intravitalmikroskopische Beobachtung der pialen Gefäße im Tierexperiment. *Zentralbl. Neurochir.* **38,** 175–181.

Auer L. M. (1978a) The pathogenesis of hypertensive encephalopathy. *Acta Neurochir.* **27** (suppl.), 1–111.

Auer L. M. (1978b) Die normalen Gefäße der Pia mater im Intravitalmikroskop. *Acta Chir. Austriaca* **23** (suppl.), 1–12.

Auer L. M. (1978c) A method for continuous monitoring of pial vessel diameter changes and its value for dynamic studies of the regulation of cerebral circulation. A preliminary report. *Pflügers Arch.* **373,** 195–198.

Auer L. M. (1978d) Pial arterial reactions to hypo- and hypercapnia, a dynamic experimental study in cats. *Eur. Neurol.* **17,** 351–362.

Auer L. M. (1984) Regulation Zerebraler Arterien und Venen. *Proc. Tagung des Tübinger Arbeitskreises fur Gefäßerkrankungen, Haigerloch.*

Auer L. M. (1986) Sympathetic control of pial vessels under in vivo conditions, in *Neural Regulation of Brain Circulation* (Owman C. and Hardebo J. E., eds.) Elsevier Science, Amsterdam.

Auer L. M. and Gallhofer B. (1981) Rhythmic activity of cat pial vessels in vivo. *Eur. Neurol.* **20,** 448–468.

Auer L. M. and Haydn F. (1985) Quantitative Measurement of CBF in Single Pial Vessels—Comparison of Techniques, in *Cerebral Blood Flow and Metabolism Measurement* (Hartmann A. and Hoyer S., eds.) Springer, Berlin, Heidelberg.

Auer L. M. and Haydn F. (1979) Multichannel videoangiometry for continuous measurement of pial microvessels. *Acta Neurol. Scand.* **60** (suppl. 72), 208–209.

Auer L. M. and Heppner F. (1978) Intravitalmikroskopische Untersuchungen an pialen Gefäßen. *Leitz Mitt. Wiss. U. Techn.* **7,** 97–102.

Auer L. M. and Ishiyama N. (1986) Pial vascular behavior during bilateral and contralateral cervical sympathetic stimulation. *J. Cereb. Blood Flow Metab.* **6,** 298–304.

Auer L. M., Ishiyama N., and Pucher R. (1987a) Cerebrovascular response to intracranial hypertension. *Acta Neurochir.* **84,** 124–128.

Auer L. M., Ishiyama N., Hodde K., Kleinert R., and Pucher R. (1987b) Effect of intracranial pressure on bridging veins in rats. *J. Neurosurg.* **67,** 263–268.

Auer L. M. and Johansson B. B. (1983) Extent and Time-Course of Pial Venous and Arterial Constriction to Cervical Sympathetic Stimulation in Cats, in *The Cerebral Veins. An Experimental and Clinical Update* (Auer L. M. and Loew F., eds.) Springer, Wien, New York.

Auer L. M. and Johansson B. B. (1980) Pial venous constriction during cervical sympathetic stimulation in the cat. *Acta Physiol. Scand.* **110,** 203–205.

Auer L. M. and MacKenzie E. T. (1984) Physiology of the Cerebral Venous System, in *The Cerebral Venous System and Its Disorders* (Kapp J. P. and Schmidek H. M., eds.) Grune and Stratton, New York.

Auer L. M. and Mokry M. (1986) Effect of topical nimodipine versus its ethanol-containing vehicle on cat pial vessels. *Stroke* **17,** 225–228.

Auer L. M., Johansson B. B., and MacKenzie E. T. (1980) Cerebral venous pressure during actively induced hypertension and hypercapnia in cats. *Stroke* **11,** 180–183.

Auer L. M., Johansson B. B., and Lund ST. (1981) Reaction of pial arteries and veins to sympathetic stimulation in the cat. *Stroke* **12,** 528–531.

Auer L. M., Oberbauer R. W., and Schalk H. V. (1983) Human pial vascular reactions to intravenous nimodipine-infusion during EC-IC bypass surgery. *Stroke* **14,** 210–213.

Auer L. M., Johansson B. B., Sayama I., and Haydn F. (1985a) Vasospasm in Hypertensive Rats Induced by Dihydralazine (Nepresol R), in *Cerebral Vascular Spasm* (Voth D. and Glees P., eds.) Walter de Gruyter, Berlin, New York.

Auer L. M., Leber K., and Sayama I. (1985b) Effect of serotonin and its antagonist ketanserin on pial vessels. *J. Cereb. Blood Flow Metab.* **5,** 517–522.

Baez S. (1966) Recording of microvascular dimensions with an image-splitter television microscope. *J. Appl. Physiol.* **21,** 299–301.

Barer R. (1960) A new micrometer microscope. *Nature* **188,** 398–399.

Beickert P. (1953) Das Verhalten der Piaarterien nach einseitiger Stellatumblockade und auf einseitigen vegetativen Reiz im Tierversuch. *Dtsch. Z. Nervenheilk.* **170,** 285–295.

Birch J., Branemark P. I., and Nilsson K. (1968a) The vascularization of a free full thickness skin graft. III. An infrared thermographic study. *Scand. J. Plast. Reconstr. Surg.* **3,** 18–22.

Birch J., Branemark P. I., Nilsson K., and Lundskog J. (1968b) Vascular reactions in an experimental burn studied with infrared thermography and microangiography. *Scand. J. Plast. Reconstr. Surg.* **2,** 97–103.

Brandt L., Ljunggren B., Andersson K. E., Edvinsson L., MacKenzie E., Tamura A., and Teasdale G. (1983) Effects of topical application of a calcium antagonist (nifedipine) on feline cortical pial microvasculature under normal conditions and in focal ischemia. *J. Cereb. Blood Flow Metab.* **3,** 44–50.

Brånemark P. I. (1965a) Capillary form and function. The microcirculation of granulation tissue. *Bibl. Anat.* **7,** 9–28.

Brånemark P. I. (1965b) Experimental biomicroscopy. *Bibl. Anat.* **5,** 51–55.

Brånemark P. I. (1962) Vitalmikroskopie. Eine Methode fur gleichzeitiges mikroskopisches Studium von Struktur and Funktion. *Leitz Mitt. Wiss.U.Techn.* **3,** 73–85.

Brånemark P. I. (1966) Intravital microscopy: Its present status and its potentialities. *Med. Biol. Illus.* **16,** 100–121.

Brånemark P. I. and Jonsson I. (1963) Determination of the velocity of corpuscules in blood capillaries. A flying spot device. *Biorheology* **1,** 143–146.

Busija D. W., Heistad D. D., and Marcus M. L. (1981) Continuous measurement of cerebral blood flow in anesthetized dogs and cats. *Am. J. Physiol.* **241,** H228–H234.

Busija D. W., Marcus M. L., and Heistad D. D. (1982) Pial artery diameter and blood flow velocity during sympathetic stimulation in cats. *J. Cereb. Blood Flow Metab.* **2,** 363–367.

Chorobski J. and Penfield W. (1932) Cerebral vasodilator nerves and their pathway from the medulla oblongata. *Arch. Neurol. Psychiatry* **28,** 1257–1289.

Clark E. R. and Wentsler N. E. (1938) Pial circulation studied by long continued direct inspection. *Res. Publ. Assoc. Res. Nerv. Ment. Dis.* **18,** 218–228.

Collmann H., Wüllenweber R., Sprung C., and Diusberg R. (1978) Early Changes of the Spinal Cord Blood Flow Regulation in the Surrounding Area of an Experimental Injury, in *Cerebrospinal Microcirculation* (Cervos-Navarro J., ed.). Raven, New York.

Donders F. C. (1849) De bewegingen der hersenen en de veranderingen der vaatvulling de pia mater, ook bij gesloten onnitzetberen schedel regtstreeks onderocht. *Onderzoek. Ged. Inh. Physiol. Lab. Utrecht Hoogeoch* **2,** 97–128.

Dora E. and Kovach A. G. B. (1981) Metabolic and vascular volume oscillations in the cat brain cortex. *Acta Physiol. Acad. Sci. Hung.* **57,** 261–275.

Dyson J. (1960) Precise measurement by image splitting. *J. Opt. Soc. Am.* **50,** 754–757.

Edvinsson L. and MacKenzie E. T. (1976) Amine mechanisms in the cerebral circulation. *Pharmacol. Rev.* **28,** 275–348.

Edvinsson L., Nielsen K. C., Owman Ch., and West K. A. (1974) Adrenergic innervation of the mammalian choroid plexus. *Am. J. Anat.* **139,** 299–308.

Eke A. (1982) Reflectometric mapping of microregional blood flow and blood volume in the brain cortex. *J. Cereb. Blood Flow Metab.* **2,** 41–53.

Eke A. (1983) Integrated Microvessel Diameter and Microregional Blood Content as Determined by Cerebrocortical Video Reflectomy, in *The Cerebral Veins. An Experimental and Clinical Update* (Auer L. M. and Loew F., eds.) Springer, Wien New York.

Finesinger J. E. (1932) Cerebral circulation. Effect of caffeine on cerebral vessels. *Arch. Neurol. Psychiat.* **28,** 1290–1395.

Finesinger J. E. (1933) Cerebral circulation. Action on the pial arteries of the convulsants caffeine, absinth, camphor and picrotoxin. *Arch. Neurol. Psychiat.* **30**, 980–1002.

Florey H. (1925) Microscopical observations on the circulation of the blood in the cerebral cortex. *Brain* **48**, 43–64.

Fog M. (1937) Cerebral circulation. The reaction of the pial arteries to a fall in blood pressure. *Arch. Neurol. Psychiatry* **37**, 351–364.

Fog M. (1939a) Cerebral circulation. I. Reaction of pial arteries to epinephrine by direct application and by intravenous injection. *Arch. Neurol. Psychiatry* **41**, 109–118.

Fog M. (1939b) Cerebral circulation. II. Reaction of pial arteries to increase in blood pressure. *Arch. Neurol. Psychiatry* **41**, 260–268.

Forbes H. S. (1937a) Cerebral circulation. XLIV. Vasodilatation in the pia following stimulation of the vagus, aortic and carotid sinus nerves. *Arch. Neurol. Psychiatry* **37**, 334–350.

Forbes H. S. (1937b) Cerebral circulation. XLV. Vasodilatation in the pia following stimulation of the geniculate ganglion. *Arch. Neurol. Psychiatry* **37**, 776–781.

Forbes H. S. (1938) Vasomotor control of cerebral vessels. *Brain* **61**, 221–233.

Forbes H. S. (1954) Study of blood vessels on cortex of living mammalian brain—description of technique. *Anat. Rec.* **120**, 309–315.

Forbes H. S. and Wolff H. G. (1928) Cerebral circulation. III. The vasomotor control of cerebral vessels. *Arch. Neurol. Psychiatry* **19**, 1057–1086.

Forbes H. S., Finley K. H., and Nason G. I. (1933) Cerebral circulation. XXIV. A. Action of epinephrine on pial vessels. B. Action of pituitary and pitressin on pial vessels. C. Vasomotor response in the pia and the skin. *Arch. Neurol. Psychiatry* **30**, 957–979.

Forbes H. S., Sohler T., and Lothrop G. N. (1941) The pial circulation of normal, monanesthetized animals. *J. Pharmacol. Exp. Ther.* **71**, 325–330.

Gotoh F., Muramatsu F., Fukuuchi Y., Okayasu H., Tanaka T., Suzuki N., and Kobari M. (1982) Video camera method for simultaneous measurement of blood flow velocity and pial vessel diameter. *J. Cereb. Blood Flow Metab.* **2**, 421–428.

Gottschewski G. (1953) Eine neue Methode zur fluoreszenzmikroskopischen Darstellung der Nierengefäßfunktion im Tierversuch. *Arzneimittelforsch.* **7**, 345–346.

Grubb R. J., Jr., Raichle M. E., Phelps M. E., and Ratcheson R. A. (1975) The Effects of Increased Intracranial Pressure upon Cerebral Blood Volume, Blood Flow, and Oxygen Utilization, in *Blood Flow and Metabolism in the Brain* (Harper M., Jennett B., Miller D., and Rowan J., eds.) Churchill Livingstone, London, New York.

Haws C. W., Gourley J. K., and Heistad D. D. (1983) Effects of nimodipine on cerebral blood flow. *J. Pharmacol. Exp. Ther.* **225,** 24–28.

Hedges T. R. and Weinstein J. D. (1964) Cerebrovascular response to increased intracranial pressure. *J. Neurosurg.* **21,** 292–297.

Heine H. (1932) Der Ultropak. *Zh. Wiss. Mikroskop.* **48,** 459–472.

Heiss W. D. and Podreka I. (1978) Assessment of pharmacological effects on cerebral blood flow. *Eur. Neurol.* **17** (suppl. 1), 135–143.

Heppner F. (1959) Die Pathophysiologie des sogenannten Gefäβkopfschmerzes. *Neurochirurgia* **1,** 158–171.

Heppner F. (1961) Kapillarmikroskopische Beobachtungen an den Piagefäβen des Groβhirns. *Acta Neurochir.* **7** (suppl.) 303–310.

Heppner F. and Marx J. E. (1958) Tierexperimentelle Untersuchungen uber die Histaminwirkung auf die Endstrombahn des Groβhirns. *Acta Neurovegetat.* **19,** 33–40.

Holt J. P. (1941) The collapse factor in measurement of venous pressure. *Am. J. Physiol.* **134,** 292.

Hossmann K. A., Hossmann V., and Takagi S. (1977) Blood Flow and Blood Brain Barrier in Acute Hypertension, in *Cerebral Vascular Disease* (Meyer J. S., Lechner H., and Reivich M., eds.) Excerpta Medica, Amsterdam-Oxford.

Illig L. (1961) *Die Terminale Strombahn.* Springer, Berlin, Gottingen, Heidelberg.

Intaglietta M. (1971) Pulsatile Velocity Components in the Omental Microvasculature, in *Proceedings of the 6th European Conference on Microcirculation, Aalborg 1970* Karger, Basel, New York.

Intaglietta M. and Tompkins W. R. (1973) Microvascular measurements by video image shearing and splitting. *Microvasc. Res.* **5,** 309–312.

Johnston I. H., Rowan J. D., Harper A. M., and Jennett W. B. (1972) Raised intracranial pressure and cerebral blood flow. 1. Cisterna magna infusion in primates. *J. Neurol. Neurosurg. Psychiat.* **35,** 285–296.

Kety S. S. (1950) Circulation and metabolism of the human brain in health and disease. *Am. J. Med.* **8,** 205–217.

Kontos H. A. and Wei E. P. (1982) Effects of Sympathetic Nerves on Pial Arterioles During Hypercapnia and Hypertension, in *Cerebral Blood Flow: Effects of Nerves and Neurotransmitters* (Heistad D. D. and Marcus M. L., eds.) Elsevier North Holland, New York.

Koyama T., Horimoto M., Mishin H., and Asakura T. (1982) Measurement of blood flow velocity by means of a laser Doppler microscope. *Optik* **61,** 411–426.

Kuschinsky W. and Wahl M. (1974) Alpha-receptor stimulation by endogenous and exogenous norepinephrine and blockade by phentolamine in pial arteries of cats. *Circ. Res.* **37,** 168–174.

Kuschinsky W. and Wahl M. (1978) Local chemical and neurogenic regulation of cerebral vascular resistance. *Physiol. Rev.* **58,** 656–689.

Kuschinsky W., Wahl M., Bosse O., and Thurau K. (1972) Perivascular potassium and pH as determinants of local pial arterial diameter in cats. *Circ. Res.* **31,** 240–247.

Langfitt T. W., Weinstein J. D., and Kassell N. F. (1965a) Cerebral vasomotor paralysis produced by intracranial hypertension. *Neurology* **15,** 622–641.

Langfitt T. W., Kassell N. F., and Weinstein J. D. (1965b) Cerebral blood flow with intracranial hypertension. *Neurology* **15,** 761–773.

Langfitt T. W., Weinstein J. D., Kassell N. F., Gagliardi L. J., and Shapiro H. H. (1966) Compression of cerebral vessels by intracranial hypertension. I. Dural sinus pressures. *Acta Neurochir.* **15,** 212–222.

Lassen N. A. (1968) Brain extracellular pH: The main factor controlling cerebral blood flow. *Scand. J. Clin. Lab. Invest.* **22,** 247–251.

Lassen N. A. and Agnoli A. (1973) The upper limit of autoregulation of cerebral blood flow—on the pathogenesis of hypertensive encephalopathy. *Scand. J. Clin. Lab. Invest.* **30,** 113–116.

Levasseur J. E., Wei E. P., Raper A. J., Kontos H. A., and Patterson J. L. (1975) Detailed description of a cranial window technique for acute and chronic experiments. *Stroke* **6,** 308–317.

MacKenzie E. T., Strandgaard S., Graham D. I., Jones J. V., and Harper A. M. (1976) Effects of acutely induced hypertension in cats on pial arteriolar caliber, local cerebral blood flow, and the blood–brain barrier. *Circ. Res.* **39,** 33–41.

McCulloch J., Edvinsson L., and Watt P. (1982) Comparison of the effects of potassium and pH on the calibre of cerebral veins and arteries. *Pflügers Arch.* **393,** 95–98.

Minard D. (1954) The lucite calvarium for direct observation of the brain in monkeys. *Anat. Rec.* **120,** 317–330.

Morii S., Ko K. R., Ngai A. C., and Winn H. R. (1985) Reactivity of Rat Pial Arterioles and Venules to Adenosine and Theophylline, An Adenosine Receptor Blocker, In Situ. *Proceedings of the International Symposium on Cerebral Blood Flow and Metabolism, Lund, 1985.*

Munro P. A. G. (1966) Methods for Measuring the Velocity of Moving Particles Under the Microscope, in *Advances in Optical and Electron Microscopy* vol. 1 (Barer R. and Cosslett V. E., eds.) Academic, New York.

Nakagawa Y., Tsuru M., and Yada M. (1974) Site and mechanism for compression of the venous system during experimental intracranial hypertension. *J. Neurosurg.* **41,** 427–434.

Nielsen K. C. and Owman C. (1967) Adrenergic innervation of pial arteries related to the circle of Willis in the cat. *Brain Res.* **6,** 773–776.

Noordergraaf A. (1978) *Circulatory System Dynamics* Academic, New York.

Okayasu H., Gotoh F., Muramatsu F., Fukuuchi Y., Amano T., and Tanaka K. (1979) A method for continuous measurement of pial vessel diameter by means of a vidicon camera system. *Acta Neurol. Scand.* **60,** 72 (suppl.), 256–257.

Owman C., Edvinsson L., and Nielsen L. C. (1974) Autonomic neuroreceptor mechanisms in brain vessels. *Blood Vess.* **11,** 2–31.

Peters T. (1955) Apparatur und Technik zur Mikroskopie an lebenden Säugetierorganen in situ in gewöhnlichem Licht und Fluoreszenzlicht. *Zh. Wiss. Mikrskop.* **67,** 348–367.

Pool J. L. (1934) Cerebral circulation. XXII. Effect of stimulation of the sympathetic nerve on the pial vessels in the isolated head. *Arch. Neurol. Psychiatry* **32,** 916–923.

Ranson S. W. and Clark S. L. (1959) *The Anatomy of the Nervous System* W. B. Saunders, Philadelphia, Pennsylvania.

Raper A. J., Kontos H. A., and Patterson, Jr., J. L. (1971) Response of pial precapillary vessels to changes in arterial carbon dioxide tension. *Circ. Res.* **28,** 518–523.

Reivich M. (1964) Arterial PCO_2 and cerebral hemodynamics. *Am. J. Physiol.* **206,** 25–35.

Röckemann W. and Plesse G. J. (1973) Registration of Erythrocyte Velocity by Selecting Wavelength and Superposition of Oscillations. *Proceedings of the 7th European Conference on Microcirculation, Aalborg 1972* Karger, Basel, New York.

Rosenblum W. I. (1969) Fluorescence angiography of cerebral microcirculation. *Am. J. Pathol.* **59,** 9.

Rosenblum W. I. (1970) Effects of blood pressure and blood viscosity of fluorescein transit time in the cerebral microcirculation in the mouse. *Circ. Res.* **27,** 825–833.

Rosenblum W. I. and Zweifach B. W. (1963) Cerebral circulation in the mouse brain. *Arch. Neurol.* **9,** 414–423.

Sato S., Miyahara Y., Dohmoto Y., Kawase T., and Toya S. (1983) Cerebral Microcirculation in Experimental Sagittal Sinus Occlusion in Dogs, in *The Cerebral Veins. An Experimental and Clinical Update* (Auer L. M. and Loew F., eds.) Springer, Wien, New York.

Schmidt H. W. (1956) Experimentelle Untersuchungen zur Frage der Gefäßspasmen in den extrakraniellen Carotis-Anteilen bei der cerebralen Angiographie. *Dtsch. Z. Nervenheilk.* **174,** 173–176.

Sercombe R., Lacombe P., Aubineau P., Mamo H., Pinard E., Reynier-Rebuffel A. M., and Seylaz J. (1979) Is there an active mechanism limiting the influence of the sympathetic system on the cerebral vascular bed? Evidence for vasomotor escape from sympathetic stimulation in the rabbit. *Brain Res.* **164,** 81–102.

Shelden C. H., Pudenz R. H., Restonska J. S., and Craig W. M. (1944) The lucite calvarium—a method for direct observation of the brain. I. The surgical and lucite processing techniques. *J. Neurosurg.* **1,** 67–69.

Shulman K. and Verdier G. R. (1967) Cerebral vascular resistance changes in response to cerebrospinal fluid pressure. *Am. J. Physiol.* **213,** 1084–1088.

Skinjoe E. and Strandgaard S. (1973) Pathogenesis of hypertensive encephalopathy. *Lancet* **1,** 461–462.

Sokoloff L. (1959) The action of drugs on the cerebral circulation. *Pharmacol. Rev.* **11,** 1–58.

Starling E. H. (1918) The law of the heart. The Linacre lecture, delivered at St. John's College, Cambridge, 1915. Longmans, Green, London.

Strandgaard S., Olesen J., Skinhoj E., and Lassen N. A. (1973) Autoregulation of brain circulation in severe arterial hypertension. *Br. Med. J.* **1,** 507–510.

Strandgaard S., MacKenzie E. T., Sengupta D., Rowan J. O., Lassen N. A., and Harper A. M. (1974) Upper limit of autoregulation of cerebral blood flow in the baboon. *Circ. Res.* **24,** 435–440.

Suzuki T. (1975) Response of pial microcirculation to changes in arterial carbon dioxide tension and to reduction in systemic blood pressure. *J. Japan. Coll. Angiol.* **15,** 171–178.

Suzuki T., Tominaga S., Strandgaard S., and Nakamura T. (1975) Fluorescein Cineangiography of the Pial Microcirculation in the Rat in Acute Angiotensin-Induced Hypertension, in *Blood Flow and Metabolism in the Brain* (Harper A. M., Jennett W. B., Miller J. D., and Rowan J. O., eds.) Churchill Livingstone, Edinburgh.

Ulrich K., Auer L. M., and Kuschinsky W. (1982) Cat pial venoconstriction to topical microapplication of norepinephrine. *J. Cereb. Blood Flow Metab.* **2,** 109–111.

Villaret M. (1939) Les repercussions vasulaires tardives de l'embolie cérébrale (en pathologie expérimentale). *Press Med.* **47,** 267–271.

Wahl M., Deetjen P., Thurau K., Ingvar D. H., and Lassen N. A. (1970) Micropuncture evaluation of the importance of perivascular pH for the arteriolar diameter on the brain surface. *Pflügers Arch.* **316,** 152–163.

Wahl M., Kuschinsky W., Bosse O., Olesen J., Lassen N. A., Ingvar D. H., Michaelis J., and Thurau K. (1972) Effect of *l*-norepinephrine on the diameter of pial arterioles and arteries in the cat. *Circ. Res.* **31,** 248–256.

Wahl M., Unterberg A., and Baethmann A. (1983) Effects of Bradykinin on Permeability and Diameter of Cerebral Vessels, in *The Cerebral Veins. An Experimental and Clinical Update* (Auer L. M. and Loew F., eds.) Springer, Wien, New York.

Wayland H. and Johnson P. C. (1967) Erythrocyte velocity measurement in microvessels by a two-slit photometric method. *J. Appl. Physiol.* **22,** 333–337.

Wiederhelm C. A. (1966) Photospectrum analysis technic. *Meth. Med. Res.* **11,** 212–216.

Winn H. R., Rubio G. R., Curnish R. R., and Berne R. M. (1981) Changes in regional cerebral blood flow (rCBF) caused by increases in CSF concentrations of adenosine and 2-chloradenosine (CHL-ADO). *J. Cereb. Blood Flow Metab.* **1,** (suppl. 1), S401–S402.

Wolff H. G. (1929) The cerebral circulation. XIa. The action of acetylcholine. *Arch. Neurol. Psychiatry* **22,** 686–699.

Wolff H. G. and Lennox W. G. (1930) The cerebral circulation. XII. The effect on pial vessels of variations in the oxygen and carbon dioxyde content of the blood. *Arch. Neurol. Psychiatry* **23,** 1097–1120.

Wright D. R. (1928) Experimental observations on increased intracranial pressure. *Aust. N. Z. J. Surg.* **7,** 215–235.

Yada K., Nakagawa Y., and Tsuru M. (1973) Circulatory disturbance of the venous system during experimental intracranial hypertension. *J. Neurosurg.* **39,** 723–729.

Zweifach B. W. (1957) General principles governing the behaviour of the microcirculation. *Am. J. Med.* **23,** 684–696.

Zweifach B. W. (1954) Direct observation of the mesenteric circulation in experimental animals. *Anat. Rec.* **120,** 277–291.

Autoradiography and Cerebral Function

Joel H. Greenberg

1. Introduction

The development of the Kety-Schmidt technique for the measurement of cerebral blood flow and metabolism (Kety and Schmidt, 1948) led to a significantly increased interest in the cerebral circulation. The ability to make quantitative measurements, especially in humans, provided investigators with the tools for collecting data not obtainable by previous means. As a tremendous amount of data began to be collected, it soon became obvious that the brain, unlike many other organs, is extremely heterogeneous in nature. Physiologic stimuli that, intuitively, would be expected to produce an alteration in cerebral blood flow and/or metabolism (and have subsequently been shown to produce regional changes) do not appear to cause any measureable changes in global cerebral blood flow or metabolism. This includes sleep (Mangold et al., 1955) and intellectual effort (Sokoloff et al., 1955). It is this heterogeneous nature of brain function that has prompted the development of techniques yielding regional or local information on cerebral hemodynamics and metabolism. Principal among these are autoradiographic methods for visualizing local function. Autoradiography, the production of an image from the radioactive decay occurring in a specimen, is only a tool for determining tissue radionuclide levels, and as such is not specific to the measurement of any particular parameter.

This chapter will discuss a variety of techniques for the measurement of certain functional parameters in the brain. The one thing that all of these techniques have in common is the use of autoradiography for the measurement of local tissue radionuclide levels. As will become evident, the tissue radionuclide levels are necessary to calculate the local value of the particular parameter of interest (e.g., blood flow, metabolism, and so on). One of the advantages of autoradiography is that not only does the autoradiogram yield quantitative information about radionuclide levels, it also produces an anatomically intact image of the tissue, allowing the investigator to easily visualize functional distribution.

2. Local Cerebral Blood Flow

One of the earliest uses of autoradiography in the study of the cerebral circulation was that of the measurement of local cerebral blood flow (lCBF) developed by a group at the National Institutes of Health (Landau et al., 1955; Freygang and Sokoloff, 1958). This method, as originally proposed, utilizes an inert gas as the tracer, and as such is based on the principles that determine the exchange of inert gases between blood and tissue (Kety, 1951).

2.1. Theory

There are a number of assumptions inherent in the technique for the measurement of lCBF using radiolabeled inert substances (Kety, 1951; Freygang and Sokoloff, 1958). If these assumptions and conditions are satisfied, then rather straightforward equations can be written to define the behavior of these substances in the tissue. The first assumption, and the one that, as we shall discuss below, appears to be the most difficult to satisfy is that the substance that we choose to use as a blood flow tracer be freely diffusible in the tissue. In other words, the substance must be flow-limited and not diffusion-limited such that the venous blood is in diffusional equilibrium with the tissue region from which it drains. Second, the tracer must be inert. Third, arteriovenous shunting must be either nonexistent or negligible. Finally, the tissue region must be homogeneous with respect to flow, and this flow must be constant during the period of measurement. This final requirement can usually be satisfied by limiting the measurement to very small tissue regions.

If the above assumptions are satisfied, then the conservation of mass equation can be used to describe the behavior of the tracer in the tissue (Kety, 1951). Define the symbols as follows (Fig. 1):

F = blood flow through tissue element (mL/min)
f = blood flow per unit mass (mL/g/min)
Q_i = quantity of tracer in tissue element i (μCi)
M_i = mass of tissue element i (g)
C_i = concentration of tracer in tissue element i (μCi/g)

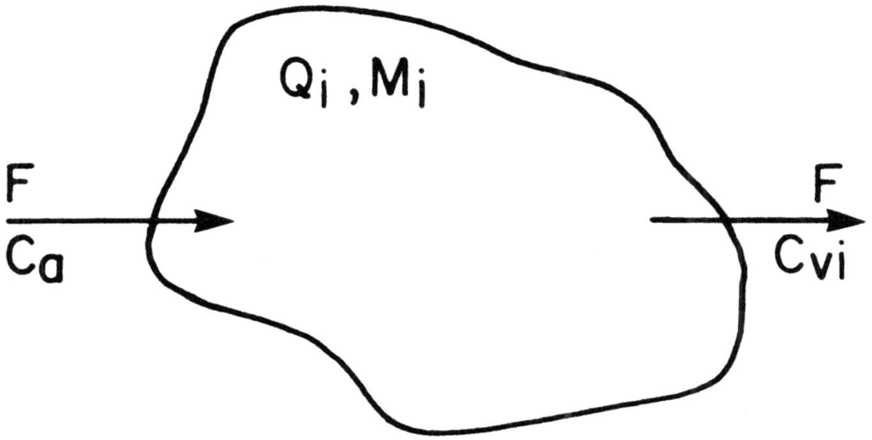

Fig. 1. Schematic representation of a tissue element, i, with mass M_i. The blood flow into the element is F, the concentration of the tracer in the blood entering into the tissue element is C_a, the concentration of the tracer in the blood leaving the element is C_{vi}, and the quantity of tracer in the tissue element is Q_i.

C_a = concentration of tracer in arterial blood (μCi/mL)
C_{vi} = concentration of tracer in venous blood of element i (μCi/mL)
λ = brain–blood equilibrium partition coefficient for tracer (mL/g)
k = f/λ (1 per min)
m = constant representing nonequilibrium between tissue element and blood perfusing tissue element

By the Fick principle (Kety, 1951), the rate of change of the tracer in the tissue element is simply the difference between the rate the tracer is delivered to the tissue from the arterial blood and the rate at which the tracer is removed from the tissue by the venous blood.

$$dQ_i/dt = FC_a - FC_{vi} \tag{1}$$

Substituting the expression $Q_i = C_iM_i$ into Eq. (1) yields:

$$d(C_iM_i)/dt = F(C_a - C_{vi}) \tag{2}$$

Since the mass of the tissue element is invariant in time, it can be removed from the differentiation, and both sides can be divided by M:

$$dC_i/dt = f(C_a - C_{vi}) \quad (3)$$

If the tissue element is homogeneous in blood flow (a reasonable assumption for very small tissue regions), then the following expression relating the arteriovenous and the arterio-tissue tracer concentration can be derived (Kety, 1951):

$$(C_a - C_{vi}) = m(C_a - C_i/\lambda) \quad (4)$$

where m is a constant between 0 and 1 that represents the degree of diffusional equilibrium of the tracer. This factor combines arteriovenous shunting, capillary permeability limitations, and diffusion limitations, and for a tracer with negligible limitations, m approaches unity. If the venous blood draining the tissue element is in equilibrium with the tissue ($m = 1$), then Eq. (4) reduces to:

$$C_v = C_i/\lambda \quad (5)$$

Substituting Eq. (5) into Eq. (3) and rearranging,

$$dC_i/dt + FC_i/\lambda = FC_a(t) \quad (6)$$

This differential equation can be solved using integration by parts. Substituting $k = f/\lambda$:

$$C_i(T) = \lambda k \int_0^T C_a(t)\, e^{-k(T-t)}dt \quad (7)$$

Equation (7) is the operational equation of the technique for the measurement of local cerebral blood flow autoradiographically. It states that local cerebral blood flow (f) can be determined with the following information: (1) the time course of the concentration of the tracer in the arterial blood from the start of the administration of the tracer ($t = 0$) until time T [$C_a(t)$], (2) the tracer concentration in the tissue element at a given time T [$C_i(T)$], and (3) the partition coefficient of the tracer in the tissue (λ). Its validity is entirely dependent on the validity of the assumptions discussed above, upon which it is based. Equation (7) is usually utilized to measure lCBF according to a fixed protocol (to be described below), although the equation is rather general, so that a variety of experimental designs can be used. It should be noted that unless $C_a(t)$ is monotonic, Eq. (7) will have multiple solutions. For example, if the tissue concentration, C_i, is measured some time following a bolus injection of the tracer, then a specific tissue level can correspond to two blood flow values—a low flow that would represent tissue whose tracer level is still rising and a high flow that would represent tissue whose tracer level was falling. The effect of a variety of infusion schedules on the accuracy of the technique will be discussed in section 2.6.1.

2.2. Experimental Technique

The experimental procedure for the determination of lCBF in animals autoradiographically is actually quite straightforward. The radiolabeled tracer is generally infused into a systemic vein over a relatively short time period—usually 1 min. During this time arterial blood samples must be obtained in order to define the time course of the arterial blood tracer concentration as required in the operational equation [Eq. (7)]. The actual blood concentration that is required is that in the cerebral arteries, but since the loss of the tracer through the walls of the large arteries is negligible, obtaining blood samples from any large artery is sufficient to define cerebral arterial concentrations. At the end of the study ($t = T$), the perfusion to the brain is rapidly stopped, and the brain is quickly removed from the skull and frozen in Freon XII or isopentane (cooled with dry ice) at –40 to –60°C.

The brain must then be sectioned in a cryostat (–18 to –22°C), and the sections placed in apposition to film in order to produce images of the radiolabeled tracer distribution. In order to quantitate the concentration of the tracer on the autoradiogram, standards are placed on the film along with the sections. These standards, available commercially (for ^{14}C and ^{3}H), are usually made from methylmethacrylate that is imbedded with the radionuclide of interest (^{14}C, ^{3}H, and so on) and are calibrated to be tissue-equivalent (Reivich et al., 1969). The thickness of the sections is usually determined by the ease of sectioning in combination with the energy of the radionuclide. The standards, however, must be calibrated to the section thickness, which for ^{14}C autoradiography is usually 20 μm. For short-lived radionuclides, standards are generally not commercially available, and the investigator must make standards separately for each study (or for a closely timed group of studies) from tissue homogenates. The standard film for most routine ^{14}C autoradiography is Kodak single-coated, blue-sensitive Medical X-ray Film SB5 (Eastman Kodak Co.). This film has been chosen as a fast film that produces acceptable images with a 5–7-d exposure (with an injected dose of 100 μCi/kg body weight). Slightly sharper images can be obtained with finer grain films (mammographic films: Kodak MR1, Dupont Lo-Dose; panchromatic: Kodak Pan-X, Kodak Plus-X), but at significantly longer (2–4 times) exposure times.

The radioactivity in the blood samples is quantitated by counting the samples in an appropriate (beta or gamma) counter cross-calibrated to the standards used for autoradiography. Knowledge of the arterial time course of tracer activity and the partition coefficient allows a flow table to be constructed [Eq. (7)] relating tissue concentrations to lCBF.

2.3. Diffusible Tracers

2.3.1. ^{131}I-Trifluoroiodomethane

The original method for the measurement of lCBF autoradiographically employed ^{131}I-trifluoroiodomethane ($CF_3{}^{131}I$) as the radiolabeled tracer (Landau et al., 1955; Freygang and Sokoloff, 1958). This substance was chosen since it was inert and known to be freely diffusible. In addition, it could be labeled with a radionuclide whose blood activity could be conveniently counted and that would produce reasonably good images on film. $CF_3{}^{131}I$ as a tracer suffered from a number of problems that limited its widespread use in the laboratory. In fact, except for the initial elegant and very carefully conducted experiments described by its developers, it has not been used since. For this reason, details of the techniques necessary to use this tracer will not be discussed.

The main drawback of the use of $CF_3{}^{131}I$ for lCBF measurement is its volatility. In order to obtain an autoradiogram from the tissue, it was necessary to cut relatively thick sections from the brain with a band saw and place these tissue slabs onto film for approximately 1 d, keeping the tissue *frozen* during the exposure. Even at the dry ice temperatures used for the exposure, there was the possibility of volatilization of the gas from the surface of the section, leading to errors in quantitation of tissue radionuclide activities. Other drawbacks of $CF_3{}^{131}I$ include: (1) its lack of commercial availability, necessitating that it be synthesized for each experiment; (2) the requirement for thick sections, making it difficult to obtain closely spaced sections (and hence images); (3) the high energy of the beta particles of ^{131}I yields images of quite low resolution; and (4) the solubility of $CF_3{}^{131}I$ in blood varies with hematocrit, so the partition coefficient will vary among animals and, in addition, is different for gray matter and white matter.

These limitations did not prevent this group from collecting data that dramatically demonstrated the heterogeneity of blood flow in the brain (Landau et al., 1955; Freygang and Sokoloff, 1958). In addition, they were able to show the regional effects on cerebral blood flow of thiopental anesthesia, hypercapnia, moderate hypoxia, and visual stimulation. These studies, in particular the visual stimulation experiments in which light was flashed into both eyes of the cat, presaged the results of regional glucose utilization studies two decades later.

2.3.2. Antipyrine

The technical difficulties of trifluoroiodomethane autoradiography described above prompted Reivich and coworkers to search for another tracer for measuring local cerebral blood flow (Reivich et al., 1969). The main goal was to find a tracer that was nonvolatile at room temperature. This would permit the film exposure to be undertaken with tissue that was not frozen and, in addition, would minimize any loss of tracer from the surface of the section. The compound that they chose was antipyrine labeled with ^{14}C. With this compound the brain could be cut into thin sections (usually 20 μm), so that blood flow in all of the brain tissue could be measured. These sections could be brought to room temperature and exposed to the film in readily available X-ray cassettes. Besides its nonvolatility, ^{14}C-antipyrine had a number of advantages over $CF_3{}^{131}I$. The lower energy of ^{14}C (mean energy 55 keV) produces autoradiograms with a significantly better resolution than ^{131}I, which has a beta emission at 608 keV along with a gamma emission. The long half-life of ^{14}C meant that permanent standards could be made (and reused), and the longer exposure times made possible by the long half-life permitted less radionuclide to be used.

These properties of ^{14}C-antipyrine simplified the procedure appreciably, making this technique a practical tool for the measurement of lCBF in the laboratory. The nonvolatility of the tracer permitted, as discussed above, thin sections to be cut and then exposed to film at room temperature. It also simplified measurement of the time course of the arterial concentration of the tracer since the blood samples did not have to be handled anaerobically. Reivich et al. (1969) recognized the difficulty in obtaining accurate arterial blood samples in many experimental preparations. Blood sampled from a catheter of any appreciable length will tend to be smeared and time-delayed in the catheter. They therefore developed a smearing correction for blood sampling and outlined a clever technique for determining this correction in any experimental sampling arrangement. Correcting for smearing, however, has generally been supplanted with the use of a very short (2–3 cm) arterial sampling catheter whose dead space is adequately cleared prior to each blood sample. This is usually done by allowing a few drops of blood to drop from the catheter prior to taking a blood sample.

2.3.3. Iodoantipyrine

As will be discussed in the next section, shortly after the development of ^{14}C-antipyrine as an agent for the measurement of lCBF, evidence began to accumulate that antipyrine was more diffusion-limited than originally thought. Consequently, interest turned to other agents that would have fewer diffusion limitations. Sakurada et al. (1978) showed that ^{14}C-iodoantipyrine was a more appropriate blood flow agent in that it could diffuse more freely through the blood–brain barrier, probably because of its appreciably higher lipid solubility. They compared the partition coefficients of iodoantipyrine and antipyrine between mixtures of two organic solvents (chloroform, which is least polar, and *n*-butanol, which is highly polar) and water. Over the range of from 100% chloroform to 100% *n*-butanol, the partition coefficient of iodoantipyrine was approximately five times that of antipyrine. They found that the partition coefficients for both iodoantipyrine and antipyrine were uniform in brain tissue and were 0.80 and 0.91, respectively.

Using the partition coefficients for these compounds measured in brain tissue, they calculated cerebral blood flow from two parallel series of studies in the rat using ^{14}C-antipyrine and ^{14}C-iodoantipyrine, respectively. The results of these studies are shown in Fig. 2A. It can be seen that the blood flows obtained with ^{14}C-iodoantipyrine are significantly higher than the blood flows in identical structures measured with ^{14}C-antipyrine. The only structure in which the blood flow using ^{14}C-antipyrine was not significantly lower was the corpus callosum, a white matter structure whose blood flow is quite low and hence a region in which a diffusion limitation is less of a problem. lCBF measured with ^{14}C-iodoantipyrine in the cat was compared to data from the literature under similar conditions in which lCBF was measured using ^{131}I-trifluoroiodomethane (Landau et al., 1955) or ^{14}C-antipyrine (Eckman et al., 1975). The lCBF values obtained with ^{14}C-iodoantipyrine were very close to the flow values measured with ^{131}I-trifluoroiodomethane. Only five out of 26 structures exhibited different blood flows, and in four of these the lCBF values were higher with ^{14}C-iodoantipyrine than with the ^{131}I-labeled gas tracer

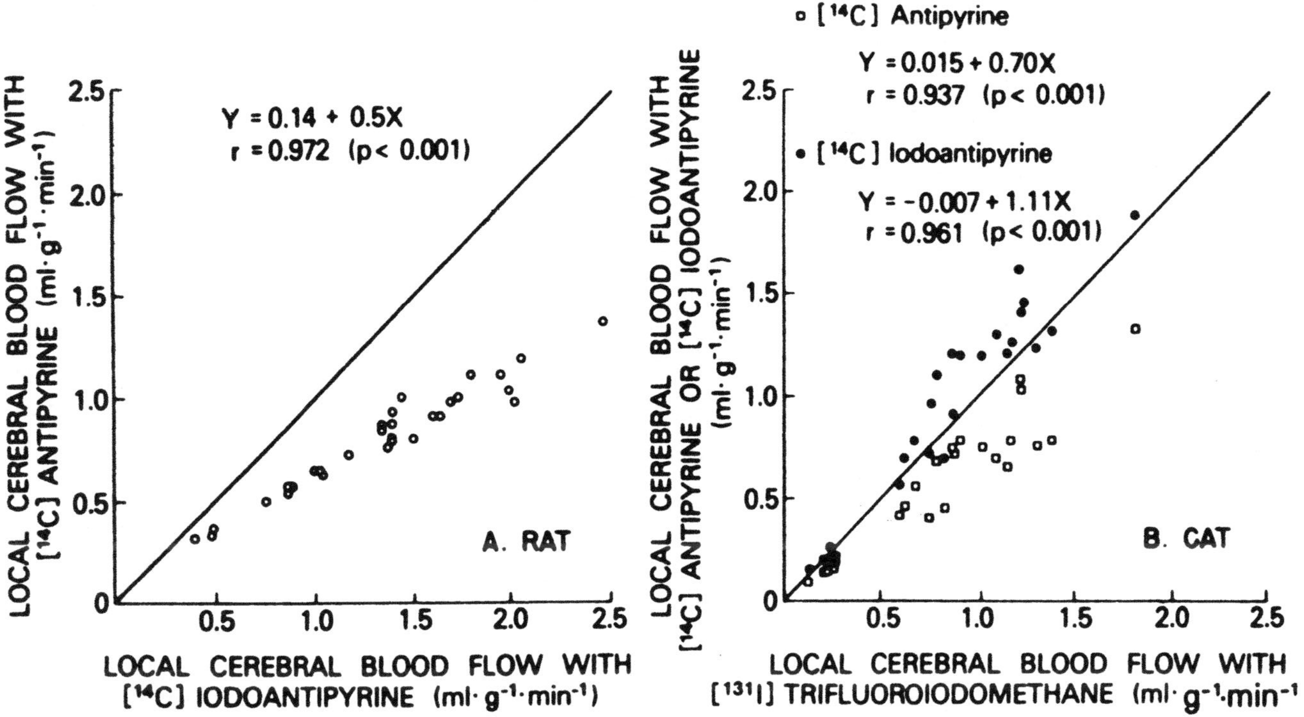

Fig. 2. Comparison between local cerebral blood flow measured with antipyrine, iodoantipyrine, and trifluoroiodomethane in the rat (A) and the cat (B). At blood flows above 0.5 mL/g/min, antipyrine underestimates blood flow, whereas iodoantipyrine agrees well with blood flow measured with trifluoroiodomethane, which is assumed not to be diffusion-limited. The solid lines are lines of identity (from Sakurada et al., 1978).

(Fig. 2B). The flows obtained with ^{14}C-antipyrine in the cat were, as observed with the rat data, lower than with the other two tracers.

The better diffusion capacity of iodoantipyrine over antipyrine has led to its present widespread use as the tracer of choice in the autoradiographic measurement of lCBF.

2.3.4. Unsymmetrical Alkyl Aryl Thiourea Compounds

A class of compounds, the unsymmetrical alkyl aryl thioureas, has been proposed by Goldman et al. (1980) as better tracers than ^{14}C-iodoantipyrine for measuring cerebral blood flow. These compounds were investigated because of a suspicion that iodoantipyrine was diffusion-limited at high flow rates. They synthesized a number of unsymmetrical alkyl aryl thiourea compounds and characterized these by measuring the octanol-water partition coefficients (to determine lipid solubility), the single-pass brain extraction, and the brain uptake index. The description of these techniques for the measurement of blood–brain transport, along with Goldman's data, is contained in section 2.4. Goldman et al. (1980) concluded, based on these studies, that one of these unsymmetrical alkyl aryl thiourea compounds, 1-butyl-3-phenylthiourea (BPTU), was a significantly better substance for quantitating local cerebral blood flow at high flow rates (as in the rat) because of its greater diffusibility than iodoantipyrine in cerebral tissue. In spite of these data, BPTU has not received widespread use as a blood flow tracer, and iodoantipyrine continues to be considered the tracer of choice.

2.4. Compound Characterization and Technique Limitations

As alluded to in section 2.3.2, diffusible tracers for the measurement of lCBF, and in particular ^{14}C-antipyrine, have been subjected to intense scrutiny. The major thrust of these investigations concerns diffusion limitations of the tracer. Eklof et al. (1974) compared regional cerebral blood flow in the rat measured with four different tracers—^{14}C-antipyrine, ^{14}C-ethanol, ^{3}H-water, and ^{133}Xe. This study grew out of an observation that the ^{14}C-antipyrine technique underestimated CBF in comparison to data obtained with the Kety-Schmidt technique (Kety and Schmidt, 1948). Cortical blood flow was measured with each of these tracers using the standard techniques discussed above, with the following modifications: (1) tissue activity was measured by counting the tissue directly instead of using autoradiography and (2) the blood and tissue samples were handled appropriately when using volatile tracers. These studies revealed that at relatively low flow rates (<1.0 mL/g/min) the blood flow obtained with these tracers only slightly underestimated CBF, compared to that obtained with the Kety-Schmidt technique; the exception was ethanol, which slightly overestimated blood flow (Fig. 3). When CBF was increased with the administration of CO_2, all of the tracers significantly underestimated blood flow. The amount by which these tracers underestimated CBF appeared to increase as the time of infusion increased. For example, the underestimation of ^{14}C-antipyrine was 56% at 30 s, 67% at 60 s, and 72% at 120 s, all compared to the blood flow obtained with the Kety-Schmidt technique.

Eklof et al. (1974) discuss three possible explanations for the consistent underestimation of blood flow using these tracers, and in particular using ^{14}C-antipyrine as an autoradiographic tracer. The first possibility concerns the inhomogeneity of blood flow. The derivation of the equations in section 2.1 [and in particular Eq. (4)] assumes that blood flow is homogeneous throughout the tissue element. Since the protocol of Eklof et al. (1974) involved tissue sampling of volumes that were not homogeneous with respect to flow, an error will result. Eklof et al. (1974) modeled this inhomogeneity and showed that this error increases as the time of infusion increases, but even at long infusion times (120 s), it is only 7% and thus cannot explain the significant underestimation of cerebral blood flow obtained with these tracers in comparison to the Kety-Schmidt technique.

The second explanation for the significant underestimation of CBF concerns diffusion limitations. As blood flow increases, the time for diffusion into the tissue decreases along with the capillary transit time. The assumption about instantaneous equilibrium becomes more and more tenuous, and Eq. (5) no longer holds. This diffusion limitation can potentially explain all of the underestimation of blood flow observed in this study. The higher blood flow measured with ethanol as compared to water is consistent with the observation that water itself is diffusion-limited (Eichling et al., 1974). Using an external detection system, Eichling et al. (1974) measured the extraction of ^{15}O-labeled water following a bolus internal carotid injection in the Rhesus monkey. They found that, even at flows of 0.3 mL/g/min, water does not freely equilibrate with the exchangeable water in the brain and, at a flow of 0.5 mL/g/min, only 90% of the water is extracted. These studies have

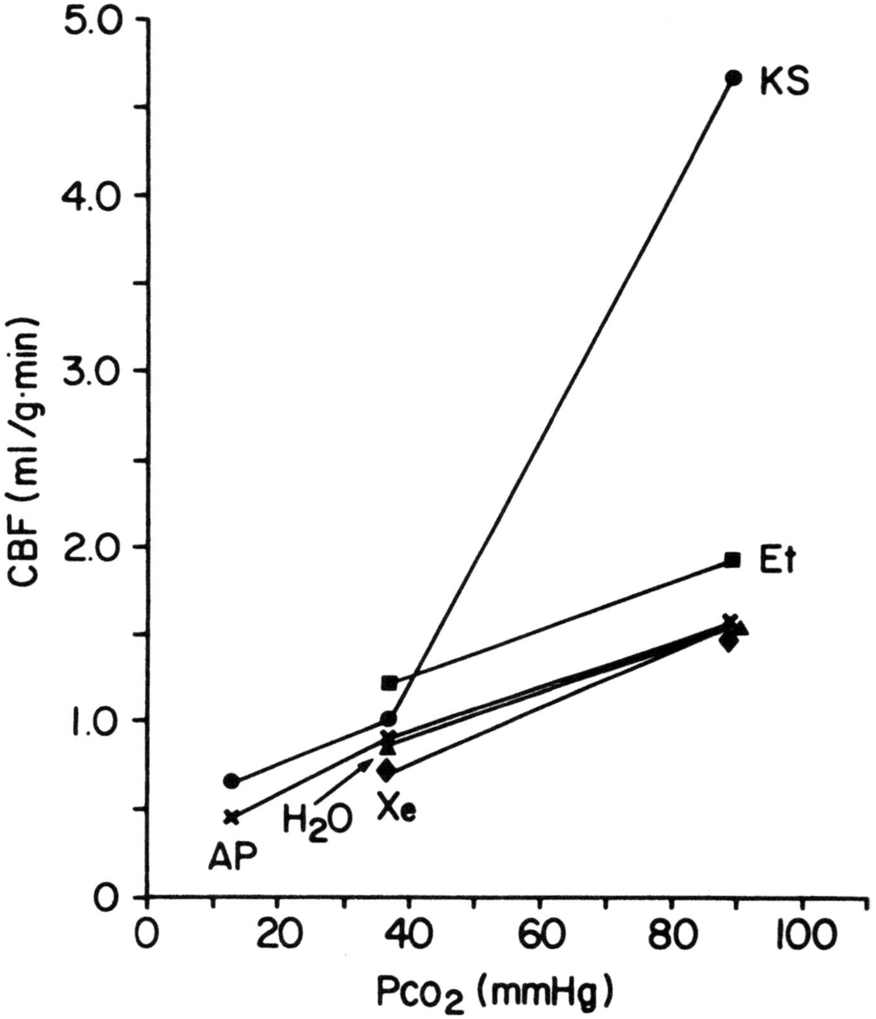

Fig. 3. Cortical blood flow CBF measured using a variety of tracers: ^{14}C-antipyrine (AP), ^{133}Xenon (Xe), ^{14}C-ethanol (Et), and H_2O-^3H-water (H_2O) compared to cerebral blood flow measuring using the Kety-Schmidt (KS) technique. The cortical blood flow was calculated using Eq. (7) (data from Eklof et al., 1974).

been extended to the extraction of ^{11}C-labeled alcohols (methanol, ethanol, isopropanol), which, at a blood flow of 0.5 mL/g/min, are 93, 97, and 99% extracted, respectively (Raichle et al., 1976). Thus, even the alcohols, which owe their increased brain permeability to higher lipid solubilities, are diffusion-limited.

The discrepancy noted by Eklof et al. (1974) between blood flow as determined by the Kety-Schmidt technique and that using the diffusible tracers with tissue sampling cannot, however, be explained totally by diffusion limitations. Xenon was used as the agent for measuring blood flow with the Kety-Schmidt technique as well as with the tissue-sampling technique. At high blood flows, the two techniques yielded appreciably different flow values. Since there is no reason to suspect that xenon is diffusion-limited (McHenry et al., 1969), there may have been a systematic error in the use of one or both of these techniques in this study. Use of the Kety-Schmidt technique at high flow rates is subject to appreciable error, which may have led to an overestimation of blood flow. It is plausible, however, that in the process of tissue sampling some ^{133}Xe may have been lost from the tissue; this would be more consistent with lower flows using tissue sampling than with the clearance measurements utilized in the Kety-Schmidt technique.

A more theoretical analysis of the diffusion limitation of the diffusible-indicator method was undertaken by Eckman et al. (1975). They reexamined the equations for diffusible tracers (*see* Eqs. 1–7), but did not assume that m, which represents the degree of diffusional equilibrium of the tracer, was equal to unity. In essence, k in Eq. (7) becomes mf/λ, instead of f/λ. Kety (1951) has derived an expression that, with some approximations, relates the factor m to the capillary permeability coefficient (P), capillary surface area (S), and blood flow:

$$m = 1 - e^{-PS/f} \tag{8}$$

For m to be close to 1, as was assumed for the derivation of Eq. (7), it is necessary for PS/f to be quite large, i.e., greater than 2–3. Since PS is a constant for the tracer (e.g., antipyrine) in the tissue, this will only occur for flows 2–3 times less than PS. The permeability–surface area product has been measured for antipyrine in brain and found to be approximately 0.02 $cm^3/g \cdot s$ (1.2 $cm^3/g \cdot min$) (Crone, 1963). Thus m is only within 5–10% of unity for local blood flows less than 0.4–0.6 mL/g·min. Eckman set up a computer simulation to evaluate the effects of permeability limitation on the measurement of local blood flow. Figure 4 shows some of the results of this computer simulation. The factor $\lambda k/f$ (where f is the true blood flow rate) equals m, so that this figure indicates how m deviates from unity as a function of local blood flow. If PS is assumed to be 0.02 $cm^3/g \cdot s$ as has been measured for antipyrine (Crone, 1965), then m is significantly less than 1 for flows above 0.2 mL/g·min. This simulation also allowed these investigators to determine the value

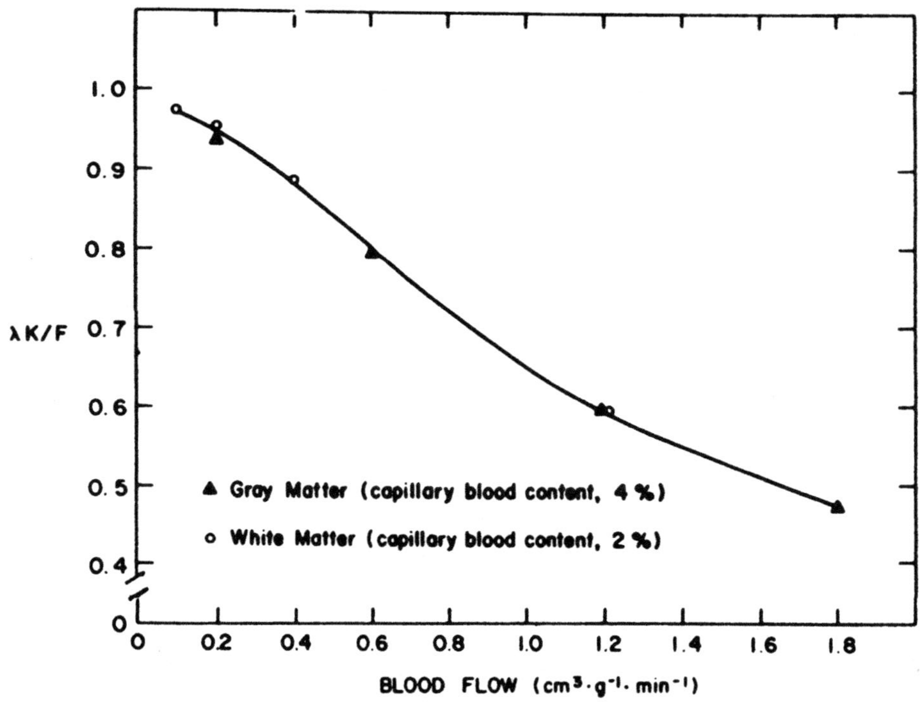

Fig. 4. Relationship between m (= $\lambda K/F$), which represents the extent to which diffusion equilibrium of the tracer is achieved as it passes through the tissue, and blood flow. This data is the result of a computer simulation with a *PS* comparable to that for antipyrine (0.02 mL/g/s). The value of m, although not different between gray and white matter, is not constant, but decreases appreciably as flow increases (from Eckman et al., 1975).

of PS necessary to be able to quantitative blood flow at the normal rate of perfusion. In order for the measured flow to be within 5% of the true flow, it is necessary for PS to be above 0.12 $cm^3/g \cdot sec$, which is significantly above the measured value for antipyrine. A simulation of $\lambda k/f$ as a function of time shows that this factor is not equal to $1 - e^{-PS/f}$, but instead asymptotically approaches this factor, indicating that m is time-dependent. The apparent discrepancy between Kety's derivation of $m = 1 - e^{PS/f\lambda}$ (Kety, 1951) and the expression $m = \lambda k/f$ is caused by the simplifying assumptions that Kety made in his derivation. Kety, in his treatment of the exchange between capillary blood and tissue, considered the changes in capillary concentration along the length of the capillary, but not changes in capillary concentrations with time. In any case, the simulations of Eckman et al. (1975) confirm the observations of Eklof et al. (1974) that antipyrine is too diffusion-limited to be used as a quantitative local cerebral blood flow tracer.

This diffusion limitation is in agreement with studies by Gjedde et al. (1975) and Bradbury et al. (1975) who measured the brain uptake index (BUI) of antipyrine and compared it to a variety of compounds including the alcohols, lactate, and nicotine. BUI is defined as the ratio of the brain concentration of a test substance relative to that of water following an intracarotid bolus of a mixture of the two substances (Oldendorf, 1970). The animal (usually rat) is decapitated approximately 15 s following the tracer injection, and so a single pass extraction is obtained. Gjedde et al. (1975) measured a BUI of 68.4 for antipyrine and 137.5 for ethanol (i.e., higher than water). Bradbury obtained similar values—67.9 for antiypyrine and 110.0 for isopropanol. Nicotine, which is also minimally metabolized by the brain, had a BUI of 125.0, comparable to the alcohols. Nicotine, however, is not a freely diffusible substance in the central nervous system since it binds to brain tissue (Schmiterlow et al., 1967), so that the equations derived in section 2.1 are not applicable. Nicotine is, however, a potential blood flow tracer that with an appropriate model for its behavior in the tissue could be utilized for the quantitative measurement of cerebral blood flow (Ohno et al., 1979).

As part of an investigation of the unsymmetrical alkyl aryl thiourea compounds, Goldman et al, (1980) measured both single-pass extraction and BUI of these compounds and compared the values to those obtained with antipyrine, iodoantipyrine, and n-butanol. Single-pass brain extraction is determined by injecting a mixture of the radiolabeled compound and ^3H-inulin (which is not diffusible and so remains in the vascular space, therefore becoming a measure of dilution in the blood) into the carotid artery, while blood samples are taken from a catheter in the sinus confluens (cerebral venous blood) (Crone, 1963). The operational equation used for extraction is $E = 1 - C_t/C_r$, where C is a ratio of the counts in the cerebral venous blood to the counts in the injectate of the test (t) and reference (r) tracers at the peak of the extraction curve, usually taken to be 2–8 s following the carotid injection. The single pass extraction, *E*, is identical to the factor *m* defined by Kety (1951), as discussed above. They obtained values for *E* of 0.45 for antipyrine, 0.55 for iodoantipyrine, 0.74 for *n*-butanol, and 0.73 for BPTU. The respective BUI values obtained were 78.5 (antipyrine), 148.2 (iodoantipyrine), 125.5 (*n*-butanol), and 144.3 (BPTU). The BUI for antipyrine agrees reasonably well with the data of Eklof et al. (1974) and Bradbury et al. (1975). If the extraction data obtained in this investigation are correct, iodoantipyrine apparently underestimates lCBF by approximately 50%, whereas *n*-butanol and BPTU underestimate lCBF by 27%. It is difficult to reconcile this significant underestimation of blood flow using iodoantipyrine with the data of Sakurada et al. (1978) if it is assumed that trifluoroiodomethane is not diffusion-limited in the brain. In the report of Sakurada et al. (1978), blood flow as measured with ^{14}C-iodoantipyrine was compared to local blood flow measured with $CF_3{}^{131}I$, and excellent agreement was found for all regions examined at flows of as high as 1.8 mL/g/min.

2.5. Indicator Fractionation Technique

Some of the methodological limitations of the tissue equilibration technique discussed above can be obviated with a technique that is a modification of both the tissue equilibration technique and the indicator fractionation technique first discussed by Goldman and Sapirstein (1973). In the indicator fractionation technique, the indicator is assumed to be completely extracted by the tissue with negligible indicator backflux. This is the basis of the radiolabeled microsphere technique, in which microspheres are totally trapped in the capillary bed with negligible loss after trapping (Rudolph and Heyman, 1967). The indicator fractionation technique has been extended for use with diffusible tracers to measure regional cerebral blood flow (Van Uitert and Levy, 1978; Gjedde et al., 1980; Sage et al., 1981). The mass balance equation is rewritten here [*see* Eq. (3)]:

$$dC_i / dt = f(C_a - C_v) \tag{9}$$

If the experiment time is sufficiently short, then the venous outflow will be negligible, and Eq. (8) can be significantly simplified:

$$dC_i / dt = fC_a \tag{10}$$

The solution of this equation does not require knowledge of the cerebral venous tracer concentration and can be simply written:

$$f = C_i(T) / \int_0^T C_a(t)\, dt \tag{11}$$

In this expression, cerebral blood flow, f, is a linear function of tissue tracer concentration divided by the integrated history of the arterial blood tracer concentration. In practice this denominator is obtained by constantly withdrawing blood from an artery following an intravenous bolus injection of the radiolabeled tracer. It is important that the arterial blood withdrawal terminate at the time the animal is sacrificed (T). This is done by cutting the arterial sampling line as the animal is being decapitated. The integrated arterial blood sample can be expressed in terms of the rate of blood withdrawal, R, and the total activity in the withdrawn blood sample, Q_A,

$$\int_0^T C_a(t)\, dt = Q_A/R \tag{12}$$

The validity of this modified indicator-fractionation technique rests on two assumptions. First, it is necessary that there be no backflux of the tracer from the tissue during the course of the study. Otherwise the assumption that C_v is negligible does not hold. Second, it is necessary that the tracer be completely extracted from the blood during the time interval 0 to T. Van Uitert and Levy (1978) used this technique to study the regional cerebral blood flow in the conscious gerbil utilizing ^{14}C-butanol as the tracer. They found that at normocapnia, close to 96% of the butanol was extracted at 10 s (i.e., $\int (C_a - C_v)dt = \int C_a dt$). The same research group (Sage et al., 1981) found a smaller apparent extraction in the rat (=90%) that decreases precipitously as blood flow increases above 1.5 mL/g/min. A more complete discussion of the errors involved with this technique is contained in section 2.6.2. The studies of Van Uitert and Levy (1978) and Sage et al. (1981) were undertaken with butanol, making autoradiography cumbersome if not impossible. This technique can be used autoradiographically by the substitution of a less volatile tracer, such as iodoantipyrine.

2.6. Error Analysis of Blood Flow Measurement

2.6.1. Equilibration Technique

As with any technique, blood flow measurement with radiolabeled tracers is subject to errors from a variety of sources. Examination of Eq. (7) indicates that with the equilibration technique an error in blood flow can arise from errors in the measurement of tissue indicator concentration, inaccuracies in the brain–blood partition coefficient (λ), timing errors, as well as errors in the measurement of indicator concentrations in the arterial blood samples. Errors in the calculated blood flow caused by permeability limitations has been discussed above. A theoretical treatment of the first three of these factors has been undertaken by Patlak and coworkers (Patlak et al., 1984). The analysis of the errors in flow (f) caused by an incorrect assessment of the tissue tracer concentration [$C_i(T)$] was made by examining the partial derivative of f with respect to $C_i(T)$. The results of this analysis are shown in Fig. 5, where the relative error in blood flow with respect to the relative error in the determination of $C_i(T)$ is plotted for four different input arterial curves: (1) a bolus input, (2) a constant blood concentration, (3) a linearly increasing blood tracer concentration, and (4) a quadratically increasing blood concentration. As can be seen from these data, the error is largest for the bolus injection followed in order by the constant arterial tracer concentration, the linearly increasing arterial tracer concentration curve, and the quadratically increasing arterial curve. This implies that the error will be reduced by forcing the arterial curve to increase with time, the error reduction being greatest with a rapidly increasing arterial input. Note that the abcissa in Fig. 5 is Tf/λ, so that the error increases with flow as well as the length of the experiment (T). The increase in error for larger flows can be easily seen when flow is plotted as a function of tissue concentration for a typical arterial curve produced by a constant infusion of isotope (Fig. 6). This curve becomes steeper at higher flows so that small errors in the tissue tracer concentration lead to quite large errors in calculated blood flow.

The errors in calculated blood flow caused by errors in the partition coefficient (λ) were also calculated by Patlak et al. (1984). Similar to the errors caused by inaccuracies in the tissue concentration, the errors are greatest for the constant arterial curve and least for the rapidly rising curve, with the relative error in flow increasing appreciably with increases in blood flow (Fig. 7). The partition coefficient essentially expresses the partitioning of the compound

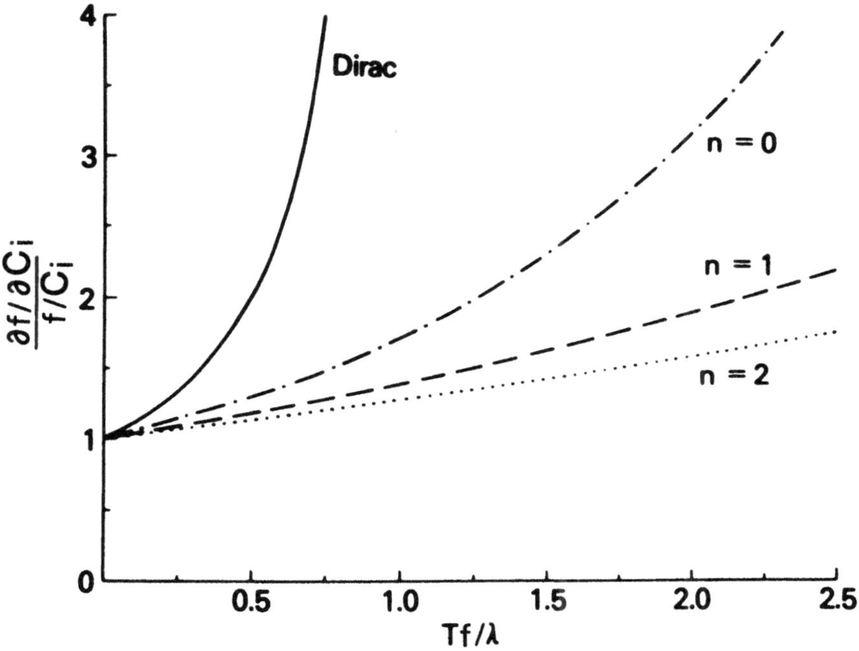

Fig. 5. The relative error in calculated blood flow (*f*) with respect to the relative error in the tissue activity (C_i) as a function of Tf/λ. As the time of death (*T*) or blood flow increases, so does the error. The various curves represent different arterial curve functions: $n = 0$ is a constant arterial curve, $n = 1$ is a linearly increasing arterial curve, whereas $n = 2$ is a quadratically increasing arterial concentration. The solid line is the result of a Dirac δ blood concentration (what would occur with an ideal bolus with no recirculation) (from Patlak et al., 1984).

between the blood and the tissue. Errors in λ will be greater in the situation in which more of the tracer is introduced to the tissue toward the beginning of the experiment (and has longer to redistribute) and least when it is introduced toward the end (such as for the parabolically rising arterial input function). Again, errors caused by inaccuracies in the partition coefficient are reduced by shortening the length of the experiment.

When Patlak et al. (1984) examined the effect of errors in the experiment length (*T*) on the errors in calculating blood flow, they found that for a constant arterial curve the error was independent of flow and was quite minimal (Fig. 8). The error is larger, however, for arterial curves that rise more rapidly. Additionally, as both flow

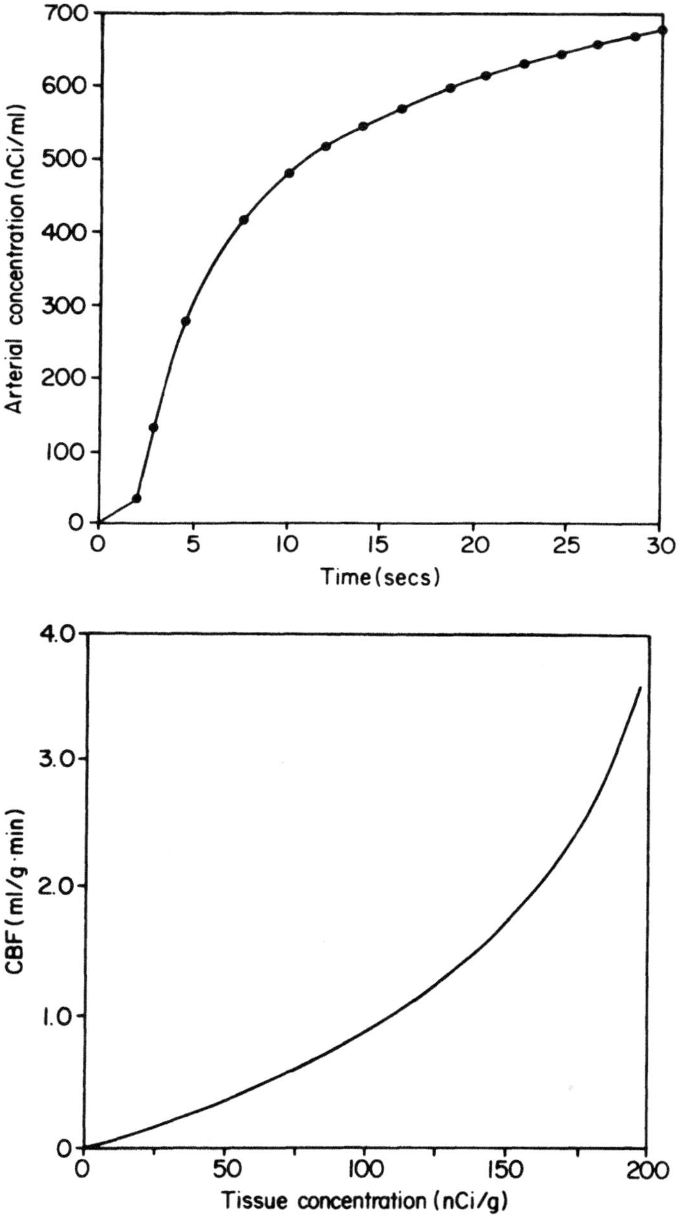

Fig. 6. A typical arterial concentration curve for a 30-s infusion of ^{14}C-iodoantipyrine (top). The relationship between tissue tracer concentration and calculated blood flow shown in panel (bottom) was obtained using this arterial time course. This curve is nonlinear, so that at higher tissue concentrations flow increases precipitously.

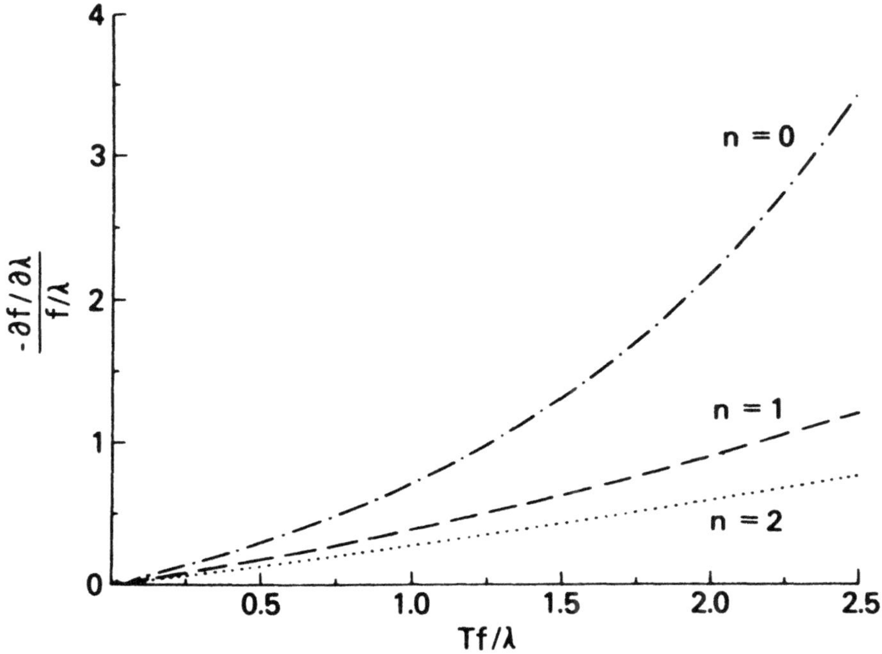

Fig. 7. The relative error in blood flow (*f*) with respect to errors in the partition coefficient (λ) as a function of *Tf*/λ for the tissue equilibration technique. The error in blood flow is greatest for the constant arterial curve time course ($n = 0$) and least when the curve changes quatratically ($n = 2$). The dashed line represents a linearly increasing arterial curve ($n = 1$) (modified from Patlak et al., 1984).

and the experiment time increase, so does the error in blood flow. Note that this is exactly the opposite to the case for inaccuracies in both tissue tracer concentrations and partition coefficient, where the flow errors were minimized by the use of a rapidly increasing arterial input. The data in Fig. 8 (as well as in Figs. 5 and 7) suggest that the flow errors will be minimized with very short experimental times. As the experiment time is reduced, errors in timing of the arterial samples with respect to the tissue sampling (end of study) become very appreciable, making studies shorter than, say, 30 s very error-prone. When all of these sources of error are combined, the total error can become appreciable. For a 30-s study with a ramp function as the arterial input function and modest errors in the tissue concentration, the partition coefficient, and the timing, the potential error in flow is ±13% at a flow of 0.4 mL/g/min and ±20% at a flow of 1.6 mL/g/min (Patlak et al., 1984).

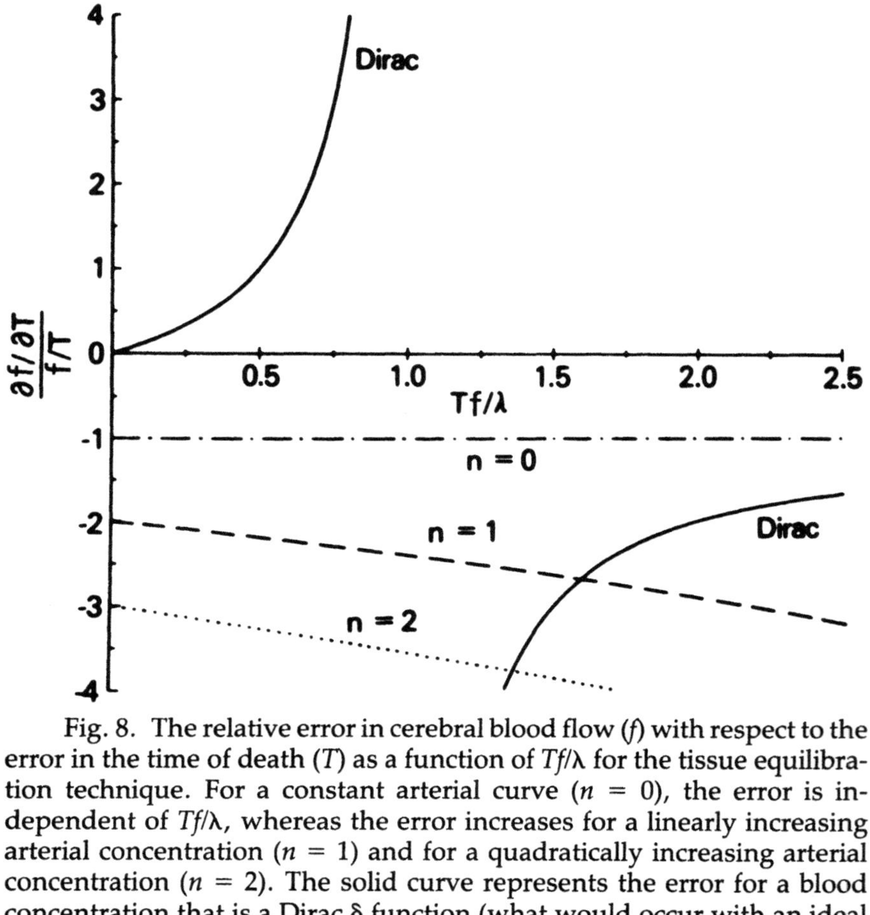

Fig. 8. The relative error in cerebral blood flow (f) with respect to the error in the time of death (T) as a function of Tf/λ for the tissue equilibration technique. For a constant arterial curve ($n = 0$), the error is independent of Tf/λ, whereas the error increases for a linearly increasing arterial concentration ($n = 1$) and for a quadratically increasing arterial concentration ($n = 2$). The solid curve represents the error for a blood concentration that is a Dirac δ function (what would occur with an ideal bolus with no recirculation) (from Patlak et al., 1984).

2.6.2. Indicator Fractionation

A similar theoretical error analysis has also been made for the indicator fractionation technique (Patlak et al., 1984). Errors caused by inaccuracies in the determination of the tissue tracer concentration and errors caused by timing were examined (*see* section 2.6.1). The partition coefficient of the tracer does not appear in the formulation of the model for the indicator fractionation technique, but the potential for loss of the tracer from the tissue (backflux) does exist, and so the backflux error was included in the error

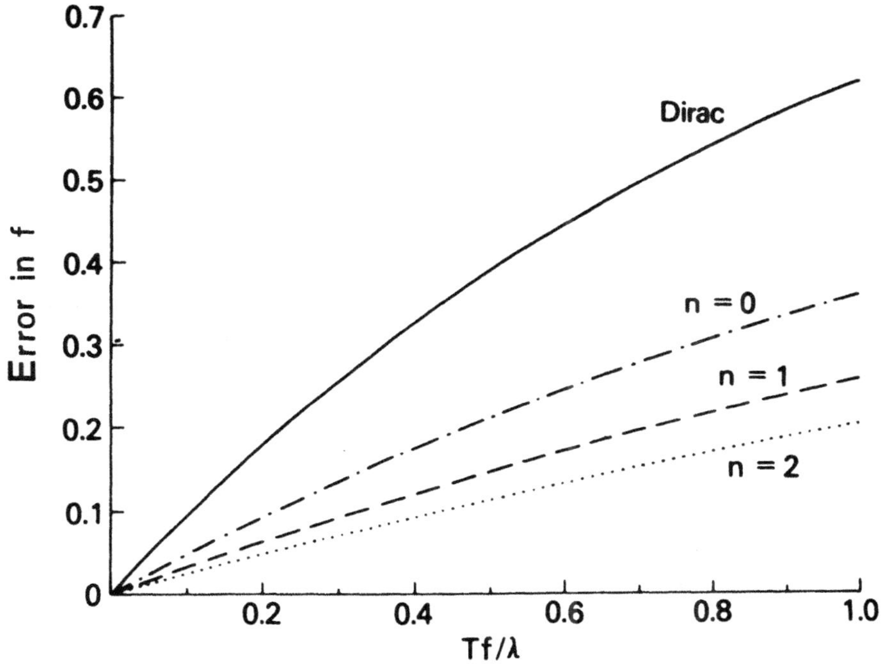

Fig. 9. The relative error in blood flow (f) caused by indicator backflux in the indicator fractionation technique. This error increases with flow and study duration (T). It is lowest for a quadratically increasing arterial concentration ($n = 2$) and increases progressively if the arterial curve increases linearly ($n = 1$) or is a constant ($n = 0$). When the arterial curve is a Dirac δ, the error is largest (from Patlak et al., 1984).

analysis. As one might expect, this error becomes appreciable when either the experiment time increases or the blood flow increases (Fig. 9). Examination of Fig. 9 indicates that when more of the isotope is delivered to the tissue at the beginning, as for the bolus injection (represented as the Dirac δ function), the greater the opportunity for backflux from the tissue. For more sharply rising arterial input curves in which the isotope is delivered later, the loss of the tracer from the tissue becomes less of a problem. This situation is directly analogous to shortening the study duration. If the experimental time becomes too short, however, timing errors can become significant. A 5–10-s study appears to be a good compromise (Sage et al., 1981). It should be noted that the backflux error decreases for increasing partition coefficient; since the partition coefficient is a measure of the extraction of the compound from the blood by the tissue, this result simply states that the backflux error decreases as the extraction increases. Substances, such as microspheres, with close to 100% extraction do not produce any backflux error. For substances that do not stick in the tissue like microspheres, the apparent extraction increases for "short" times (Sage et al., 1981); that is why the backflux error diminishes with short experiment times.

The errors in measuring blood flow caused by timing errors vary with both the shape of the arterial tracer input as well as the experiment duration (Patlak et al., 1984). If most of the input occurs early in the experiment, then the arterial curve will be relatively small toward the end of the experiment, and errors in timing between the tissue sampling (decapitation) and the arterial blood will be small. In the case of a bolus input, the arterial concentration is close to zero 8–10 s into the study, and the error caused by inaccurate timing is only a few percent. Therefore, in order to minimize the errors caused by timing, a bolus injection is advantageous. In practice, this is the input most frequently used in this technique. Although lengthening the experiment time decreases the effect of timing errors, it also appreciably increases the backflux error (which is significantly more important) and, hence, is not advisable.

The error in measuring tissue tracer concentrations leads to an error in calculated blood flow that is independent of the shape of the arterial curve and is not altered by either flow or the study duration (Patlak et al., 1984). Other than attempting to quantitate the tissue activity as accurately as possible, no change in experimental design will reduce or minimize this error.

As discussed above, the errors involved with both the equilibration technique and the indicator fractionation technique are comparable, making it difficult to choose between them. The sources of the largest errors in each of the techniques are, however, different. The tissue equilibration technique is susceptible to errors in λ, so that if there is potential for significant alterations in λ in a particular study, the equilibration technique would probably not be the technique of choice. In addition, the tissue equilibration technique is more susceptible to timing errors than is the indicator fractionation technique, so this technique should only be used if timing can be done accurately. On the other hand, the backflux error can be very appreciable with the indicator fractionation technique, introducing appreciable error at high flow rates. Investigators should be aware of the potential sources of error in the technique they will be using so they can execute their study in such a manner as to minimize these errors.

3. Local Cerebral Glucose Utilization

3.1. Introduction

Local cerebral glucose utilization is probably a better window into the function of the central nervous system than is local cerebral blood flow. Until quite recently, however, there did not exist any techniques for measuring cerebral metabolism regionally. In the early 1970s, Sokoloff and coworkers (Reivich et al., 1971; Kennedy et al., 1975; Sokoloff et al., 1977) developed a technique for the measurement of local cerebral glucose utilization (lCGU). This technique has been the subject of many reviews and monographs (Sokoloff, 1981, 1982, 1985a; Greenberg and Reivich, 1983). The introduction to the basis of this technique as discussed by Sokoloff (1985a) is excellent in that it relates the method to the more general concept of biochemical reaction rates, and so shall be briefly summarized here. In this approach, the classical expression for the measurement of biochemical reactions using radiolabeled tracers is considered. If labeled substrate is added to a system, then the reaction rate is equal to the quantity of labeled product formed (or the quantity of substrate consumed) over a time interval divided by the time integral of the specific activity of the precursor. As discussed by Sokoloff (1985a), measuring the quantity of labeled product formed is generally easier than measuring the quantity of substrate consumed, since the change in the amount formed is usually appreciably larger. Measurement of the specific activity is necessary since it permits a correction for the concentration difference between the total species and the labeled species. Sometimes, in the measurement of reaction rates, instead of using a labeled precursor, a labeled analog of the precursor is used, necessitating the introduction of a correction factor, which Sokoloff calls the isotope effect correction factor. The next section contains a brief review of the theoretical basis of the deoxyglucose technique. As will become apparent (*see* Fig. 10), the operational equation of the deoxyglucose technique can be placed into the format for the calculation of the rates of a biochemical reaction.

$$R_i = \frac{\overbrace{\underbrace{C_i^*(T)}_{\text{Total }^{4}\text{C in Tissue at Time, T}} - \underbrace{k_1^* e^{-(k_2^*+k_3^*)T} \int_0^T C_p^* e^{(k_2^*+k_3^*)t}\,dt}_{^{14}\text{C in Precursor Remaining in Tissue at Time, T}}}^{\text{Labeled Product Formed in Interval of Time, 0 to T}}}{\underbrace{\left[\frac{\lambda \cdot V_m^* \cdot K_m}{\Phi \cdot V_m \cdot K_m^*}\right]}_{\text{Isotope Effect Correction Factor}} \underbrace{\left[\underbrace{\int_0^T \left(\frac{C_p^*}{C_p}\right) dt}_{\text{Integrated Plasma Specific Activity}} - \underbrace{e^{-(k_2^*+k_3^*)T} \int_0^T \left(\frac{C_p^*}{C_p}\right) e^{(k_2^*+k_3^*)t}\,dt}_{\text{Correction for Lag in Tissue Equilibration with Plasma}}\right]}_{\text{Integrated Precursor Specific Activity in Tissue}}}$$

Fig. 10. Operational equation for the calculation of the local rate of cerebral glucose utilization. R is the rate of glucose utilization in tissue i; $C_p{}^*$ and C_p are the arterial plasma ^{14}C-deoxyglucose and glucose concentrations; T is the time of sacrifice; $C_i{}^*$ is the total ^{14}C concentration in tissue i at time T; t is the variable time; $k_1{}^*$ and $k_2{}^*$ are the rate constants for the transport of ^{14}C-deoxyglucose phosporylation by hexokinase; λ is the ratio of the tissue distribution space for ^{14}C-deoxyglucose to that of glucose; ϕ is a constant representing the steady-state fraction of glucose, which once phosphorylated, continues down the glycolytic pathway for further metabolism; and $K_m{}^*$, $V_m{}^*$, K_m, and V_m are the Michaelis-Menten constants and the maximal velocities of hexokinase for deoxyglucose and glucose, respectively. The latter six constants can be combined into one constant ($\lambda V_m{}^* K_m / \phi V_m K_m{}^*$), which has been termed the lumped constant (from Sokoloff, 1979).

3.2. *Theoretical Basis*

The natural first choice for a labeled precursor for the measurement of cerebral glucose metabolism would be labeled glucose (e.g., [^{14}C]glucose). This was considered by Sokoloff et al. (1977) to be unsuitable, however, because of the rapid conversion of glucose to carbon dioxide, which is rapidly cleared from the cerebral tissue. They therefore chose 2-deoxy-D-[^{14}C]glucose (2DG), which is a labeled analog of glucose, as a potentially useful tracer for the measurement of local cerebral glucose metabolism. This compound differs from glucose only in that on the second carbon atom of the molecule a hydrogen atom replaces the hydroxyl group that is present in glucose. It has, however, a number of properties that make it a very useful metabolic tracer. Deoxyglucose is transported across the blood–brain barrier by the same saturable carrier-transport system that transports glucose, and once in the tissue it competes with glucose for hexokinase and is phosphorylated to deoxyglucose 6-phosphate (Sols and Crane, 1954; Bidder, 1968; Bachelard et al., 1971; Horton et al., 1973). The major property of the deoxyglucose that makes it a useful tracer is that further metabolism does not occur, since the deoxyglucose 6-phosphate cannot be isomerized to fructose 6-phosphate (Sols and Crane, 1954; Wick et al., 1957; Tower, 1958; Bachelard, 1971). Additionally, the deoxyglucose 6-phosphate is not a substrate for glucose 6-phosphate dehydrogenase, the first step in the hexose-monophosphate shunt (Sols and Crane, 1954; Tower, 1958). There is a potential for loss of the deoxyglucose 6-phosphate since deoxyglucose 6-phosphate can be hydrolyzed back to free deoxyglucose by glucose 6-phosphatase, which, although in very low concentrations in mammalian brain, does exist. This will be discussed in section 3.6.4.

The theoretical basis of the deoxyglucose technique can best be discussed with reference to Fig. 11. This three-compartment model schematically describes the transport and metabolism of both glucose and deoxyglucose in the brain. These substances share and compete for a common carrier for the transport from the blood plasma compartment into the cerebral tissue where they initially exist as free glucose and deoxyglucose, respectively. The constants k_1 and k_1^* represent the rate constants for carrier-mediated transport of glucose and deoxyglucose, respectively. Once in this precursor pool, they compete either for the carrier for transport back from brain to plasma (k_2 and k_2^* are the respective

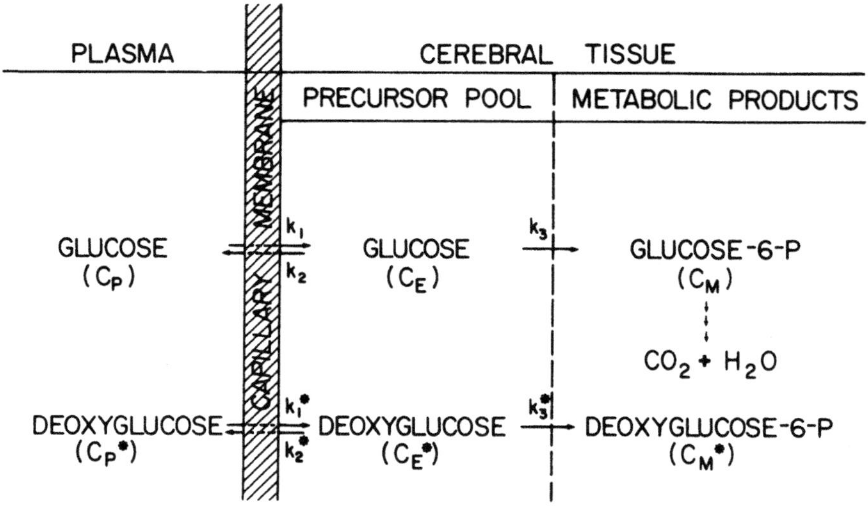

Fig. 11. Theoretical model for transport and metabolism of glucose and deoxyglucose in the brain. C_p and C_p* represent the concentration of glucose and ^{14}C-deoxyglucose in the arterial plasma, respectively; C_E and C_E* represent their respective concentrations in the tissue pools that serve as substrates for hexokinase; and C_M and C^*_M represent the concentrations of glucose 6-phosphate and ^{14}C-deoxyglucose 6-phosphate in the tissue. The constants k_1*, k_2*, and k_3* represent the rate constants for carrier-mediated transport of ^{14}C-deoxyglucose from plasma to tissue, from tissue back to plasma, and for phosphorylation by hexokinase, respectively. The constants k_1, k_2, and k_3 are the equivalent rate constants for glucose (from Reivich et al., 1982).

rate constants for glucose and deoxyglucose, respectively) or for the enzyme hexokinase, which phosphorylates them to their respective hexose 6-phosphates (represented by the rate constants k_3 and k_3*). As mentioned above, the deoxyglucose 6-phosphate that is formed is essentially trapped (except for the small amount that is dephosphorylated by glucose 6-phosphatase). The glucose 6-phosphate, however, is further metabolized, eventually to carbon dioxide and water. Using this model and a number of assumptions, a set of equations can be written to describe the concentration of the various species in the three compartments (Sokoloff et al., 1977). The equations used are mass balance equations, intercompartmental transport equations, and the classical Michaelis-Menten equation modified for the presence of competitive substrates.

The principal assumptions used in the derivation of the equations are as follows (Sokoloff et al., 1977):

1. Each region is homogeneous with respect to blood flow and glucose metabolism as well as the concentrations of the chemical species (hexoses and hexose 6-phosphates). Additionally, the rate of transport of the species between compartments is homogeneous.
2. Glucose (and deoxyglucose) metabolism is in a steady state over the course of the measurement.
3. The deoxyglucose is administered in tracer quantities only. Otherwise the deoxyglucose 6-phosphate will competitively inhibit glucose 6-phosphate metabolism to such an extent as to influence glucose metabolism.
4. The concentration of deoxyglucose in the capillary plasma can be represented by the concentration in the arterial plasma. Since the extraction ratios of both glucose and deoxyglucose are low, this is a reasonable assumption.

Utilizing these assumptions, Sokoloff and coworkers have derived an operational equation for the deoxyglucose technique. This equation is shown in Fig. 10, which describes each term of the equation in terms of the general equation for the calculation of the rate of a biochemical reaction. The numerator is the quantity of labeled product formed from time 0 to time T (the time duration of the study). Unfortunately, it is difficult to differentiate between the amount of labeled product formed and the amount of labeled substrate that has yet to enter into the biochemical reaction. Autoradiographically, all that can be measured is the total amount of label (radioactivity) in the tissue. It is necessary to subtract from this total tissue radioactivity the amount of free (unphosphorylated) deoxyglucose in the tissue. This is estimated from the arterial plasma activity level (which will be free deoxyglucose) and knowledge of the constants of transport of deoxyglucose between the three compartments (k_1^*, k_2^*, k_3^*). Thus the numerator of the operational equation is the total tissue radioactivity measured directly from the density of the autoradiogram minus the amount of free deoxyglucose in the tissue calculated from the arterial curve and the transport constants. As discussed above, the denominator of the equation for biochemical reaction rates is the integrated precursor specific activity in the tissue. In simplistic terms, this is just the first term in the expression in the denominator of Fig. 10. Because the plasma is not in instantaneous equilibrium with the tissue, a correction term is necessary to account for the lag between the tissue and the arterial plasma. This is the second term in the denominator.

The term in the denominator labeled "isotope effect" correction factor is necessary because an analog of glucose, deoxyglucose, and not glucose, is used as the tracer. Our interest is in measuring glucose metabolism, not deoxyglucose metabolism, in the brain, and it is therefore necessary to correct for the use of this analog. In the derivation of the equations for glucose metabolism using deoxyglucose, a number of "constants" appear, primarily as a result of the use of the Michaelis-Menten equation for competitive inhibition. In the operational equation these constants appear together and have become known as the lumped constant (Sokoloff et al., 1977). The lumped constant can be considered as the product of four terms: $1/\phi$, λ, $K_m/K_m{}^*$, and $V_m{}^*/V_m$. (ϕ) is the fraction of glucose, which, once phosphorylated, continues down the glycolytic pathway and is generally considered to be unity. Lambda (λ) is the ratio of the distribution volume of deoxyglucose to that of glucose and K_m, V_m, $K_m{}^*$, and $V_m{}^*$ are the Michaelis-Menten constants and the maximal velocities of the phosphorylation reaction for glucose and deoxyglucose, respectively. The techniques for measuring the lumped constant and the conditions under which it may change will be discussed in section 3.4.

3.3. Determination of Rate Constants

The operational equation shown in Fig. 10 contains rate constants for deoxyglucose describing the transport from plasma to brain tissue and back ($k_1{}^*$ and $k_2{}^*$) along with the rate constant for deoxyglucose phosphorylation by hexokinase ($k_3{}^*$). These constants cannot be determined in each experimental animal, but instead must be measured in a separate series of studies and then applied to cerebral glucose metabolism studies. (In positron emission tomography, in which tissue concentrations can be followed in time following the injection of the glucose analog, the rate constants can be measured directly in each subject. This has formed the basis of an alternate approach for the calculation of regional cerebral glucose metabolism (Phelps et al., 1979).) Theoretically, they may vary as a function of the experimental conditions—species, type of pathology, anesthetic, and so on. And, in actual fact, they obviously are not constants. k_3^* is a phosphorylation rate that must certainly be different under different circumstances and even in the same animal must vary depending upon the region of the brain. The equation (Fig. 10) is set up in such a manner so that, operationally, the calculated rate of glucose utilization is relatively insensitive to alterations in the rate constants. This will be discussed in detail in section 3.6.1. Consequently, the rate constants have been determined in the conscious rat (Sokoloff et al., 1977) and used unaltered in practically all published reports that have utilized the deoxyglucose technique. The technique for the determination of the rate constants derives from the model of Fig. 11. Using the mass balance equation for deoxyglucose, the following equation can be written relating the time course of the ^{14}C concentration in the brain tissue [$C_i(T)$], the time course of the arterial plasma ^{14}C concentration ($^{14}C_p$), and the rate constants (k_1^*, k_2^*, and k_3^*):

$$C_i(T) = k_1^* \exp[-(k_2^* + k_3^*)T] \int_0^T C_P^* \exp[(k_2^* + k_3^*)t]\mathrm{d}t + k_1^* k_3^* \int_0^T \left\{ \exp[-(k_2^* + k_3^*)t] \times \int_0^t C_P^* \exp[(k_2^* + k_3^*)\tau]d\tau \right\} dt \tag{13}$$

If $C_i(T)$ could be measured at a number of time points in an animal, then the rate constants could be determined directly by a least-squares, iterative technique by measuring $C_i(T)$ and $C_p(t)$ over time following administration of labeled deoxyglucose. This has been accomplished using positron emission tomography in humans (Phelps et al., 1979; Reivich et al., 1979), but is not practical in laboratory animals since there are regional differences in the rate constants (at least gray matter is significantly different from white matter), and the resolution is not sufficient to resolve these differences. Consequently, in order to determine k_1^*, k_2^* and k_3^*, it is necessary to use a series of animals in which each animal receives a constant infusion of ^{14}C-deoxyglucose and blood samples are obtained for the measurement of $C_p(t)$ at frequent enough intervals so as to define the arterial plasma time course of activity. At various times following the start of the ^{14}C-deoxyglucose infusion (T = 5,

10, 20, 30, 45 min), the animals are sacrificed and the brain removed and processed for quantitative autoradiography to determine $C_i(T)$. These data provide the time course of the tissue ^{14}C-deoxyglucose concentration which, utilizing Eq. (13), allows for the determination of k_1^*, k_2^*, and k_3^* by a nonlinear iterative process.

The kinetic constants of deoxyglucose have been measured in 18 regions of the normal conscious albino rat (Sokoloff et al., 1977). No significant differences have been found among the gray matter structures or the white matter structures, although there was a significant difference in the values of these parameters between gray and white matter (Table 1). They have not been measured in any other laboratory animal, although some preliminary data are available in the nonhuman primate, indicating that they are very similar (Kennedy et al., 1978). Table 1 also lists, for comparison, the values in humans obtained by positron emission tomography.

3.4. Determination of the Lumped Constant

As discussed above, the lumped constant appears as a multiplicative factor in the denominator of the operational equation and accounts for the difference in the steady-state rates of deoxyglucose and glucose phosphorylation. The parameters that comprise the lumped constant, namely (1) a factor proportional to the amount the glucose 6-phosphatase activity (ϕ), (2) the ratio of the distribution volumes of glucose and deoxyglucose in the brain (λ), (3) the ratio of the maximal velocities of phosphorylation of deoxyglucose and glucose (V_m^*/V_m), and (4) the ratio of the Michaelis-Menten constants (K_m/K_m^*), very likely are uniform throughout the brain (Sokoloff et al., 1977). Most of these factors are ratios between glucose and deoxyglucose, so that although the parameter may be different in different regions of the brain, the ratio is most probably constant. It is also a reasonable assumption that the lumped constant will not be altered under normal physiological conditions. There are measured differences, however, in the lumped constant between species (Table 2), and under pathophysiological conditions there may be significant alterations in this parameter, particularly when the glucose supply becomes rate-limiting. Preliminary evidence exists that indicates that in ischemia the lumped constant may increase appreciably (Ginsberg and Reivich, 1979), although estimates of the lumped constant in patients with stroke showed only small variations between ischem-

Table 1
Deoxyglucose Rate Constants

	Rat[a]		Monkey[b]		Human[c]	
Constant	Gray matter	White matter	Gray matter	White matter	Gray matter	White matter
k_1^*, per min	0.189[d] ± 0.012	0.079 ± 0.008	—	—	0.090 ± 0.006	0.057 ± 0.004
k_2^*, per min	0.245 ± 0.040	0.133 ± 0.046	—	—.	0.221 ± 0.018	0.109 ± 0.015
k_3^*, per min	0.052 ± 0.010	0.020 ± 0.020	—	—	0.105 ± 0.009	0.078 ± 0.021
Half-life of precursor pool ln $2/(k_3^* + k_3^*)$, min	2.38 ± 0.40	4.51 ± 0.90	2.0	4.81	2.37 ± 0.20	3.29 ± 0.40
Distribution volume mL/g	0.647 ± 0.073	0.516 ± 0.171	—	—	0.238 ± 0.029	0.234 ± 0.033

[a]From Sokoloff et al. (1977).
[b]From Kennedy et al. (1978).
[c]From Reivich et al. (1982).
[d]Values are means ± SE.

Table 2
Values of the Lumped Constant

Species	Condition	Mean ± SEM
Deoxyglucose		
Albino rat[a]	Conscious	0.464 ± 0.026
	Barbiturate	0.512 ± 0.039
	Combined	0.481 ± 0.023
Cat[b]	Anesthetized	0.411 ± 0.005
Dog[c]	Conscious puppy	0.558 ± 0.031
Sheep[d]	Conscious newborn	0.382 ± 0.012
Human[e]	Conscious	0.56 ± 0.043
Fluorodeoxyglucose		
Cat[f]	Barbiturate	0.44 ± 0.012
Human[e]	Conscious	0.52 ± 0.028

[a]From Sokoloff et al. (1977).
[b]From Sokoloff (1986).
[c]From Duffy et al. (1982).
[d]From Abrams et al. (1984).
[e]From Reivich et al. (1985).
[f]From Matsuda et al. (1987).

ic and nonischemic regions of the brain (Hawkins et al., 1981). The only other pathophysiological condition in which the lumped constant has been examined is in hypoglycemia, in which a large increase in the lumped constant has been measured (Suda et al., 1981), and in hyperglycemia, in which a small, but still significant, decrease in the lumped constant occurs (Schuier et al., 1981).

The lumped constant is measured as a global parameter. The technique of measurement is based on the model equations that can be manipulated into a form in which the lumped constant can be expressed in terms of measurable quantities (Sokoloff et al., 1977). If the arterial plasma deoxyglucose concentration can be made to follow a step change, then the equations simply state that the lumped constant asymptotically approaches the ratio of the extraction fraction for deoxyglucose (E^*) and glucose (E) multiplied by the ratio of their specific activities in arterial blood and plasma:

$$LC = (E^*/E)(C_A^*/C_A)/(C_P^*/C_P) \qquad (14)$$

This equation assumes that the plasma deoxyglucose concentration is a constant over the study period and that the measurements of extraction fractions are made at a sufficiently long time after the plasma deoxyglucose attains a steady value, approximately 40–50 min. This time duration is dictated by the derivation of Eq. (14), which holds only for large times in relation to the rate constants. The plasma deoxyglucose is made constant (a step change is usually sought) by use of an infusion schedule determined from knowledge of the plasma response to an impulse (bolus injection) and assumptions of the linearity and time invariancy of the physiological system.

3.5. Determination of lCGU

The operational equation discussed in section 3.2 basically defines the procedure by which rates of local cerebral glucose utilization are measured. The equation in Fig. 10 states that the lCGU in a particular region equals the total ^{14}C concentration in the region minus the amount of free 2DG (estimated from the plasma time course of the ^{14}C-2DG and knowledge of the rate constants) divided by the integrated precursor specific activity of the 2DG in the tissue (estimated from the integrated precursor specific activity in the plasma, again using the plasma time course and the rate constants), all corrected for the differences between glucose and deoxyglucose as represented by the lumped constant. Since the rate constants cannot be measured in the same animal as is lCGU (*see* section 3.3), and since the operational equation is relatively insensitive to variations in the rate constants (*see* section 3.6.1), rate constants determined in a separate series of animals (basically from the literature) are used in the calculations. Similarly, the value for the lumped constant determined for the particular species of interest is used in the equation. Consequently, the only parameters that need to be measured for the calculation of lCGU are: (1) the tissue ^{14}C concentration at the end of the study, (2) the time course of the arterial plasma ^{14}C concentration, and (3) the plasma glucose concentration (assumed constant throughout the study). (If the plasma glucose concentration does change during the study, it is necessary to use a modified equation that accounts for these changes—*see* section 3.6.3.)

The tissue concentration of ^{14}C at the end of the study is determined autoradiographically in an identical manner as described for ^{14}C-iodoantipyrine (section 2.2). The plasma ^{14}C-deoxyglucose concentration is determined from arterial blood samples drawn frequently enough over the course of the study (usually 45 min) to adequately define the time course. In order to minimize the amount of free unmetabolized deoxyglucose in the tissue, the ^{14}C-deoxyglucose is administered as a bolus. Consequently, the plasma deoxyglucose concentration changes rapidly following the administration of 2DG, necessitating more frequent sampling at the beginning and less frequent sampling in the later phase of the study. Collection of samples every 15 s is sufficient for the first minute; samples are then collected every minute until approximately 5 min into the study, at which point blood sampling every 5–10 min can adequately define the plasma deoxyglucose curve. These blood samples are immediately centrifuged to separate the plasma, which is placed on ice until measurements of radioactivity and glucose concentration can be obtained. The activity in these samples is determined with liquid scintillation counting, and lCGU is calculated from the operational equation using a computer program.

3.6. Error Analysis

3.6.1. Effect of Errors in Rate Constants

The procedure by which lCGU is measured using deoxyglucose has been defined in such a manner so as to minimize errors in the calculated metabolic rate. In particular, the rate constants, which must certainly vary among animals as well as cerebral structures, cannot be measured in each animal. The operational equation, however, is of a form such that for reasonably long experimental times (in excess of 30 min), the error in calculated metabolic rate caused by inaccuracies in the rate constants is quite small. In fact, with sufficient time, the terms containing the rate constants approach zero. Very long study times, however, are impractical since, first, a steady state in the study condition cannot usually be maintained for a long period of time, and, second, there is a small amount of glucose 6-phosphatase activity in the brain, causing some of the label to leave the tissue (section 3.6.4). As can be seen in Fig. 12, as the study duration is lengthened, the errors in the calculated metabolic rate decreases, so that by 45 min (the time duration of most studies) the error becomes quite small. The exception is the error caused by k_4, which can become appreciable for long study durations. Thus, unless there is reason to believe that the rate constants may be very significantly altered, such as in some pathological conditions, the errors in the calculated metabolic rate for glucose caused by inaccurately known rate constants is small.

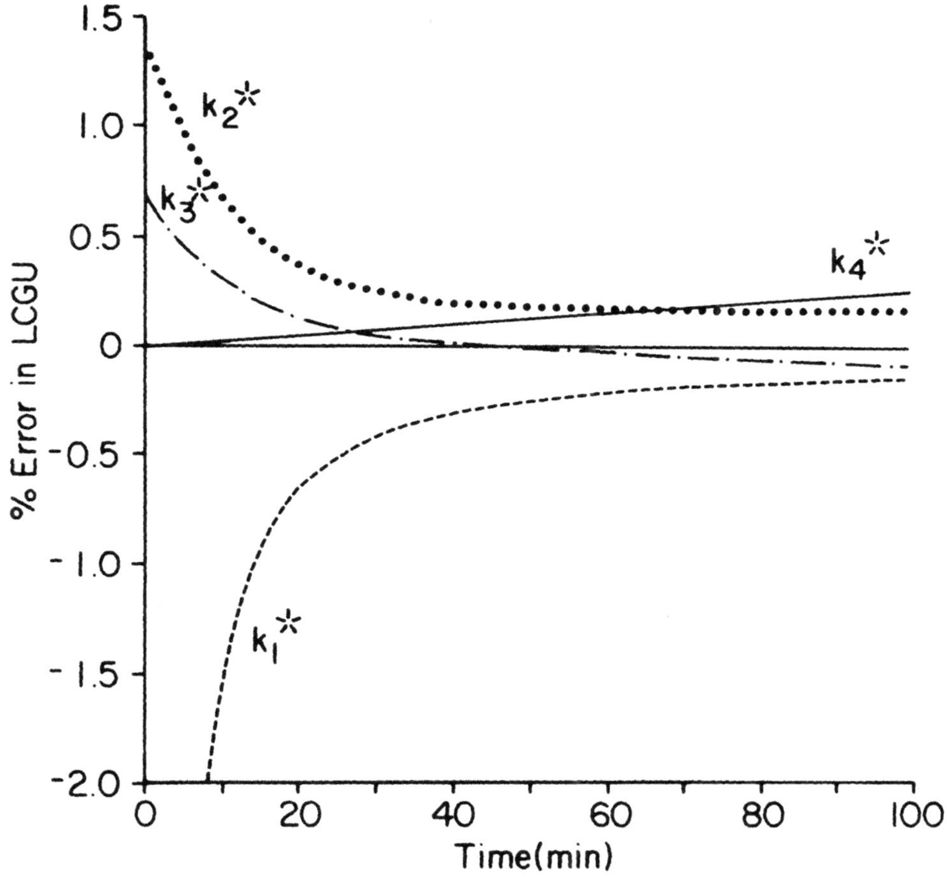

Fig. 12. The error in calculated metabolic rate caused by a 1% error in each of the rate constants as a function of time duration of study. The dashed line is the error when k_1^* is altered, the dotted line when k_2^* is altered, the dotted-dashed line when k_3^* is altered, and the solid line when k_4^* is altered, all by 1%. Note the large error in the first 15–20 min following the administration of the deoxyglucose. These errors were calculated using a typical arterial concentration time course.

In pathological conditions such as ischemia or severe hyperglycemia, however, the rate constants may vary enough so as to influence the calculated metabolic rates for glucose. The effect of ischemia on the rate constants has been examined in humans (Hawkins et al., 1981). It was found that in a series of six patients with clinical and computerized transmission tomography findings consistent with a nonhemorrhagic stroke, the rate constants are generally lower in ischemic regions compared to homologous regions of the contralateral hemisphere; the values were more depressed in the patients with the most severe ischemic insult. Using the measured rate constants for ischemia, the model generated more accurate estimates of local cerebral glucose metabolism than when the rate constants from normal young adults were used. When the latter were applied to ischemic regions, the local metabolic rate was significantly underestimated, and temporal instability of the model was observed. Thus in abnormal conditions such as these, it may be necessary to determine the rate constants in each situation.

3.6.2. Effect of Errors in Lumped Constant

The lumped constant appears in the operational equation (Fig. 10) as a multiplicative factor so that, for example, a 10% error in the lumped constant leads to exactly a 10% error in the calculated metabolic rate. Although, as discussed above (section 3.4), the lumped constant is not expected to change under physiological conditions, the significant sensitivity of the method to alterations in the lumped constant makes it important to be aware of potential changes in the lumped constant under the particular study conditions of interest. The original measurement technique for the lumped constant requires a measurement of the whole brain extraction fraction of glucose and deoxyglucose, necessitating for the measurement of the lumped constant in a particular experimental condition that the entire brain be in that experimental condition. In other words, measurement of the lumped constant in ischemia required that the entire brain be approximately equally ischemic. Gjedde has recently introduced a technique for the measurement of the regional lumped constant by measuring the regional brain glucose content utilizing labeled 3-*O*-methylglucose (Gjedde, 1982; Gjedde et al., 1985; Gjedde and Diemer, 1983). Use of this technique allows for the measurement of the lumped constant in a number of pathophysiological conditions that are focal in nature. Initial studies indicate that the lumped constant in normal animals is reasonably uniform throughout the brain, as was originally assumed by Sokoloff et al. (1977).

3.6.3. Effect of Varying Plasma Glucose Concentrations

The operational equation of Fig. 10 is based on the assumption that the arterial plasma glucose concentration remains constant during the period of measurement. This condition can frequently be too restrictive, since in a number of experimental conditions plasma glucose may vary, and, since the calculated metabolic rate for glucose is directly proportional to the plasma glucose, significant errors can result. If plasma glucose is assumed to vary during the study, then the original model equations can be used to derive a new operational equation (Savaki et al., 1980), as follows:

$$R_i = \frac{C_i^*(T) - k_1^* \exp[-k_{23}^*T] \int_0^T C_P^* \exp[k_{23}^*t]dt}{\dfrac{\lambda V_m^* K_m}{\phi V_m K_m^*} \displaystyle\int_0^T \left[\frac{k_{23}^* \exp[-k_{23}^* \tau] \int_0^\tau C_P^* \exp[k_{23}^*t]dt}{C_P(0) \exp[-k_{23}\tau] + k_{23} \exp[-k_{23}\tau] \int_0^\tau C_P \exp(k_{23}t)dt} \right] d_\tau} \tag{15}$$

where $C_p(0)$ is the arterial plasma glucose concentration at zero time, $k_{23}^* = (k_2^* + k_3^*)$, and $k_{23} = (k_2 + k_3)$ is the turnover rate constant of the free glucose pool in the brain. All of the other symbols are the same as those in Fig. 10. An estimate of the half-life of the free glucose pool has been found to be 1.2 and 1.8 min in normal conscious and anesthetized rats, respectively (Savaki et al., 1980). Although it varies with plasma glucose concentration, the equation is relatively insensitive to the value of the half-life of the glucose pool in the physiological range. It should be noted that in the case of significant alterations in plasma glucose, other factors have to be considered. As discussed above, significant hypo- and hyperglycemia affect the value of the lumped constant.

3.6.4. Phosphatase Activity

A great deal of controversy has existed during the past few years regarding the presence of glucose 6-phosphatase in the brain (Hawkins and Miller, 1978; Huang and Veech, 1982; Sacks et al., 1983). This enzyme is capable of converting deoxyglucose 6-phosphate back to free deoxyglucose, which would then be capable of leaving the brain. Since the original model does not include a term to represent this loss, if appreciable glucose 6-phosphatase exists, the operational equation would lead to a significant underestimation of lCGU. A detailed discussion of this topic has been published by Sokoloff (1985b, 1986). The absence of any appreciable phosphatase activity is supported by a number of observations. First, measurements of lCGU in rats sacrificed at either 20, 30, or 45 min yield almost identical values. If appreciable phosphatase activity existed, the calculated metabolic rates would decline at the later time points. Second, measurement of lCGU using positron emission tomography in the same subjects over several hours indicates that the glucose 6-phosphatase activity, although present, is very small since the introduction of a term representing the loss of deoxyglucose (k_4^* = 0.0055/min) is sufficient to account for the decline in tissue activity over time (Phelps et al., 1979). Detailed examination of the intracellular distribution of deoxyglucose 6-phosphate and phosphatase indicates that although the deoxyglucose 6-phosphate is formed in the cytosol, the phosphatase is located on the inner surface of the endoplasmic reticulum (Sokoloff, 1985b). Consequently, the deoxyglucose 6-phosphate must first be transported to the endoplasmic reticulum prior to any hydrolysis. The absence (or at least shortage) of any specific carrier in the brain requires that the dexoyglucose 6-phosphate diffuse to the site of the enzyme prior to hydrolysis. This slow process means that over the first 30–40 min a negligible amount of deoxyglucose 6-phosphate will be hydrolyzed to free deoxyglucose, and so a negligible error will result if it is assumed that all of the deoxyglucose 6-phosphate is irreversibly trapped (Fig. 13).

4. Resolution and Tritiated Autoradiography

4.1. Resolution

A number of factors influence the resolution of autoradiography (Rogers, 1979). These include section (source) thickness, film (emulsion) type, thickness of antiabrasive coat on the emulsion, energy of the isotope, and movement of tracer during section preparation.

As seen in Fig. 14, a point source positioned on a film will exposure grains not only immediately below the source, but also some distance away from the source because of the finite penetration of the radionuclide tangential to the film. The density distribution of the resulting image from the center outward (Fig. 15) defines the resolution of the image. If the source is not infinitely thin, as in a point source, but has a thickness of, for example, 20 μm, then the tissue component that is now elevated above the film will, by virtue of the increased solid angle, lead to a broadening of the

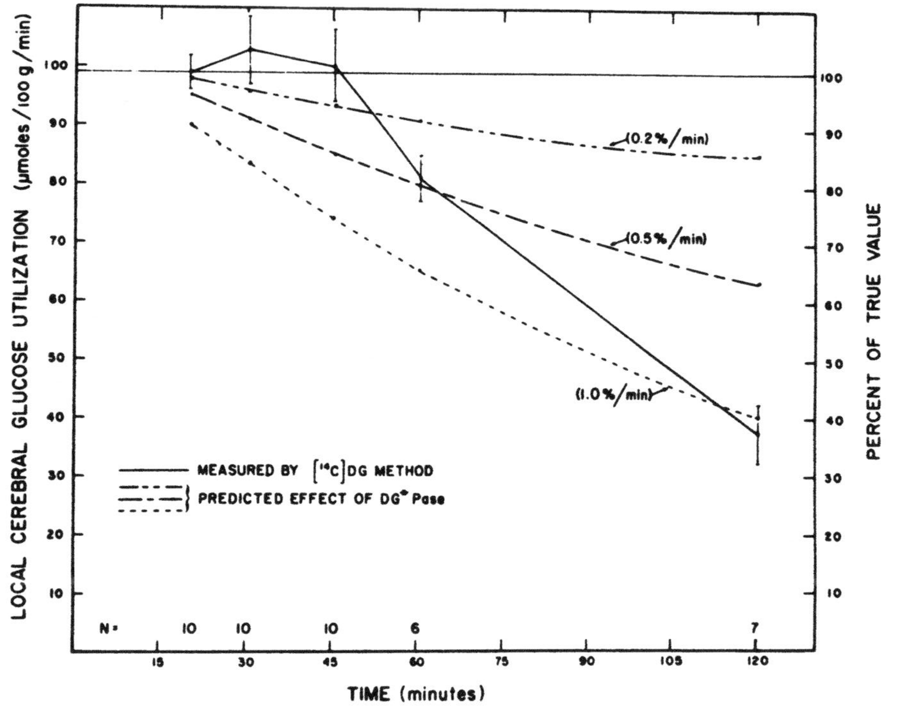

Fig. 13. Measured metabolic rates for five study durations (20, 30, 45, 60, and 120 min) calculated using the operational equation from Fig. 10, which assumes that there is no dephosphorylation of the deoxyglucose 6-phosphate. The straight lines (0.2, 0.5, and 1.0%/min) are calculated metabolic rates using the equation in Fig. 10 and assuming that ^{14}C is lost from the tissue at the rates listed. The straight line assumes no loss of ^{14}C from the tissue. Note that if the study duration is 45 min or less, the metabolic rate calculated using an equation without k^*_4 is accurate (from Sokoloff, 1982).

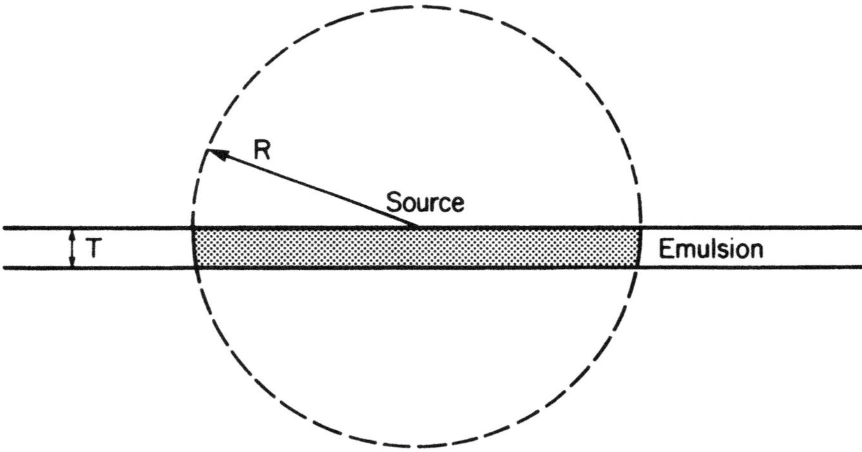

Fig. 14. Schematic representation illustrating the interaction of a point source with an emulsion with a thickness (*T*) small in relation to the track length (*R*) of the radionuclide. The image of the point source has a diameter 2*R*.

range of emissions striking the film. Although the tissue closest to the film emulsion will make a greater contribution to the image, the presence of tissue separated from the film (albeit by other tissue) will lead to a degraded image with a poorer resolution. Similarly, the presence of the coat that is placed on the emulsion for abrasion protection will have the identical effect—it will increase the distance between the source and the emulsion, increasing the solid angle of exposure and decreasing the resolution of the image. As the energy of the source increases, the resolution decreases, caused directly by the larger range of the more energetic particles in the emulsion. For example, in a nuclear track emulsion, ^{14}C will expose grains as far away as 40 μm, whereas ^{3}H will expose grains up to only 3 μm distant (Rogers, 1979).

The type of emulsion will also play a role in the ultimate resolution of the image. The major determinant of the sensitivity of a film (the relationship between the incident exposure and the resultant image optical density) is grain size. When the silver halide crystals, or grains, are larger, they have a greater probability of being "hit" by the beta particle, with the result that for the same incident radiation, more crystals will be reduced. With larger crystals, however, the size of the grain (up to 3 μm) can start affecting the resolution of the image.

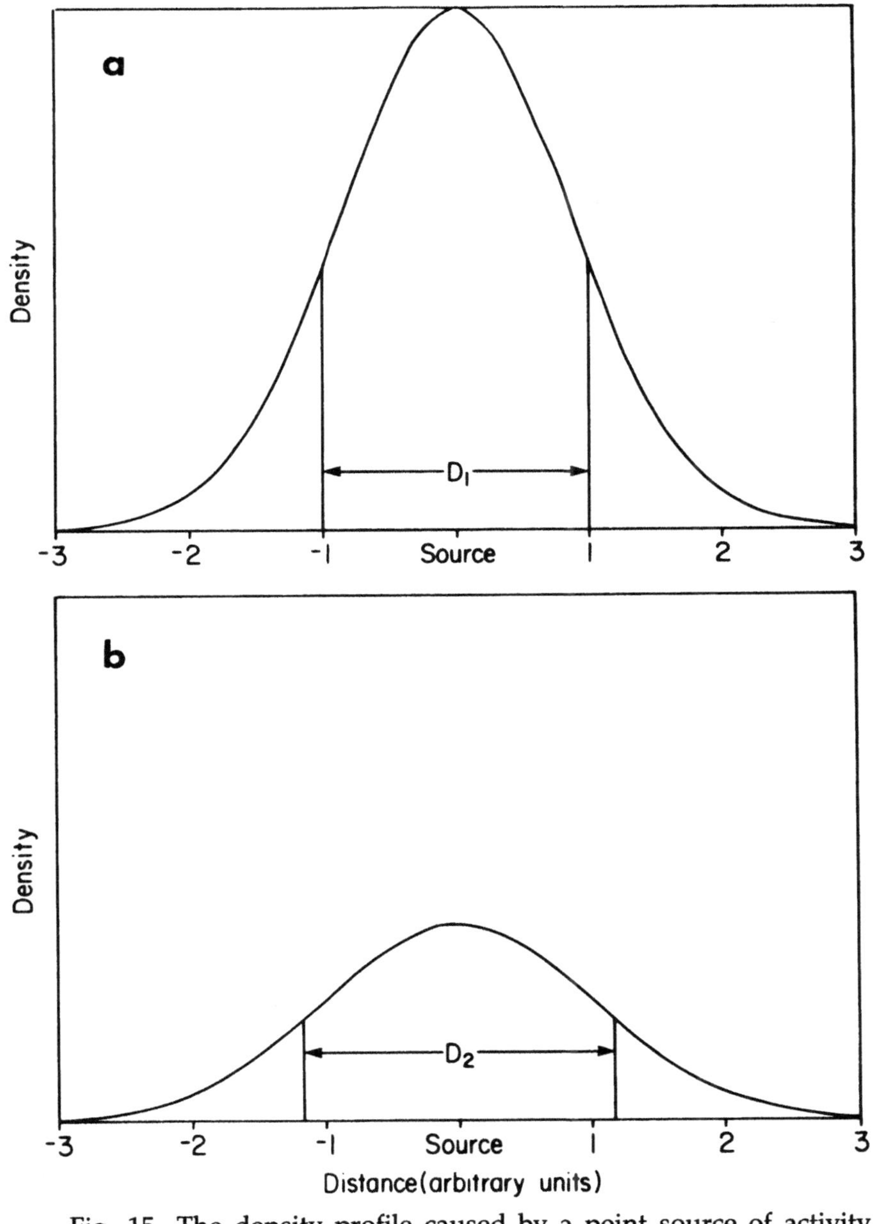

Fig. 15. The density profile caused by a point source of activity placed on film as indicated in Fig. 14. This profile changes as a function of the distance between the source and the emulsion. The distribution is less intense and broader when the source is elevated from the emulsion (b), as would occur for the upper portion of the source farthest from the film, in comparison to the line spread function of the source closest to the film (a).

Gallistel and Nichols (1983) utilized a resolution phantom to empirically examine some of these resolution-limiting factors. Gratings consisting of alternating equal-width radioactive and nonradioactive bars were produced using microcircuit lithography. The gratings were made with either ^{14}C or ^{3}H and varied from 10 to 500 μm. Different types of films were exposed to these gratings, and the resulting images were scanned with either a microdensitometer or a video digitizer (Fig. 16). These studies indicate that the film is not the resolution-limiting factor; the images on a variety of films are similar, although some films produce a grainier image than do other films. This is probably because the contribution from the grain size, generally 0.5–3 μm, is relatively small in relation to the other factors. Likewise, varying the source thickness does not appreciably change the quality of the image, although the resolution is degraded. The primary factor in improving the resolution is the power of the isotope, with ^{3}H producing an image of significantly better resolution than ^{14}C. With ^{14}C, the limit of resolution is 10 lines/mm, whereas with ^{3}H, 25 lines/mm could be resolved. Thus if resolution higher than that obtained in conventional ^{14}C autoradiography is desired, using thinner sections and/or a finer grain film will not improve the ultimate resolution of the image. This is the fundamental reason, when high resolving power is required, that some investigators have been using ^{3}H-labeled 2-deoxyglucose in their studies of cerebral glucose metabolism. These studies do not address the loss in resolution caused by any diffusion of the isotope prior to the production of the autoradiogram. By examining autoradiograms from brain sections in a manner similar to their analysis of the grating images, Gallistel and Nichols (1983) conclude that the diffusion of the isotope may be the resolution-limiting factor since the sharpest transition they found in their brain images was a factor of six less than the transitions in the images of the gratings. The very sharp images obtained with ^{3}H-labeled 2-deoxyglucose autoradiography using dipping emulsions in which the tissue is not thawed suggest that diffusion may occur during the thawing process (after the cells have been ruptured by the freezing) (Gallistel and Nichols, 1983).

The amount of isotopic diffusion in conventional 2-deoxyglucose autoradiography was examined by Smith (1983). This study was designed in order to duplicate the procedures normally used in ^{14}C-2-deoxyglucose autoradiography so that any discrepancies noted between this study and the study of Gallistel and Nichols (1983) can be attributed to the tissue preparation procedure. A hole drilled into an unlabeled rat brain was filled with a plug of brain tissue that had been labeled with ^{14}C-2DG and uniformly mixed prior to insertion. The resultant autoradiographs were scanned with an Optronics rotating drum digitizing scanner with the aperature and raster both set at 25 μm. A plot of the decrease in optical density from the "hot/cold" edge into the "cold" tissue indicated that the resolution of the autoradiograph was approximately 200 μm. In the study of Gallistel and Nichols (1983), the bar width at which there was a 50% loss in contrast between the optical density over the bar and the optical density between the bar was 50 μm. Thus the difference determined from these two experiments is a measure of the diffusion of the label during section preparation and exposure and represents the gain in resolution that potentially could be achieved with significant modification of the tissue preparation procedure along with the use of ^{3}H.

4.2. ^{3}H Autoradiography

4.2.1. Dipping Emulsions

By substituting ^{3}H ($E_{max} = 18.5$ keV) for ^{14}C ($E_{max} = 155$ keV), it was hoped that the ability of the technique to image smaller elements would be appreciably improved, the goal being the achievement of cellular resolution. Standard X-ray film used for ^{14}C autoradiography could not be used with ^{3}H since ^{3}H cannot penetrate the protective layer normally covering the emulsion. The availability of suitable films for autoradiographic detection of ^{3}H limited investigators to the use of dipping emulsions, where it was necessary to pick frozen sections up onto cold glass plates that had previously been coated with a nuclear emulsion, keeping this preparation frozen during exposure (Sharp, 1976). This process had to be done in the dark, making it rather tedious. The developed slide was counterstained, and the labeling in the region of interest was obtained by grain counting. The advantage of this procedure, aside from the improved spatial resolution that it offered, was that the autoradiogram and the histological stain were on the same plate, thereby allowing precise anatomical localization of the labeling. This technique was used by Sharp (1976) in a study of the localization of labeling in the perikarya and the neuropil in the resting and swimming rat. He found that the grain density in the neuronal perikarya was approximately equal to that in the surrounding neuropil, implying that the majority of glucose utilization in gray matter was caused by the neuropil, since the gray

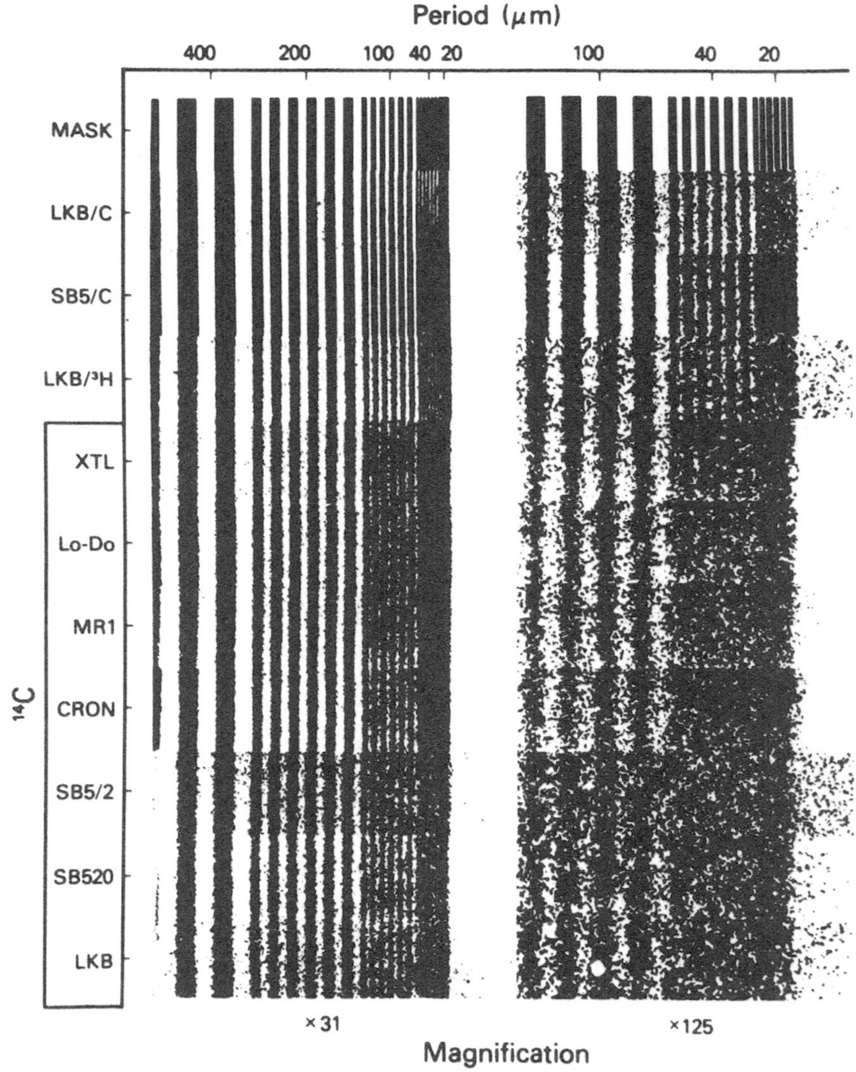

Fig. 16. Images produced by microgratings doped with either ^{14}C or ^{3}H on a variety of films at two levels of magnification. The gratings consist of lines whose width and spacing equal one-half of the period. MASK, the mask from which the microgratings were produced; LKB/C and SB5/C, contact print images of the mask on LKB film and SB5 film, respectively; LKB/^{3}H, image on LKB film of the grating impregnated with ^{3}H; the

matter is predominately neuropil. Wagner et al. (1979) used the dipping emulsion technique to examine the magnocellular neurosecretory cells of the hypothalamus during osmotic stress. They did not, however, find a preferential labeling of these cells during water deprivation. Other studies utilizing the ^{3}H-labeled deoxyglucose technique include those investigating the effect of unilateral basilar papilla removal on glucose uptake in second- and third-order auditory nuclei in the chick (Lippe et al., 1980) and the metabolic changes accompanying denervation and reinnervation of the dentate gyrus (Steward and Smith, 1980).

A slight modification of this technique involves freeze-drying tissue and impregnating it with a resin (Pilgrim and Wagner, 1981). Following tissue sectioning, it can then be placed onto predipped slides, exposed, developed, and grain counting performed, as discussed above. This has the advantage that the tissue does not have to remain frozen during the exposure period, but may produce a more serious problem in that the impregnating procedure may potentially wash out some of the labeled deoxyglucose, thereby interfering with quantification. A similar problem may exist with a resin-embedding technique developed by Des Rosiers and Descarries (1978) in which animals were perfused with a sequence of 3.5% gluteraldehyde (phosphate-buffered), 1% OsO_4 (phosphate-buffered) acetone, and three acetone-Epon mixtures of increasing concentrations of resin. Very thin sections (1 μm) of the Epon-embedded brain tissue produce very fine autoradiograms at the cellular level that are adquately preserved for light and electron microscopy. Other types of fixatives have been investigated with equal success—excellent autoradiographic resolution has been obtained, but at the expense of loss of an appreciable percentage of the label (Durham et al., 1981). It is thought, however, that the loss is such that the original relative distribution is preserved; if so, then the resulting image is still proportional to glucose metabolism.

←

bottom seven images are from a grating impregnated with ^{14}C on a variety of films. SB5, XTL, and MR1 are films made by Kodak, each with different emulsion thicknesses, anti-abrasion layer thicknesses, and grain sizes, whereas Cronex 4 (CRON) and Lo-Dose (Lo-Do) are made by Dupont. SB5/2 indicates that the grating thickness was 2.25 μm, whereas SB520 and the remainder of the images are produced by gratings that were 20 μm thick (from Gallistel and Nichols, 1983).

A fixation technique has also been used in in vitro studies in which the tissue under study was incubated with ^{3}H-2DG while being appropriately stimulated according to the experimental design (Basinger et al., 1979). The fixation was very similar to that of Des Rosiers and Descarries (1978), which, although it produces autoradiograms of excellent resolution, washes out an appreciable amount of the deoxyglucose 6-phosphate and free deoxyglucose, making quantification impossible.

There has been considerable interest in developing appropriate freeze drying and freeze substitution methods for the processing of 2-deoxyglucose autoradiograms. This interest is enhanced by the opportunity of obtaining sections that not only have excellently preserved histology, but have the potential of being quantitatively analyzable in that the washout of the tracer can be minimized. Some of these techniques are discussed in a review by Smith (1983), in which it is concluded that, with proper care and the appropriately prepared standards, the 2-deoxyglucose technique can achieve cellular resolution.

4.2.2. Tritium-Sensitive Film

With the development of a film suitable for ^{3}H autoradiography (LKB Ultrafilm), it became possible to process ^{3}H autoradiograms in an identical fashion to ^{14}C autoradiograms (Fig. 17c). This film lacks the protective coating normally found on X-ray film, and so the ^{3}H is capable of reaching the emulsion. The film has been utilized for ^{3}H autoradiography with deoxyglucose (Goldberg et al., 1980; Alexander et al., 1981; Faraco-Cantin et al., 1980), although few quantitative studies have been undertaken. For quantitation it is necessary to obtain ^{3}H standards and calibrate them for the standard 20μm tissue sections. These standards can be made by adding a tritiated substance to a methyl methacrylate monomer, adding a methylmethacrylate styrene copolymer and a methacrylate polymer, and allowing the mixture to harden (Alexander et al., 1981). The standards are calibrated identically to the technique used for calibration of ^{14}C standards (Reivich et al., 1969). Alexander et al. (1981) have quantitatively measured local cerebral glucose utilization rates in rats with ^{3}H-deoxyglucose and compared the data to those obtained with ^{14}C-deoxyglucose. The agreement was excellent. One advantage of ^{3}H autoradiography, besides the improvement in spatial resolution, is that variations in section thicknesses do not lead to inaccuracies in quantitation. This derives from the fact that 20-μm thick sections are infinitely thick for ^3H, so that the ^3H is absorbed by the tissue from all layers of the tissue greater than 5 μm from the emulsion (Alexander et al., 1981). Although this self-absorption eliminates section thickness problems, it does complicate the quantification in that different regions of the brain may have different self-absorption properties, thereby necessitating separate calibration of the standards for these regions. In particular, there is an appreciable difference in self-absorption between gray and white matter for the most part because of the different water contents of the two tissue types (Hawkins et al., 1979; Alexander et al., 1981; Kuhar and Unnerstall, 1985). Gray matter has a water content of approximately 80–85%, whereas the water content in white matter is closer to 70–75%. Although this difference may not appear appreciable, if the sections are thaw-mounted and hence dehydrated, as is the usual procedure, then what is left on the film is the tissue solid, which is more than 50% greater for white matter than for gray matter. Many cerebral structures, especially subcortically, contain varying concentrations of fiber tracts with resulting different self-absorption characteristics. It should be noted that the energy of ^{14}C is sufficiently low so that conventional ^{14}C autoradiography is not immune to self-absorption artifacts. Although the effects are very much smaller than with ^3H, they can lead to errors of the order of 10%. Most laboratories do not account for these differences.

A more practical consideration concerning autoradiography with ^3H-deoxyglucose is the large amount of radionuclide that must be used to produce high-quality autoradiograms on the film in a reasonable length of time, making ^3H-labeled autoradiography more expensive than ^{14}C-labeled autoradiography. On the other hand, the specific activities achievable with ^3H labeling permit more isotope to be administered without violating the requirement that the deoxyglucose be present in tracer quantities only.

5. Multiple-Label Autoradiography

In a number of experimental protocols, it is sometimes extremely advantageous to be able to make two (or more) autoradiographic measurements in a given animal. For example, in the study of the cerebral circulation, measurements of local cerebral blood flow along with local cerebral glucose utilization in the same animal can yield much more information than that obtained by measuring these two parameters in separate animals. Another use

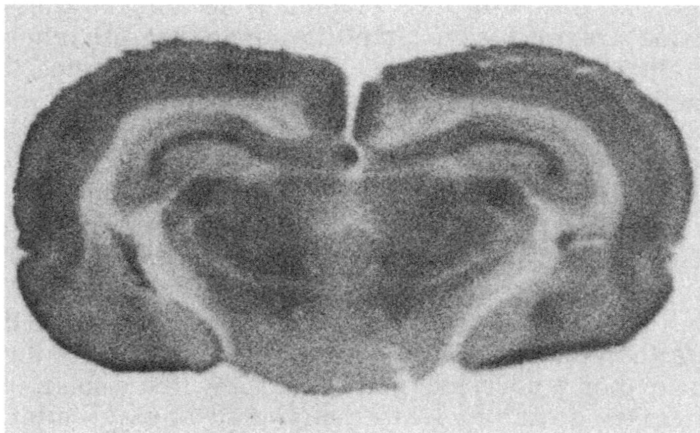

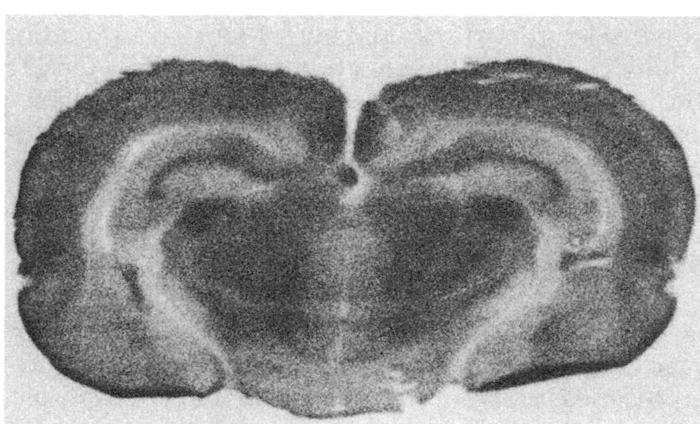

Fig. 17a (top) and b (bottom).

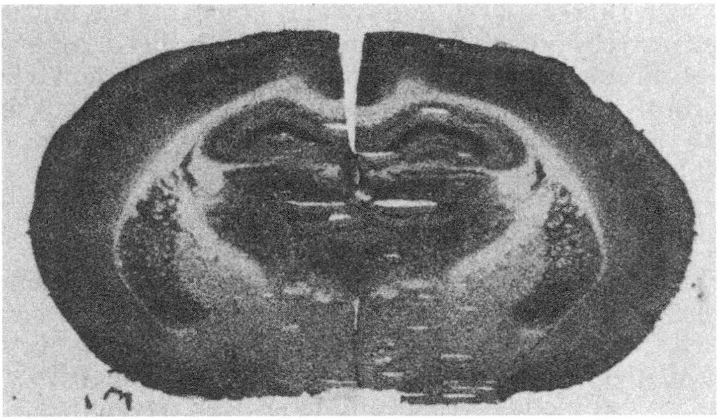

Fig. 17a–c. Double label autoradiography using ^{14}C and ^{3}H. The image produced by placing the autoradiogram on tritium-sensitive film (LKB Ultrofilm) will have contributions caused both by ^{3}H and ^{14}C (a), whereas the image produced on film with an antibrasion coating (SB5) is caused only by ^{14}C (b). With single-label tritium autoradiography, the resolution is improved, as seen in the ^{3}H-2-deoxyglucose image of panel (c).

of double-label autoradiography involves the measurement of the spatial distribution of a substance along with the local metabolic effect of that substance in different regions of the tissue.

In order to differentiate between the distribution of different compounds (i.e., deoxyglucose and iodoantipyrine) in the tissue, it is necessary to label them with different radionuclides (i.e., ^{14}C, ^{3}H, or ^{123}I). By exploiting various different properties of these radionuclides, such as differences in energy or decay constants, it is possible to obtain separate autoradiographic images of the two compounds. The general concept is to obtain multiple images from the same sections, with some parameter changing between the images. For example, for double-label autoradiography, differentiation according to energy would involve obtaining an autoradiographic image of the section which will be caused by both isotopes and then interposing a thin absorber between the section and the film capable of stopping all of the low-energy isotope, in order to obtain an image caused only by the higher-energy radionuclide. Appropriate manipulation of the images permits the concentration of each radionuclide to be obtained. Differentiation according to decay constants is similar. In this case two radionuclides with sufficiently different decay constants are utilized. An image caused by both radionuclides is first produced. After waiting for a sufficient length of time for one of the radionuclides to significantly decay, a second image of the same sections is made caused either primarily or totally (depending on the decay constants) by one of the radionuclides.

5.1. Carbon/Iodide Double Label

The radionuclides of iodide are in many ways convenient for multiple-label autoradiography. A number of different radionuclides with varying decay constants can be produced (^{123}I, ^{125}I, ^{131}I with half lives of 13.1 h, 60 d, and 8.06 d, respectively), and numerous techniques for labeling have been developed.

One of the first quantitative double-label autoradiographic techniques to be developed using isotopes of iodine was a method for the simultaneous measurement of local cerebral blood flow and local cerebral glucose utilization rates (Jones et al., 1979; Lear et al., 1981). In this technique, local cerebral blood flow was measured with ^{123}I-iodoantipyrine, and local cerebral glucose metabolism with ^{14}C-labeled 2-deoxyglucose (Fig. 18). ^{123}I has a half-life of 13.1 h, and its decay produces internal conversion electrons of energies (159 keV, 87%; 127 keV, 13%) close to ^{14}C (a spectrum with maximum energy 155 keV, mean energy 55 keV). Hence it is suitable for autoradiography. Two autoradiography images are produced. To obtain the first image, the sections are placed on the film for 12–14 h to produce an image caused primarily by ^{123}I, since the quantities of isotopes injected are such that the ^{123}I can produce a good image in this period of time, whereas ^{14}C requires approximately 1 wk for a comparable image. Thus the first image is (in normal tissue) approximately 92% caused by ^{123}I and 8% caused by ^{14}C. The sections are removed from the film for a week, at which time they are placed back on film for a period of 1 wk. By this time the ^{123}I has decayed to a negligible amount, and the second image is caused primarily by ^{14}C, with a 6–8% component in the second exposure caused by ^{125}I, a contaminant in the production of ^{123}I. Subtraction equations were derived to allow the images to be corrected so that "^{14}C only" and "^{123}I only" images could be produced (Lear et al., 1981). Because of the short half-life of ^{123}I, it is impractical to make ^{123}I standards. It was found more convenient to calibrate the ^{14}C standards for ^{123}I, i.e., to determine the relationship between how ^{123}I exposes film and how ^{14}C exposes film. In the range of exposures utilized, this ratio is a constant, thus permitting all densitometry to be expressed in ^{14}C units. The subtraction is done, therefore, in ^{14}C exposure units, not optical densities.

This technique has been used to examine the flow-metabolism couple in normal rats (Jones et al., 1981) and in an examination of the relationship between local cerebral blood flow and local cerebral glucose utilization in hemorrhagic hypotension in the rat (Greenberg et al., 1981), in focal cerebral ischemia in the cat (Choki et al., 1983), and in functional activation in the vibrissae/barrel system of the rat (Greenberg et al., 1979). It should be noted that errors arise during the subtraction process; this is in addition, of course, to the errors inherent in single-label autoradiographic procedures. In the case in which there is less than 10% cross-contamination between the images, the errors in determining the concentrations of the two radionuclides increase by a factor of 1.1, whereas when the cross-contamination is 50%, the errors increase by a factor of more than 1.7.

A modification of the $^{123}I/^{14}C$ double-label technique involves using ^{131}I instead of ^{123}I (Mies et al., 1981). The half-life of ^{131}I is 8.06 d, so that the radionuclide is more readily available and hence more conveniently utilized. Its longer half-life, however, necessitates a very long time delay between the first and second exposure in order to minimize ^{131}I contamination in the second exposure. In addition, the energy of ^{131}I (mean energy = 284 keV) is sufficient to compromise spatial resolution.

5.2. $^{14}C/^{3}H$ Double Label

Double-label autoradiographic studies have also been undertaken with ^{3}H and ^{14}C (Alexander et al., 1982). Instead of exploiting half-life differences between the isotopes (^{3}H has a half-life of 12.5 yr), the isotope distributions are separated using the large difference in energies between them (^{3}H, E_{mean} = 6 keV; ^{14}C, E_{mean} = 55 keV). An exposure is made on standard X-ray film with a protective coating that does not allow ^{3}H to expose the film. This produces a "^{14}C only" image (Fig. 17). Subsequent exposure on ^{3}H film produces an image with substantial contributions from both ^{14}C and ^{3}H (Fig. 17). Image subtraction allows the contribution for ^{3}H to be determined, although with a not insignificant error. Because of the low energy of ^{3}H and hence the poor sensitivity of film (even "^{3}H" film) to ^{3}H, it is difficult to arrange the administered doses of both ^{14}C and ^{3}H so that the image on the ^{3}H

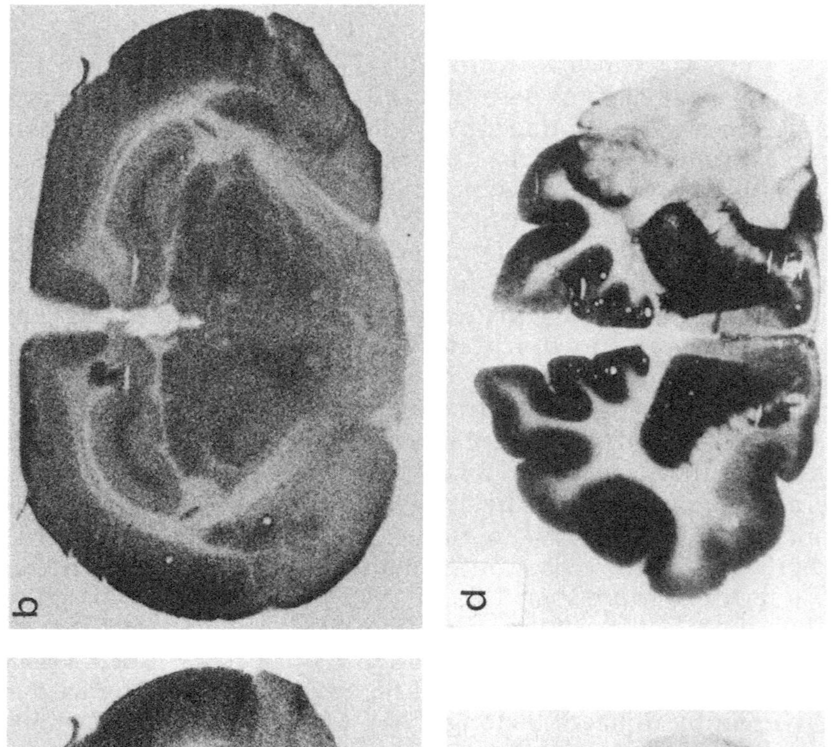
a
b
c
d

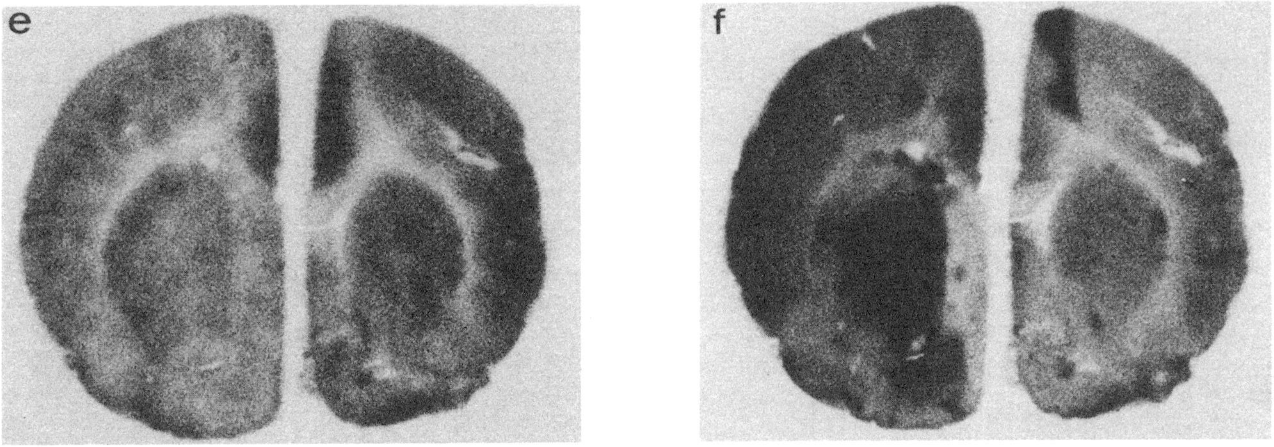

Fig. 18. Double-label autoradiograms using ^{14}C and ^{123}I. The images on the left (a), (c), and (e) are obtained from the first short (13 h) exposure and hence are caused primarily by ^{123}I-iodoantipyrine with a 7–8% contamination caused by ^{14}C. These images are thus primarily blood flow images. The images on the right were obtained after the ^{123}I had decayed and are caused primarily by ^{14}C-2-deoxyglucose, with a 6–8% contamination caused by ^{125}I, a contaminant in the ^{123}I production. Consequently the images are principally that of regional glucose metabolism. Image pairs (a,b) are from an anesthetized rat. Images (c,d) are from a cat in which the left middle cerebral artery has been occluded. The ^{14}C-2-deoxyglucose was administered 90 min after occlusion, and the ^{123}I-iodoantipyrine, 30 min later. Note that in areas of rather uniform ischemia that some regions have an increased glucose utilization, whereas other areas exhibit a significant metabolic depression. Image pairs (e,f) are from a rat at a blood pressure of 40 mm Hg. Note the dramatic mismatch between cerebral blood flow and cerebral glucose metabolism.

film (without the protective antiabrasive coating) is primarily caused by ^{3}H. In order to give equivalent exposures on LKB film, it is necessary to have tissue concentration of ^{3}H approximately 40 times greater than the ^{14}C concentration. Since the amount of ^{14}C administered to the animal is set by the requirement of producing an image on the standard X-ray film (with the protective coating) in a reasonable period of time, the only way to make the combined image (on ^{3}H film) predominately caused by ^{3}H is to inject in excess of 100 times as much ^{3}H as ^{14}C. This is usually not practical because of the limitations of tracer kinetics, not to mention expense. Another limitation of a double-label technique that uses ^{3}H as one of the labels concerns the problem of self-absorption discussed above (section 4.3.2). This must be accounted for in the subtraction procedure.

5.3. ^{14}C/^{14}C Double Label

The concept of using ^{14}C as the label on both of the compounds in a double-label procedure is novel and worth investigation primarily because of the convenience of ^{14}C autoradiography. In ^{14}C/^{14}C autoradiography, an initial exposure is made on regular X-ray film (usually Kodak SB5) that includes both tracers. The tissue sections are then washed in a solvent in order to selectively remove one of the tracers, the solvent, of course, being chosen so as wash out only one tracer. A second exposure follows, permitting the direct measurement of the concentration of the remaining tracer. Appropriate image subtraction is then used to obtain the concentration of the tracer that was washed out by the solvent. It should be noted that this subtraction must be done in exposure units; one cannot subtract optical densities. A technique for the double-label autoradiographic measurement of local cerebral blood flow and local cerebral metabolic rate for glucose using ^{14}C-IAP and ^{14}C-2DG has been described by Furlow et al. (1983). In these studies the first exposure was caused by both compounds, in approximately equal amounts. They then washed the dry sections in chloroform for 72 h, after which they placed the washed sections on film for another exposure. They reported that the cholorform extracts all of the ^{14}C-IAP from the brain section, leaving the ^{14}C-2DG (and ^{14}C-deoxyglucose-6-phosphate) intact. The validation of this procedure, that all of the IAP is washed out and all of the 2DG remains, consisted of comparing the results using this technique with data from single label studies. Since reasonably good agreement was obtained, Furlow et al. (1983) concluded that this technique can be used for quantitative measurements of lCBF and lCGU.

A more rigorous evaluation of this procedure was undertaken by Jones and Greenberg (1985). They directly measured the amount of both ^{14}C-IAP and ^{14}C-2DG that was extracted from the sections by chloroform during the wash procedure. Sections from rats administered either ^{14}C-IAP or ^{14}C-2DG were placed on film for 7 d to obtain prewash images. The same sections were then washed in chloroform for various times up to 102 h and then reexposed to film for a postwash image. The ratio of the ^{14}C in the two images is a measure of the washout of the compound from the sections. They found that although 98% of the IAP was washed out by 25 h, the 2DG also washed out, with only approximately 81% of the ^{14}C-2DG remaining after chloroform washing (Fig. 19). Not accounting for this significant washout of 2DG will lead to errors in the calculated ^{14}C-IAP concentration as well as the ^{14}C-2DG concentration of at least 20% (depending on the ratio of the two compounds in the tissue). Because of the nonlinear relationship between ^{14}C-IAP concentrations and lCBF, the errors in blood flow will be even larger.

The data of Jones and Greenberg (1985) indicate, however, that the amount of ^{14}C-2DG washed out by the chloroform is relatively reproducible, suggesting that if this washout component is included in the subtraction equations, acceptable quantification of both ^{14}C-IAP and ^{14}C-2DG can be obtained. A technique capitalizing on this reproducibility has recently been developed (Ginsberg et al., 1986). Although they attempted to measure the amount of ^{14}C-IAP remaining postwash in each study, the small magnitude of this value made this measurement prone to large errors. Utilizing optimalization techniques, it was found that a value of 2% yields data with the least error. The amount of ^{14}C-2DG that remained postwash could be measured with each study (by washing single-label sections containing ^{14}C-2DG only), but was found to be quite stable with a value of 0.6. This technique is quite promising in that ^{14}C-labeled compounds are generally quite convenient to work with. Its greatest disadvantage in comparison to some of the double-label autoradiographic techniques discussed above is the potential subtraction error generated by the fact that the first exposure contains appreciable contributions from both compounds. Theoretically this could be circumvented by giving ^{14}C-IAP in large excess over ^{14}C-2DG; this is impractical because of

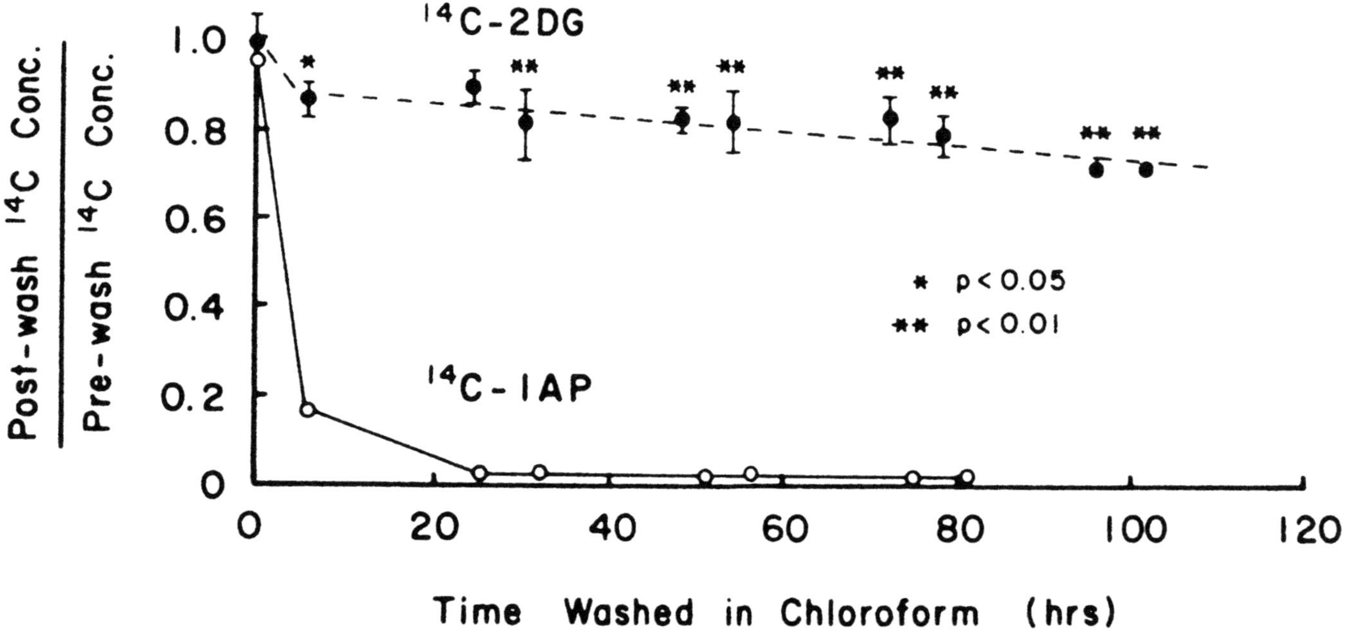

Fig. 19. The rate at which ^{14}C is removed from sections washed in chloroform. ^{14}C-Iodoantipyrine (IAP) washes out quite rapidly, whereas the ^{14}C-2-deoxyglucose (2DG) is removed from the section quite slowly. After 30 h of washing, the ^{14}C in the section after washing is significantly lower than the ^{14}C in the section prior to washing. The significance of differences from control was obtained with Dunnett's *t* test (from Jones and Greenberg, 1985).

the long exposure times that would be necessary for the second exposure. This washout technique, of course, is limited to double-label studies in which one (and only one) of the compounds can be appreciably washed out.

5.4. $^{18}F/^{14}C$ Double Label

The very short half-life of some radionuclides has been successfully utilized in double-label autoradiography. If one combines large amounts of a very short-lived radionuclide, such as ^{18}F, which has a half life of 110 min, with ^{14}C, then almost complete separation between tracers can be achieved directly on film. With the normal quantities of ^{14}C and sufficient quantities of ^{18}F, a first exposure of 2–3 h will yield an image that is almost entirely caused by ^{18}F with negligible contamination caused by the ^{14}C. Waiting a day or two prior to the second exposure will produce an image caused entirely by ^{14}C. Sako et al. (1984) utilized ^{18}F-fluorodeoxyglucose and ^{14}C-IAP to measure local cerebral glucose metabolism and local cerebral blood flow in the same animal. This technique suffers from the time constraint imposed by the short half-life of ^{18}F; the study must be well coordinated with the production of ^{18}F (which is only available to a very limited number of laboratories), and the brain must be sectioned very soon after the study is complete.

5.5. Autoradiography with More Than Two Labels

Theoretically, double-label autoradiography can be extended to the use of three or four labels. The major requirement is that the number of exposures equals the number of tracers used in the study, and that each exposure be significantly different from the previous one so that tracers can be separated and their concentrations calculated. In mathematical terms this simply means that the number of equations at least equals the number of unknowns, with redundant equations not permitted. The only reported autoradiographic studies in which more than two tracers have been used employ short-lived isotopes and require a number of images, each at (obviously) different times so that each image contains a different percentage of the various tracers. Lear et al. (1984) has derived a set of equations relating the half-lives of the tracers to the times of exposures for optimal separation of the tracers. With knowledge of the relative half-lives of the tracers, the derived nomograms permit the investigator to determine the best dose ratio for the radionuclides that will be used. Introduction of more than two tracers increases the errors involved appreciably.

6. Image Analysis

In the early days of quantitative autoradiography, the optical densities of the autoradiograms were obtained by using a simple densitometer consisting of a small light spot and a photometer. A panel meter would indicate the amount of light transmitted, and the isotope concentration would be determined by comparing this light transmission to the transmission through a set of isotopic standards. This soon became the limiting factor in the experimental studies; each animal experiment can easily produce 100 images, each with generally a very heterogeneous isotopic distribution. As autoradiography became a more common technique, and as technology became more available and less expensive, a number of tools began to be applied to the analysis of autoradiographic images. Principal among these has been the digitization of the autoradiogram with subsequent computer analysis. In the very simplest of configurations one can interface a digital computer to a single spot densitometer so that the transmitted light can be compared to transmission through calibrated standards and an isotope concentration determined instantaneously. Without extensive programing, this isotope concentration can be converted to a blood flow value or a glucose metabolic rate and, with appropriate software, coded according to location, structure name, and so on. This type of semiautomated densitometer has been in active use in our laboratory for over a decade and offers a significant advantage over previous methods in that it is fast and very accurate (10–12 bits of resolution is obtained) and can be constructed at a very modest cost.

Relatively inexpensive hardware that has become available over the past few years permits whole-image digitization. The digitized image is then displayed on a monitor and can be easily manipulated and analyzed. A variety of devices are available for digitizing the image. These include video cameras, charge-coupled cameras (two-dimensional arrays), charge-coupled linear arrays, and rotating drum scanners. Each has different characteristics with regard to signal:noise, resolution, geometric linearity, sensitivity, stability, scan time, and cost. A detailed discussion of all of these devices is beyond the scope of this chapter. The vidicon, which is the most common type of video camera, does not have a very large signal:noise ratio, so that high accuracy is not possible. It is inexpensive ($500–2500), however, and digitized images can be obtained quite quickly, making it attractive for low-cost systems. Its reasonably poor accuracy is probably not a problem for single-label studies, but would produce significant errors in multiple-label studies when image subtraction is necessary. Two-dimensional, charge-coupled cameras are significantly more expensive, but exhibit much higher signal:noise ratios with a larger dynamic range. The array may not be exactly uniform, which can potentially lead to artifacts in the image. If a normalization scan is performed first (without the film), however, this nonuniformity can be removed. They are also very fast and can digitize an image in seconds. Linear charge-couple arrays are less expensive than two-dimensional arrays. They can be incorporated into scanners by mounting the array on a track that moves mechanically across the image (Lear et al., 1985). Since fewer detectors comprise the array, they can be made quite uniform. They also have excellent signal-:noise characteristics. The rotating drum scanner is basically a single-spot, single-photomultiplier/detector system. The film is mounted on a cylindrical drum; on the outside of the drum is a light source with a selectable aperature of 25–200 μm, and on the inside sits a light guide connected to a photomultiplier tube. The light passes through the film onto the photomultiplier while the film rotates on the drum. Simultaneously, the light source/photomultiplier assembly moves along the film so that the entire film can be scanned. Everytime that the drum rotates, an air-based reference reading is obtained. This compensates for any drift that may occur in the light source-detector system. This system is accurate (8 bits), but scanning time is long (approximately 5 min/image) and the scanner, which is much more mechanically complex than the other scanners, is quite expensive. An image analysis system built around this type of scanner is described by Goochee et al. (1980).

Once the image is digitized, it is generally displayed on a monitor, and, depending upon the hardware configuration as well as the software available, it can be analyzed in a variety of ways. Most image analysis systems allow the image to be passed through a look-up table defining the relationship between optical density (or light transmission) and display gray scale or color level (Fig. 20).

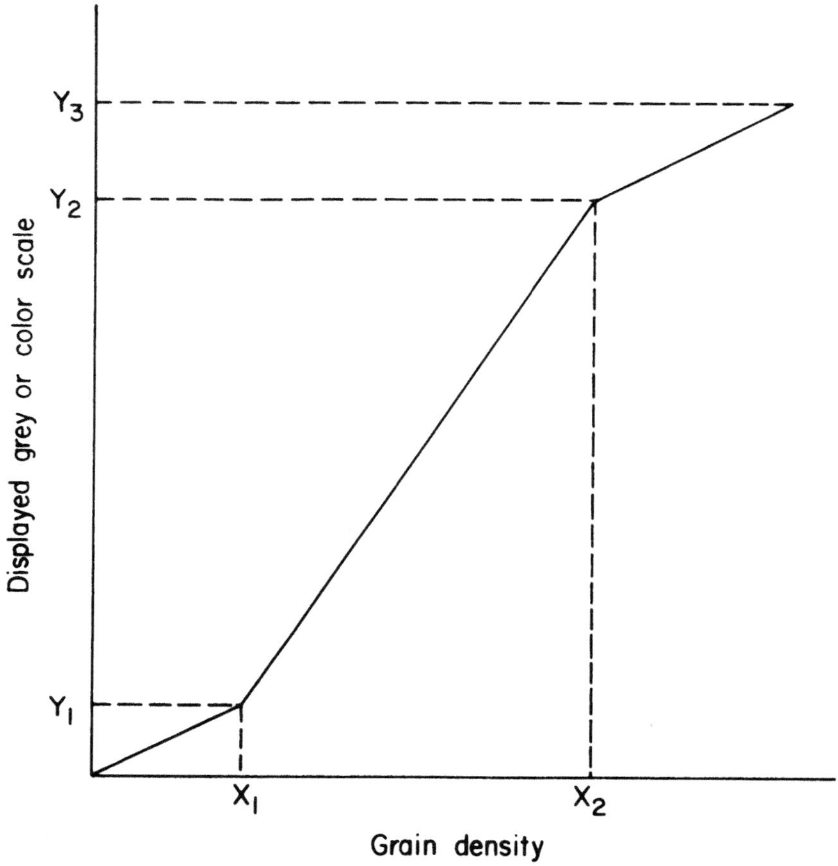

Fig. 20. Transformation between grain density on the film and the gray scale or color scale being displayed on the monitor. By varying X_1, X_2, Y_1, Y_2, and Y_3, the contrast of the image can be manipulated in such a manner as to highlight various density ranges in the original autoradiograms. For example, if the density range of interest is between X_1 and X_2, then this density range can be mapped into the entire gray scale by making $Y_1 = 0$ and $Y_2 = Y_3$.

Thus the density range on the autoradiogram (from the lightest to the darkest structure) can be mapped into a wider gray scale range, thereby increasing the contrast. This allows the investigator to more easily visualize changes in labeling that may otherwise be missed. The ability to psuedo-color the image, that is, assign different colors (and shades of colors) to different densities also enables the investigator to more easily see the isotope distribution. Psuedo-coloring can be deceptive in that, depending upon the density-to-color transformation chosen, very small density changes can be represented by very dramatic color changes. It is important that pseudo-color images be accompanied by a quantitative color scale so that the image can be properly interpreted.

Most image analysis systems have the facility to obtain average optical densities in a select portion of the image by indicating regions of interest on the image. With minimal programing, this average density can be converted to a more useful parameter such as blood flow, glucose metabolism, neuroreceptor density, and so on. Unless the relationship between optical density (or transmission) is linear, this transformation should be done on a pixel-by-pixel basis. Other functions incorporated into many image analysis systems include line profiling, multisegment display, and image alignment, subtraction, and division (necessary for multiple-label autoradiography). Line or density profiling allows the user to place a line through the image and obtain a plot of the density (flow, metabolism, and so on) along the line. This is particularly useful in examining data in which unilateral effects may exist. A plot of the line profile, e.g., through the oculomotor nuclear complex, would graphically show asymmetries caused by unilateral visual stimulation (Kennedy et al., 1976).

7. Applications

Autoradiography is, after all, only a tool by which local radioisotopic distributions can be measured. With this tool one can measure a variety of parameters, only two of which have been discussed above—cerebral blood flow and cerebral glucose metabolism. A variety of other functions can be mapped; one of the most exciting parameters of recent interest has been neuroreceptor distribution in the brain.

The use of the diffusible tracer technique for the measurement of local cerebral blood flow autoradiographically has, for the most part, been limited to measurements of local flow in a variety of pathological models, including ischemia (Ginsberg et al., 1978), seizures (Ingvar and Siesjo, 1983), and hypertension (Greenberg et al., 1985). A limited number of studies have utilized this technique to examine blood flow alterations during functional activation. These include visual activation (Sokoloff, 1961), sleep (Reivich et al., 1968), and vibrissae stimuation (Greenberg et al., 1979).

The 2-deoxyglucose technique for the measurement of local cerebral glucose utilization has found very extensive use in the neurosciences. The close relationship between local functional activity and glucose metabolism as demonstrated in studies in which animals are functionally activated (visual and auditory stimulation and deprivation, vibrissae stimulation, nerve stimulation, and so on) has led to the use of this technique for studies in which measurements of neuronal activity are sought. Even a brief review is beyond the scope of this chapter. A number of reviews exist (Greenberg and Reivich, 1983; Sokoloff, 1985a), although because of the wide use of this technique, these reviews are not current.

The autoradiographic techniques discussed above are essentially similar to techniques of emission tomography discussed elsewhere in this volume, so much so that emission tomography is frequently referred to as in vivo autoradiography. Both techniques enable the measurement of the organization of the central nervous system and the derangements that occur in pathological states. As such, they are extremely valuable tools.

References

Abrams R., Ito M., Frisinger J. E., Patlak C. S., Pettigrew K. D., and Kennedy C. (1984) Local cerebral glucose utilization in fetal and neonatal sheep. *Am J. Physiol.* **246,** R608–R618.

Alexander G., Schwartzman R. J., Bell R. D., and Bertoni J. M. (1982) Simultaneous autoradiographic quantification of cerebral blood flow and metabolism. *Neurology* **32,** A196.

Alexander G. M., Schwartzman R. J., Bell R. D., Yu., and Renthal A. (1981) Quantitative measurement of local cerebral metabolic rate for glucose utilizing tritiated 2-deoxyglucose. *Brain Res.* **223,** 59–67.

Bachelard H. S. (1971) Specificity and kinetic properties of monosaccharide uptake into guinea pig cerebral cortex in vitro. *J. Neurochem.* **18,** 213–222.

Bachelard H. S., Clark A. G., and Thompson M. F. (1971) Cerebral-cortex hexokinases. Elucidation of reaction mechanisms by substrate and dead-end inhibitor kinetic analysis. *Biochem. J.* **123,** 707–715.

Basinger S. E., Gordon W. C., and Lam D. M. K. (1979) Differential labeling of retinal neurones by ^{3}H-2-deoxyglucose. *Nature* **280,** 682–684.

Bidder T. G. (1968) Hexose translocation across the blood-brain interface: Configurational aspects. *J. Neurochem.* **15**, 867–874.

Bradbury M. W. B., Patlak C. S., and Oldendorf W. H. (1975) Analysis of brain uptake and loss of radiotracers after intracarotid injection. *Am. J. Physiol.* **229**, 1110–1115.

Choki J., Greenberg J. H., Jones S. C., and Reivich M. (1983) Correlation between blood flow and glucose metabolism in the focal cerebral ischemia in the cat: An application of double label autoradiographic method. *J. Cereb. Blood Flow Metab.* **3** (suppl. 1), S399–S400.

Crone C. (1963) The permeability of capillaries in various organs as determined by use of the "indicator difussion" method. *Acta Physiol. Scand.* **58**, 292–305.

Des Rosiers M. H. and Descarries L. (1978) Adaptation de la methode au desoxyglucose a l'echelle cellulaire: Preparation histologique du systeme nerveux central en vue de la radioautographie a haute resolution. *C.R. Acad. Sci.* Ser. D **287**, 53–56.

Duffy T. E., Cavazzuti M., Cruz N. F., and Sokoloff L. (1982) Local cerebral glucose metabolism in newborn dogs: Effects of hypoxia and halothane anesthesia. *Ann. Neurol.* **11**, 233–246.

Durham D., Woolsey T. A., and Kruger L. (1981) Cellular localization of 2-[^{3}H]deoxy-D-glucose from paraffin-embedded brains. *J. Neurosci.* **1**, 519–526.

Eckman W. W., Phair R. D., Fenstermacher J. D., Patlak C. S., Kennedy C., and Sokoloff L. (1975) Permeability limitation in estimation of local brain-blood flow with [^{14}C]antipyrine. *Am. J. Physiol.* **229**, 215–221.

Eichling J. O., Raichle M. E., Grubb R. L., and Ter-Pogossian M. M. (1974) Evidence of the limitations of water as a freely diffusible tracer in brain of the rhesus monkey. *Circ. Res.* **35**, 358–364.

Eklof B., Lassen N. A., Nilsson L., Norberg K., Siesjo B. K., and Torlof P. (1974) Regional cerebral blood flow in the rat measured by the tissue sampling technique; a critical evaluation using four indicators C^{14}-antipyrine, C^{14}-ethanol, H^3-water and xenon^{133}. *Acta. Physiol. Scand.* **91**, 1–10.

Faraco-Cantin F., Courville J., and Lund J. P. (1980) Methods for ^{3}H-2-D-deoxyglucose autoradiography on film and fine-grain emulsions. *Stain Technol.* **55**, 247–252.

Freygang W. H. and Sokoloff L. (1958) Quantitative measurement of regional circulation in the central nervous system by the use of radioactive inert gas. *Adv. Biol. Med. Phys.* **6**, 263–279.

Furlow T., Martin R., and Harrison L. (1983) Simultaneous measurement of local glucose utilization and blood flow in the rat brain: An autoradiographic method using two tracers labeled with carbon-14. *J. Cereb. Blood Flow Metab.* **3,** 62–66.

Gallistel C. R. and Nichols S. (1983) Resolution-limiting factors in 2-deoxyglucose autoradiography. I. Factors other than difussion. *Brain Res.* **267,** 323–333.

Ginsberg M. D. and Reivich M. (1979) Use of the 2-deoxyglucose method of local cerebral glucose utilization in the normal brain: Evaluation of the lumped constant during ischemia. *Acta Neurol. Scand.* **60** (suppl. 72), 226–227.

Ginsberg M. D., Budd W. W., and Welsh F. A. (1978) Diffuse cerebral ischemia in the cat. I. Local blood flow during severe ischemia and recirculation. *Ann. Neurol.* **3,** 482–492.

Ginsberg M. D., Smith D. W., Wachtel M. S., Gonzalez-Carvajal M., and Busto R. (1986) Simultaneous determination of local cerebral glucose utilization and blood flow by carbon-14 double-label autoradiography: Method of procedure and validation studies in the rat. *J. Cereb. Blood Flow Metab.* **6,** 273–285.

Gjedde A. (1982) Calculation of cerebral glucose phosphorylation from brain uptake of glucose analogs in vivo: A re-examination. *Brain Res. Rev.* **4,** 237–274.

Gjedde A. and Diemer N. H. (1983) Autoradiographic determination of regional brain glucose content. *J. Cereb. Blood Flow Metab.* **3,** 303–310.

Gjedde A., Andersson J., and Eklof B. (1975) Brain uptake of lactate, antipyrine, water and ethanol. *Acta. Physiol. Scand.* **93,** 145–149.

Gjedde A., Hansen A. J., and Siemkowicz E. (1980) Rapid simultaneous determination of regional blood flow and blood-brain glucose transfer in brain of rat. *Acta. Physiol. Scand.* **108,** 321–330.

Gjedde A., Wienhard K., Heiss W. D., Kloster G., Diemer N. H., Herholz K., and Pawlik G. (1985) Comparative regional analysis of 2-fluorodeoxyglucose and methylglucose uptake in brain of four stroke patients with special reference to the regional estimation of lumped constant. *J. Cereb. Blood Flow Metab.* **5,** 163–178.

Goldberg L., Courville J., Lund J. P., and Kauer J. S. (1980) Increased uptake of [^{3}H]deoxyglucose and [^{14}C]deoxyglucose in localized regions of the brain during stimulation of the motorcortex. *Can. J. Physiol. Pharmacol.* **58,** 1086–1091.

Goldman H. and Sapirstein L. A. (1973) Brain flow in the conscious and anesthetized rat. *Am. J. Physiol.* **224,** 122–126.

Goldman S. S., Hass W. K., and Ransohoff J. (1980) Unsymmetrical alkyl aryl thiourea compounds for use as cerebral blood flow tracers. *Am. J. Physiol.* **238,** H776–H787.

Goochee C., Rasband W., and Sokoloff L. (1980) Computerized densitometry and color coding of [^{14}C]deoxyglucose autoradiographs. *Ann. Neurol.* **7,** 359–370.

Greenberg J. H. and Reivich M. (1983) CNS Regulation of Carbohydrate Metabolism, in *Advances in Metabolic Disorders* vol. 10 (Szabo A. J., ed.) Academic, New York.

Greenberg J. H., Hand P., Sylvestro A., and Reivich M. (1979) Localized metabolic-flow couple during functional activity. *Acta Neurol. Scand.* **60** (suppl. 72), 12–13.

Greenberg J. H., Nadasy G., Kovach A. G. B., Jones S. C., and Reivich M. (1981) Local cerebral glucose metabolism during hemorrhagic hypotension. *J. Cereb. Blood Flow Metab.* **1** (suppl. 1), S254–S255.

Greenberg J. H., Burke A. N., Sladky J., and Reivich M. (1985) Regional variation in cerebral perfusion during acute hypertension. *J. Cereb. Blood Flow Metab.* **5** (suppl. 1), S455–S456.

Hawkins R. A. and Miller D. L. (1978) Loss of radioactive 2-deoxy-D-glucose-6-phosphate from brains of conscious rats: Implications for quantitative autoradiographic determination of regional glucose utilization. *Neuroscience* **3,** 251–258.

Hawkins R., Hass W. K., and Ranschoff J. (1979) Measurement of regional brain glucose utilization in vivo using [2-^{14}C]glucose. *Stroke* **10,** 690–703.

Hawkins R., Phelps M. E., Huang S. C., and Kuhl D. E. (1981) Effect of ischemia on quantification of local cerebral glucose metabolic rate in man. *J. Cereb. Blood Flow Metab.* **1,** 37–52.

Horton R. W., Meldrum B. S., and Bachelard H. S. (1973) Enzymatic and cerebral metabolic effects of 2-deoxy-D-glucose. *J. Neurochem.* **21,** 507–520.

Huang M. T. and Veech R. L. (1982) The quantitative determination of the in vivo dephosphorylation of glucose 6-phosphate in rat brain. *J. Biol. Chem.* **257,** 11358–11363.

Ingvar M. and Siesjo B. K. (1983) Local blood flow and glucose consumption in the rat brain during sustained bicuculline-induced seizures. *Acta Neurol. Scand.* **68,** 129–144.

Jones S. C. and Greenberg J. H. (1985) Evaluation of a double-tracer autoradiographic technique for the measurement of both local cerebral glucose metabolism and local cerebral blood flow. *J. Cereb. Blood Flow Metab.* **5,** 335–337.

Jones S. C., Lear J. L., Greenberg J. H., and Reivich M. (1979) A double label autoradiographic technique for the quantitative measure of cerebral blood flow and glucose metabolism. *Acta Neurol. Scand.* **60** (suppl. 72), 202–203.

Jones S. C., Fedora T., Lear J., Greenberg J. H., and Reivich M. (1981) The flow-metabolism couple in normal rat brain. *J. Cereb. Blood Flow Metab.* **1** (suppl. 1), S488–S489.

Kennedy C., Des Rosiers M., Reivich M., Sharp F., Jehle J. W., and Sokoloff L. (1975) Mapping of functional neural pathways by autoradiographic survey of local metabolic rate with [^{14}C]deoxyglucose. *Science* **187,** 850–853.

Kennedy C., Des Rosiers M. H., Sakurada O., Shinohara M., Reivich M., Jehle J. W., and Sokoloff L. (1976) Metabolic mapping of the primary visual system of the monkey by means of the autoradiographic [^{14}C]deoxyglucose technique. *Proc. Natl. Acad. Sci. USA* **73,** 4230–4234.

Kennedy C., Sakurada O., Shinohara M., Jehle J., and Sokoloff L. (1978) Local cerebral glucose utilization in the normal conscious macaque monkey. *Ann. Neurol.* **4,** 293–301.

Kety S. S. (1951) The theory and applications of the exchange of inert gas at the lungs and tissues. *Pharmacol. Rev.* **3,** 1–41.

Kety S. S. and Schmidt C. F. (1948) The effects of altered arterial tensions of carbon dioxide and oxygen on cerebral blood flow and cerebral oxygen consumption of normal young men. *J. Clin. Invest.* **27,** 484–492.

Kuhar M. J. and Unnerstall J. R. (1985) Quantitative receptor mapping by autoradiography: Some current technical problems. *Trends Neurosci.* **8,** 49–53.

Landau W. M., Freygang W. H., Roland L. P., Sokoloff L., and Kety S. S. (1955) The local circulation of the living brain: Values in the unanesthetized and anesthetized cat. *Trans. Am. Neurol. Assoc.* **80,** 125–129.

Lear J., Ackermann R., Kameyama M., Carson R., and Phelps M. (1984) Multiple-radionuclide autoradiography in evaluation of cerebral function. *J. Cereb. Blood Flow Metab.* **4,** 264–269.

Lear J. L., Jones S. C., Greenberg J. H., Fedora T. J., and Reivich M. (1981) Use of ^{123}I and ^{14}C in a double for simultaneous measurement of LCBF and LCMRgl. *Stroke* **12,** 589–597.

Lear J. L., Muth R., and Plotnick J. (1985) Digital image analyser for autoradiography. *J. Nucl. Med.* **26,** p44.

Lippe W. R., Steward O., and Rubel E. W. (1980) The effect of unilateral basilar papilla removal upon nuclei laminaris and magnocellularis of the chick examined with [^{3}]H-2-deoxy-D-glucose autoradiography. *Brain Res.* **196,** 43–58.

Mangold R., Sokoloff L., Conner E., Kleinerman J., Gherman P. G., and Kety S. S. (1955) The effects of sleep and lack of sleep on the cerebral circulation and metabolism of normal young men. *J. Clin. Invest.* **34,** 192–1100.

Matsuda H., Nakai H., Jorkar S., Diksic M., Evans A. C., Meyer E., Redies C., and Yamamoto Y. L. (1987) Alternate approach to estimate lumped constant in the deoxyglucose model: Simulation and validation. *J. Nuc. Med.* **28,** 471–480.

McHenry L. C., Jr., Jaffe M. E., and Goldberg H. I. (1969) Regional cerebral blood flow measurement with small probes. *Neurology* **19,** 1198–1206.

Mies G., Niebuhr I., and Kossman K. A. (1981) Simultaneous measurement of blood flow and glucose metabolism by autoradiographic techniques. *Stroke* **12,** 581–588.

Ohno K., Pettigrew K. D., and Rapoport S. I. (1979) Local cerebral blood flow in the conscious rat as measured with ^{14}C-antipyrine, ^{14}C-iodoantipyrine and ^{3}H-nicotine. *Stroke* **10,** 62–67.

Oldendorf W. H. (1970) Measurement of brain uptake of radiolabeled substances using a tritiated water internal standard. *Brain Res.* **24,** 372–376.

Patlak C. S., Blasberg R. G., and Fenstermacher J. D. (1984) An evaluation of errors in the determination of blood flow by indicator fractionation and tissue equalibration (Kety) methods. *J. Cereb. Blood Flow Metab.* **4,** 47–60.

Phelps M., Huang S. C., Hoffman E. J., Selin C., Sokoloff L., and Kuhl E. E. (1979) Tomographic measurement of local cerebral glucose metabolic rate in humans with (F-18)2-fluoro-2-deoxy-D-glucose: Validation of method. *Ann. Neurol.* **6,** 371–388.

Pilgrim C. and Wagner H. J. (1981) Improving the resolution of the 2-deoxy-D-glucose method. *J. Histochem. Cytochem.* **29** (suppl. 1A), 190–194.

Raichle M. E., Eichling J. O., Straatmann M. G., Welch M. J., Larson K. B., and Ter-Progossian M. M. (1976) Blood-brain barrier permeability of ^{11}C-labeled alcohols and ^{15}O-labeled water. *Am. J. Physiol.* **230,** 5423–552.

Reivich M., Alavi A., Wolf A., Fowler J., Russell J., Annett C., MacGregor R. R., Shiue C. Y., Atkins H., Anand A., Dann R., and Greenberg J. H. (1985) Glucose metabolic rate kinetic model parameter determination in humans: the lumped constants and rate constants for [18F] fluorodeoxyglucose and [14C] deoxyglucose. *J. Cereb. Blood Flow Metab.* **5,** 179–192.

Reivich M., Alavi A., Wolf A., Greenberg J. H., Fowler J., Christman D., MacGregor R., Jones S. C., London J., Shiue C., and Younekura T. (1982) Use of 2-deoxy-D-[$^{1\text{-}11}$C]glucose for the determination of local cerebral glucose metabolism in humans: Variation within and between subjects. *J. Cereb. Blood Flow Metab.* **2,** 307–319.

Reivich M., Isaccs G., Evarts E., and Kety S. S. (1968) The effects of slow wave sleep and REM sleep on regional cerebral blood flow in cats. *J. Neurochem.* **15,** 301–306.

Reivich M., Jehle J., Sokoloff L., and Kety S. S. (1969) Measurement of regional cerebral blood flow with antipyerine-^{14}C in awake cats. *J. Appl. Physiol.* **27,** 296–300.

Reivich M., Kuhl D., Wolf A., Greenberg J., Phelps M., Ido T., Casella V., Fowler J., Hoffman E., Alavi A., Som P., and Sokoloff L. (1979) The [^{18}F]fluoro-deoxyglucose method for the measurement of local cerebral glucose utilization in man. *Circ. Res.* **44,** 127–137.

Reivich M., Sano N., and Sokoloff L. (1971) Development of an Autoradiographic Method for Regional Cerebral Glucose Consumption, in *Brain and Blood Flow* (Russell R., ed.) Pitman, London.

Rogers A. W. (1979) *Techniques in Autoradiography* 2nd Ed., American Elsevier, New York.

Rudolph A. M. and Heyman M. A. (1967) Circulation of the fetus in utero. Methods for studying distribution of blood flow, cardiac output and organ flow. *Circ. Res.* **21,** 163–184.

Sacks W., Sacks S., and Fleischer A. (1983) A comparison of the cerebral uptake and metabolism of labeled glucose and deoxyglucose in vivo in cats. *Neurochem. Res.* **8,** 661–685.

Sage J. I., Van Uitert R. L., and Duffy T. E. (1981) Simultaneous movement of cerebral blood flow and unidirectional movement of substances across the blood brain barrier: theory, methods and application to leucine. *J. Neurochem.* **36,** 1731–1738.

Sako K., Kato A., Diksic M., and Yamamoto, L. (1984) Use of short lived ^{18}F and long-lived ^{14}C in double tracer autoradiography for simultaneous measurement of LCBF and LCGU. *Stroke* **15,** 896–900.

Sakurada O., Kennedy C., Jehle J., Brown J. D., Carbin G. L., and Sokoloff L. (1978) Measurement of local cerebral blood flow with iodo-[^{14}C] antipyrine. *Am. J. Physiol.* **234,** H59–H66.

Savaki H. E., Davidson L., Smith C., and Sokoloff L. (1980) Measurement of free glucose turnover in brain. *J. Neurochem.* **35,** 495–502.

Schmiterlow G. C., Hansson E., Andersson G., Applegren L.-E., and Hoffmann P. C. (1967) Distribution of nicotine in the central nervous system. *Ann. NY Acad. Sci.* **142,** 2–14.

Schuier F., Orzi F., Suda S., Kennedy C., and Sokoloff L. (1981) The lumped constant for the [^{14}C]deoxyglucose method in hyperglemic rats. *J. Cereb. Blood Flow Metab.* **1,** S63.

Sharp F. R. (1976) Relative cerebral glucose uptake of neuronal perikarya and a neuropil determined with 2-deoxyglucose in resting and swimming rat. *Brain Res.* **110,** 127–139.

Smith C. B. (1983) Localization of Activity-Associated Changes in Metabolism of the Central Nervous System with the Deoxyglucose Method: Prospects for Cellular Resolution, in *Current Method in Cellular Neurobiology* (Barker, J. L and McKelvy J. FF., eds.) John Wiley, New York.

Sokoloff L. (1961) Local Cerebral Circulation at Rest During Altered Cerebral Activity Induced by Anesthesia or Visual Stimulation, in

The Regional Chemistry, Physiology and Pharmacology of the Nervous System (Kety S. S. and Elkes J., eds.) Pergamon, Oxford.

Sokoloff L. (1981) Localization of functional activity in the central nervous system by measurement of glucose utilization with radioactive deoxyglucose. *J. Cereb. Blood Flow Metab.* **1,** 7–36.

Sokoloff L. (1982) The Radioactive Deoxyglucose Method: Theory, Procedure and Applications for the Measurement of Local Glucose Utilization in Central Nervous System, in *Advances in Neurochemistry* (Agranoff B. W. and Aprison M. H., eds.) Plenum, New York.

Sokoloff L. (1985a) Application of Quantitative Autoradiography to the Measurement of Biochemical Processes In Vitro, in *Positron Emission Tomography* (Reivich M and Alavi A., eds.) Alan R. Liss, New York.

Sokoloff L. (1985b) Basic Principles in Imaging of Regional Cerebral Metabolic Rates, in *Brain Imaging and Brain Function* (Sokoloff L., ed.) Raven, New York.

Sokoloff L. (1986) Cerebral Circulation, Energy Metabolism, and Protein Synthesis: General Characteristics and Principles of Measurement, in *Positron Emission Tomography and Autoradiography* (Phelps M. E., Mazziotta J. C., and Schelbert H. R., eds.) Raven, New York.

Sokoloff L., Mangold R., Wechsler R. L., Kennedy C., and Kety S. S. (1955) The effect of mental arithmetic on cerebral circulation and metabolism. *J. Clin. Invest.* **34,** 1101–1108.

Sokoloff L., Reivich M., Kennedy C., Des Rosiers M. H., Patlak C. S., Pettigrew K. D., Sakurada O., and Shinohara M. (1977) The [^{14}]deoxyglucose method for the measurement of local cerebral glucose utilization: Theory, procedure, and normal values in the conscious and anesthetized albino rat. *J. Neurochem.* **28,** 897–916.

Sols A. and Crane R. K. (1954) Substrate specificity of brain hexokinase. *J. Biol. Chem.* **210,** 581–595.

Steward O. and Smith L. K. (1980) Metabolic changes accompanying denervation and reinnervation of the dentate gyrus of the rat measured by [^{3}-H]2-deoxyglucose autoradiography. *Exp. Neurol.* **69,** 513–527.

Suda S., Shinohara M., Miyaika M., Kennedy C., and Sokoloff L. (1981) Local cerebral glucose utilization in hypoglycemia. *J. Cereb. Blood Flow Metab.* **1,** S62.

Tower D. B. (1958) The effects of 2-deoxy-D-glucose on metabolism slices of cerebral cortex incubated in vitro. *J. Neurochem.* **3,** 185–205.

Van Uitert R. L. and Levy D. E. (1978) Regional brain blood flow in conscious gerbil. *Stroke* **9,** 67–72.

Wagner H. J., Pilgrim C., and Zwerger H. (1979) A system of cells in the unstimulated rat brain characterized by preferential accumulation of [3-H]deoxyglucose. *Neurosci. Lett.* **15,** 181–186.

Wick A. N., Drury D. R., Nakada H. I., and Wolfe J. B. (1957) Localization of the primary metabolic block produced by 2-deoxyglucose. *J. Biol. Chem.* **224,** 963–969.

Measurement of Regional Cerebral Hemodynamics and Metabolism by Positron Emission Tomography

Peter Herscovitch

1. Introduction

The past 40 years have seen progressive advances in the techniques available for measuring the blood flow and metabolism of the human brain. These advances have culminated in the development of positron emission tomography (PET), an imaging technique that permits the noninvasive, in vivo study of regional brain physiology and biochemistry.

A brief review of the earlier methods will help put the advantages of PET into perspective. In the 1940s, the pioneering work of Kety and Schmidt led to the nitrous oxide method for measuring cerebral blood flow (CBF) (Kety and Schmidt, 1948). Measurements of arterial and cerebral jugular venous concentrations of the gas during nitrous oxide inhalation permitted the calculation of cerebral hemispheric blood flow. These CBF measurements, when combined with measurements of brain arterial-venous differences for glucose and oxygen, permitted the determination of cerebral metabolism as well. The desire to obtain regional rather than global data led to the development of external probe systems to measure CBF. These were used to measure the clearance from the brain of freely diffusible radioactive gases, such as xenon-123, that were administered either by injection into the carotid artery or by inhalation (Obrist et al., 1967; Lassen and Ingvar, 1972). Subsequently, external probe techniques using intracarotid injections of radiotracers labeled with positron-emitting radionuclides were developed for the measurement of not only CBF, but also cerebral metabolism and blood volume (Ter-Pogossian et al., 1969, 1970; Eichling et al., 1975; Raichle et al., 1975).

These techniques, however, have several drawbacks. Only global measurements of cerebral metabolism can be obtained unless invasive carotid artery injections are used. The external probes used to obtain regional measurements of CBF have certain limitations. These systems record radioactivity from a volume of brain tissue extending a variable depth beneath the probe. Thus, radioactivity measurements from heterogeneous tissue elements are superimposed, and the presence of underperfused tissue in the field of view may not be detected. Also, measurements cannot be made from deeper structures such as the basal ganglia.

These limitations provided the impetus for the development of PET. The first tomograph was developed at Washington University in St. Louis by Ter-Pogossian and colleagues in the mid 1970s (Ter-Pogossian et al., 1975). There are three components necessary for the application of PET to the study of cerebral physiology. These are: (1) tracer compounds of physiological interest that are labeled with positron-emitting radionuclides and are safe for administration to humans; (2) a positron emission tomograph to provide images from which one can accurately measure the amount of positron-emitting radioactivity and thus the amount of tracer compound throughout the brain; and (3) a mathematical model that describes the in vivo behavior of the specific radiotracer used, so that the physiologic process under study can be quantitated from the tomographic measurements of regional radioactivity.

Several positron-emitting radiopharmaceuticals and corresponding tracer models have been developed to perform a variety of regional physiologic and biochemical measurements in the brain with PET. This chapter will provide a review of the positron emission tomographic measurement of cerebral blood flow, blood volume, and metabolism of glucose and oxygen. The imaging process and radiotracers used in PET will be described so as to provide a basis for the subsequent discussion of tracer methodology. Strategies for analyzing and interpreting the quantitative physiologic images provided by PET will also be reviewed.

2. PET Detection Systems

Certain radionuclides decay by the emission of a positron, a subatomic particle that has the same mass as an electron, but with a positive charge. After its emission from the nucleus, the positron travels a few millimeters in the tissue, losing its kinetic energy. When almost at rest, it interacts with an electron. This interaction results in the annihilation of both positron and electron, so that their combined mass is converted into energy in the form of electromagnetic radiation. This radiation consists of two high-energy (511 keV) gamma photons that travel away from the annihilation site with the speed of light at 180° to each other. Detection of these annihilation photon pairs, which result from radioactive decay events, is used to measure both the amount and the location of radioactivity. The two annihilation photons can be detected by two radiation detectors that are connected by an electronic coincidence circuit (Fig 1). The circuit records a decay event only when both detectors sense the arrival of the photons almost simultaneously. A very short time window for photon arrival, typically 5–20 ns, is allowed for registration of a coincidence event. The time window, called the coincidence resolving time,

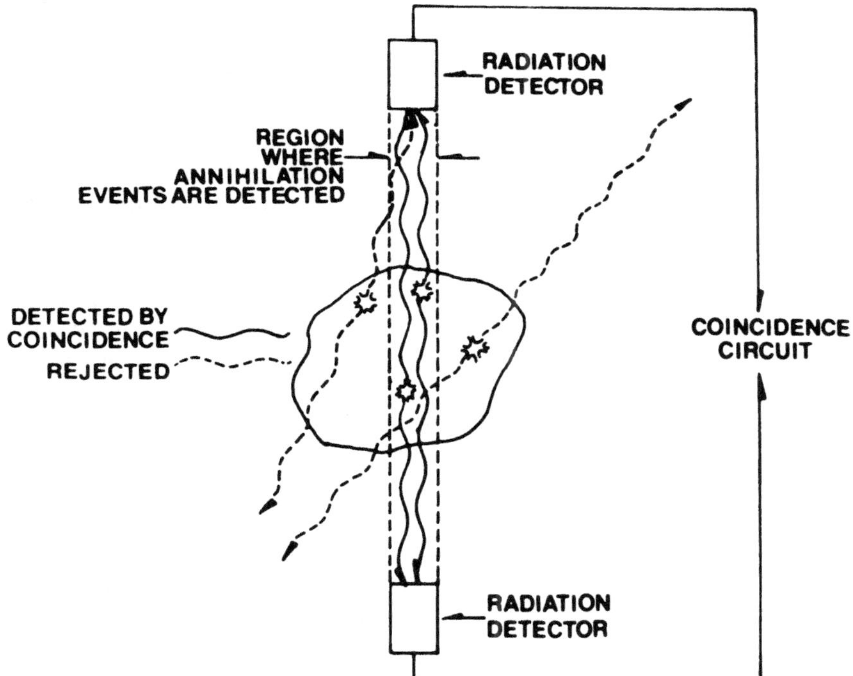

Fig. 1. Detection of annihilation photons by two radiation detectors connected by a coincidence circuit. A decay event is recorded only when both photons are detected almost simultaneously. This coincidence requirement localizes the site of the annihilation to the volume of space between the detectors (reproduced from *Sem. Nuclear Med.* **7**, 109, 1977, courtesy of Dr. M. M. Ter-Pogossian and by permission of Grune and Stratton).

depends upon the design of the detectors and the electronic coincidence circuits used in the tomograph. The coincidence requirement for photon detection localizes the site of the annihilation to the volume of space between the detector pair.

In practice, a ring of radiation detectors connected in pairs by coincidence circuits is used to surround the distribution of positron-emitting radioactivity in the body. With each decay event, the annihilation photons are detected as a coincidence line so that the number of coincidence lines recorded by any pair of detectors is proportional to the amount of radioactivity between them. A computer is used to reconstruct an image of the distribution of radioactivity from the coincidence lines recorded by the detector ring (Fig. 2). These lines are sorted into parallel groups, each group representing a profile or projection of the radioactivity distribution, viewed from a different angle. These profiles are then combined by use of the same mathematical principles developed for X-ray computed tomography (Brooks and DiChiro, 1976) to reconstruct an image of regional radioactivity. The reconstruction process requires a correction for the attenuation or absorption of annihilation photons that occurs within the object being scanned. Estimates of the amount of radiation attenuation by the head can be calculated mathematically. Actual measurement of attenuation is more accurate, however (Frackowiak et al., 1980). This measurement is obtained by performing a separate transmission scan with a ring source of radioactivity surrounding the subject's head prior to the emission scan. The amount of photon attenuation by tissue between each opposing detector pair is measured and used in the reconstruction process.

The intensity of any portion of the PET image is proportional to the amount of radioactivity in the corresponding region of tissue in the field of view of the scanner (Eichling et al., 1977). To obtain *absolute* measurements of regional radioactivity, the system must be calibrated. This is done by imaging a cylinder containing a solution of radioactivity and then measuring the amount of radioactivity in an aliquot of the solution with a calibrated well counter (Raichle et al., 1983).

A PET system consists of several components. There are one or more rings of radiation detectors mounted on a gantry. Multiple rings employed in more recently designed scanners permit several contiguous tomographic images to be obtained simultaneously. The subject rests on a special couch fitted with a head holder to prevent head movement during the scanning procedure. A com-

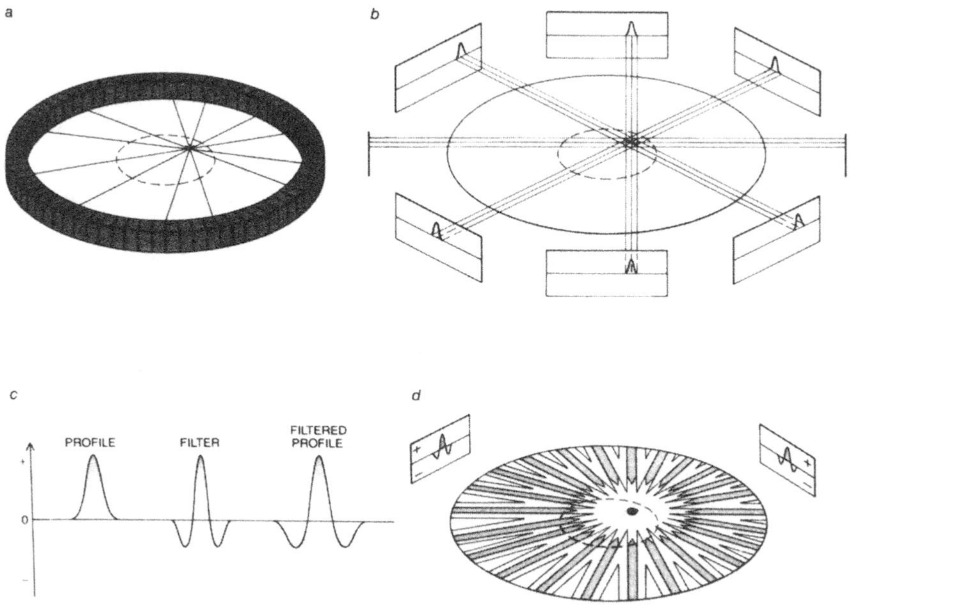

Fig. 2. The PET reconstruction process can be explained with reference to this schematic diagram depicting the imaging of a small uniform region of radioactivity. (a) Multiple coincidence lines are recorded by opposing detectors in the circular array, each line resulting from a positron emission and annihilation. (b) The lines are sorted into parallel groups, each group representing the profile of the radioactivity in the image plane, viewed from a different angle. (c) Each profile is then mathematically processed by means of a special function called the filter function. (d) The modified profiles are then combined by "back-projection" to reconstruct an image of the distribution of radioactivity in the field of view of the scanner (from Positron-Emission Tomography, M. M. Ter-Pogossian et al., Copyright © October, 1980 by Scientific American, Inc. All rights reserved. Courtesy of Dr. M. M. Ter-Pogossian and by permission of *Scientific American*).

puter is required to record the coincidence events and to reconstruct and display the tomographic images of radioactivity. The engineering design of positron emission tomographs involves several complex factors. These include selection of the shape, size, and material for the scintillation crystals in the radiation detectors, the arrangement of the detectors in the gantry, the design of coincidence electronic circuitry, and the mathematical methods used to reconstruct the tomographic image (Ter-Pogossian, 1981; Brooks et al., 1981; Hoffman, 1982; Muehllehner and Karp, 1986). All of these factors affect the performance of the tomograph. A discussion of the design of tomographs is beyond the scope of this review. Two performance characteristics, image noise and image resolution, that have a major impact on the quality of physiologic measurements made with PET, will be reviewed, however.

Statistical noise occurs in the PET image because of the random nature of radioactive decay. The disintegration rate of a radioactive sample undergoes moment-to-moment random variation that can be described by a Poisson statistical distribution. The resultant uncertainty in the measurement of the amount of radioactivity, expressed as a fraction of the number of counts recorded, N, is proportional to $1/\sqrt{N}$. Thus, counting noise decreases as N increases. Similarly with PET, the statistical reliability of a local radioactivity measurement depends upon the number of counts recorded. The situation is more complex, however, than the simple Poisson prediction. This is because the radiation measurement in any small brain region is obtained from an image reconstructed from multiple views or projections of the radioactivity distribution throughout the brain slice. Thus, the measurement noise in any individual brain region will be affected by noise in other brain regions. The resultant noise is greater than would be predicted by Poisson statistics (Budinger et al., 1978). Image noise can be decreased by increasing the number of counts collected (Fig. 3). This can be accomplished either by prolonging the duration of the PET scan data collection period or by increasing the count rate by administering more radioactivity to the subject. However, the latter approach is limited not only by considerations of radiation safety, but also by the ability of tomographs to measure radioactivity accurately at high count rates. At high count rates, random coincidences are recorded by the tomograph. Coincidence lines are erroneously detected from pairs of photons originating from two *unrelated* positron annihilations occurring within the coincidence resolving time of a detector pair. These random coincidences add noisy background to the image. Although corrections can be made to subtract the average background value of the randoms, the contribution to image noise persists (Hoffman et al., 1981). The number of randoms that are registered increases with the amount of radioactivity in the field of view of the tomograph and with its coincidence resolving time. Thus, for any given tomograph, the amount of radioactivity administered and the scan duration must

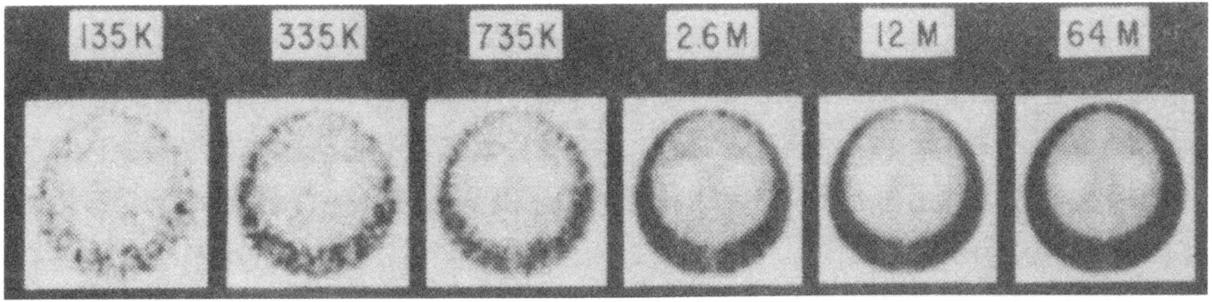

Fig. 3. Effect of varying counts on image noise. The images are of a phantom with chambers of different radiotracer concentrations. With increasing counts, the noise in the image decreases (toward the right) (from *IEEE Trans Nuclear Sci.* **26,** 2746, 1979, courtesy of Dr. M. E. Phelps, and by permission of the Institute of Electrical and Electronic Engineers, © 1979 IEEE).

be carefully selected so as to minimize overall image noise caused by counting statistics and random coincidences. These factors must be taken into account in the design of the tracer kinetic models used to analyze PET data.

The spatial resolution of a PET system is ultimately defined by how accurately the location of a positron-emitting nucleus in the field of view is recorded. This determines the accuracy with which the spatial distribution of a positron-emitting radiotracer can be measured. Two major factors that limit resolution are the physics of the positron annihilation process and the design of the tomograph itself.

The annihilation radiation detected by the tomograph does not originate in the radioactive nucleus, the location of which one wishes to measure. Rather it is produced only after the positron has traveled up to several millimeters away from the nucleus. The average range the positron travels before annihilation, which depends upon the specific radionuclide involved (e.g., 1.1 mm for ^{11}C, 2.5 mm for ^{15}O), limits the accuracy with which the location of the radionuclide can be determined (Phelps et al., 1975). A second physical factor limiting resolution is the non-colinearity of the annihilation photons. The angle between the two photons deviates slightly from 180° because of residual momentum of the positron at the time of annihilation. The combined effect of positron range and annihilation radiation non-colinearity sets a theoretical lower limit on PET resolution of about 2–5 mm. However, it is tomographic design that limits the resolution of current PET systems. The size and shape of the radiation detectors used are especially important in determining how accurately the position of the coincidence lines are recorded, with smaller crystals in general providing better resolution. The resolution of tomographs currently in use is approximately 0.5–1.5 cm.

The effect of limited resolution is visually apparent as a blurring or lack of sharpness of the tomographic images (Figs. 4 and 5). More important though, is the effect on the accuracy of measurements of regional radioactivity (Hoffman et al., 1979; Mazziotta et al., 1981). The radioactivity in a brain region will appear to be blurred or spread over a larger area. The worse the resolution, the greater this effect. Thus, in the reconstructed image, a brain region will appear to contain only a portion of the radioactivity that was in the corresponding brain structure. In addition, some of the radioactivity in surrounding brain regions will appear to be spread into the region of interest. As a result of this effect, called partial volume averaging, a regional measurement will contain a contribution from the radioactivity in the structure of interest and as well as from surrounding structures. Thus, in an area of high radioactivity surrounded by low levels of radioactivity, the measurement of radioactivity will be underestimated, whereas in a low-radioactivity area surrounded by high activity, the measurement will be overestimated. These errors will be less when the size of the area or structure of interest is large with respect to the scanner resolution. In general, though, it is not possible to obtain measurements that reflect pure gray or white matter radioactivity, especially in cortical regions. In addition, because of partial volume averaging, PET measurements will frequently include contributions from metabolically inactive cerebrospinal fluid in sulci or ventricles. This will lead to an underestimation of radioactivity, and therefore of tissue blood flow and metabolism (Herscovitch et al., 1986).

3. Radiotracers for PET

The second component required for PET is a tracer compound of physiologic interest that is labeled with a positron-emitting radionuclide. A wide variety of radiotracers has been synthesized for use with PET, not only for the measurement of cerebral blood flow and metabolism, but also for the study of neuroreceptors, tissue pH, blood–brain barrier permeability, and drug kinetics.

The commonly used positron-emitting radionuclides and their half-lives are listed in Table 1. These nuclides have several properties that make them well suited for PET. The chemical nature of ^{15}O, ^{13}N, and ^{11}C is virtually identical to that of their nonradioactive counterparts, which are the basic constituents of living matter, as well as of most drugs. Thus, they can be incorporated into radiotracers that have the same biochemical behavior as the corresponding nonradioactive compound. Although ^{18}F, ^{82}Rb, ^{68}Ga, and ^{75}Br are not nuclides of biologically significant elements, they can be used to label a variety of compounds of physiologic interest. For example, ^{18}F is used as a substitute for hydrogen to synthesize analogs of drugs or naturally occurring molecules. The half-lives of these positron-emitting radionuclides are short. Thus, relatively large amounts of radioactivity can be administered to human subjects (e.g., 75 mCi of ^{15}O, 30 mCi of ^{11}C)

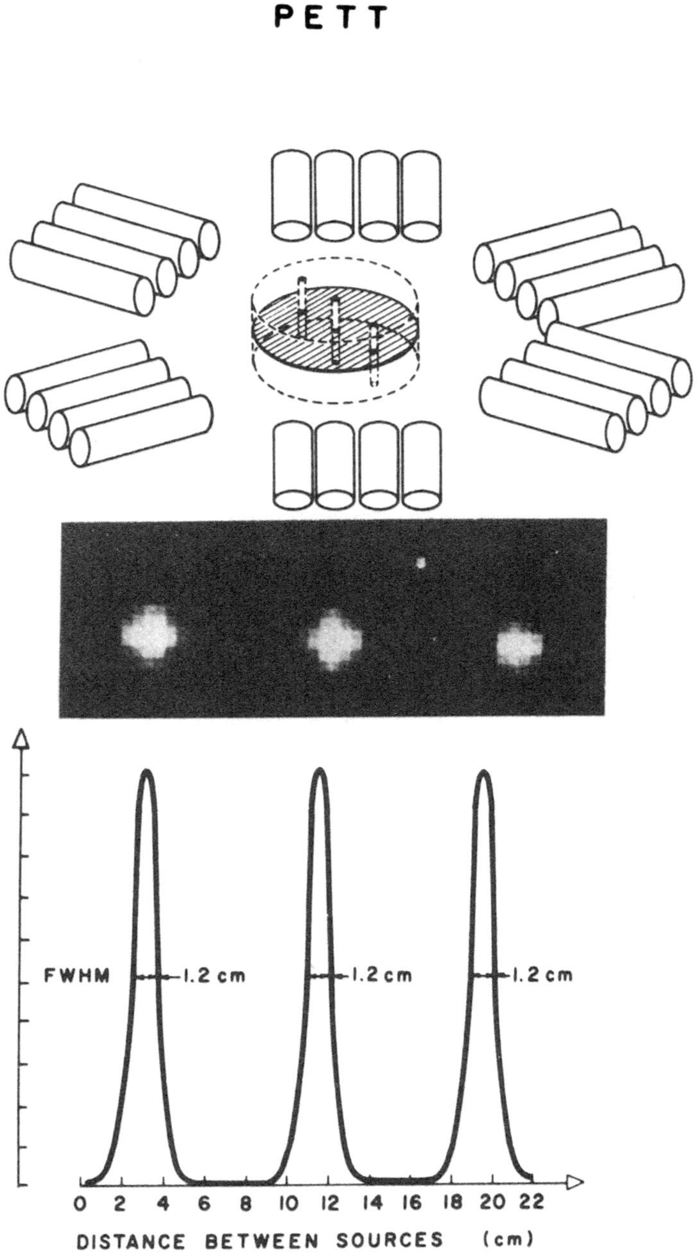
PETT
FWHM
1.2 cm
1.2 cm
1.2 cm
0 2 4 6 8 10 12 14 16 18 20 22
DISTANCE BETWEEN SOURCES (cm)

to obtain tomographic images of good statistical quality, but because of rapid decay, the radiation exposure is within acceptable limits (Kearfott, 1982). Also, the short half-life of some of these nuclides, especially ^{15}O, permits multiple repeat studies in the same subject because of rapid decay of the radiotracer after each administration.

A disadvantage of these short half-lives is the resultant difficulty in the preparation of positron-emitting radiotracers. The nuclides most commonly used for the measurement of cerebral blood flow and metabolism, ^{15}O, ^{11}C, and ^{18}F, are produced by means of a cyclotron, a complex and expensive electromechanical apparatus. To use these nuclides to label radiotracers for PET, one requires a cyclotron in the immediate medical environment. In addition, because of the short half-lives of these nuclides, fast radiochemical labeling techniques are required. These must yield radiotracers safe for human use that are sterile, nontoxic, and apyrogenic. The radiotracers must be of high enough specific activity (i.e., mCi/mass of compound) so that the administration of physiologically trace amounts of the compound will result in tissue radioactivity levels high enough to be measured with PET. Finally, the synthetic product must be radiochemically pure. Most or all of the positron-emitting radioactivity in the product should be attached to the radiotracer compound. The presence of other labeled products that behave differently from the desired product would confound the use of the corresponding tracer kinetic model. In spite of these requirements, a wide variety of positron-emitting radiopharmaceuticals has been synthesized (Wolf, 1981; Comar et al., 1982; Welch and Kilbourn, 1984).

Fig. 4. Definition and measurement of the resolution of a positron emission tomograph. Images are obtained by scanning line sources of radioactivity (upper panel). Rather than appearing as points on the reconstructed image, the radioactivity in each source appears spread over a larger area (middle panel) because of limited image resolution. Resolution is defined in terms of the amount of spreading that occurs. A plot of the image intensity (lower panel) shows that this spreading takes the form of a Gaussian-like curve, called the line spread function. The width of the line spread function at one half of its maximum amplitude (termed the full width at half maximum, FWHM) is used to quantify the resolution of the tomograph. Here the resolution is 1.2 cm (from Ter-Pogossian et al., 1975, courtesy of Dr. M. M. Ter-Pogossian and by permission of the Radiological Society of North America).

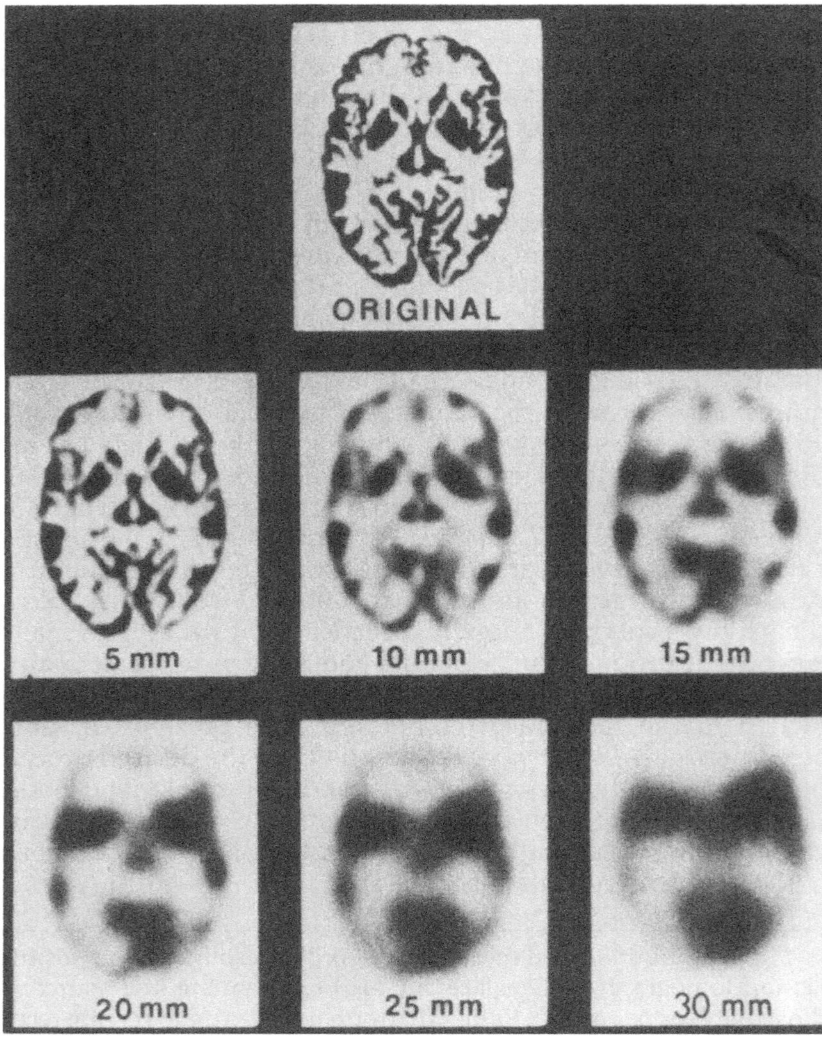

Fig. 5. Simulation of the effect of differing scanner resolutions on image quality. At upper center is an original brain image. Subsequent images simulate the effect of scanning this image with tomographs of varying resolution, from 5 to 30 mm FWHM. Note the apparent spreading or redistribution of radioactivity in the reconstructed images. This effect increases as the scanner resolution worsens (from Mazziotta et. al. 1981, courtesy of Dr. J. Mazziotta and by permission of Raven Press).

Table 1
Positron-Emitting Nuclides Used in PET

Nuclide	Half-life, min
Rubidium-82 (^{82}Rb)	1.25
Oxygen-15 (^{15}O)	2.05
Nitrogen-13 (^{13}N)	10.0
Carbon-11 (^{11}C)	20.4
Gallium-68 (^{68}Ga)	68.1
Bromine-75 (^{75}Br)	101
Fluorine-18 (^{18}F)	110

4. Tracer Techniques

Once an appropriate radiotracer has been synthesized and administered, its distribution in the brain can be determined by means of the tomograph. The final component required in the application of PET is the mathematical model describing the in vivo behavior of the specific radiotracer used. By means of such models, one can calculate the value of a specific biological variable, such as CBF, from tomographic measurements of local brain radiotracer concentration and from measurements of the radiotracer concentration in blood over time. This use of mathematical models is the fundamental step that makes PET a quantitative physiological and biochemical tool rather than only an imaging technique.

Two general approaches to modeling can be used, compartmental or distributed. Because of their relative mathematical simplicity, essentially all models currently used for PET are compartmental. These models assume that there are units or compartments in which the radiotracer concentration is uniform at any instant in time, and that have uniform properties with respect to the tracer. The compartments can be physical spaces, such as the intravascular blood pool or the extracellular space, or biochemical entities, such as the amount of neuroreceptor available for binding by a specific radiolabeled ligand. In contrast, distributed models account for the presence of time-varying concentration gradients of the radiotracer. For example, the radiotracer concentration may vary along the length of a capillary because of movement of tracer into and out of brain tissue. Although distributed models may provide a more accurate description of radiotracer behavior (Larson et al., 1987; Graham et al., 1985), their greater mathematical complexity has limited their use in PET.

The tracer kinetic model is a mathematical description of the relationship over time between the amount of tracer presented to the brain in its arterial input and the amount of radiotracer in brain tissue. It includes variables representing measurable quantities, such as brain and peripheral blood radiotracer concentrations, and variables whose value is desired, such as local blood flow or metabolism. Several factors must be considered in the development of tracer kinetic models. These include the local delivery of tracer by arterial blood to the tissue, the transport of tracer across the blood–brain barrier, the metabolic fate of the tracer in the tissue, the behavior of any labeled metabolites of the tracer formed in brain, the presence of tracer in the vascular space of the brain rather than in brain tissue, the recirculation of radiotracer and any labeled metabolites from the rest of the body to the brain, and the potential for alterations in radiotracer behavior in the presence of local pathology. In addition, there are several practical issues that must be considered. The number of unknown variables in the model equation cannot be so large that it is impossible to solve the model from the available scan data. Radioactive decay of the tracers during the scanning period must be taken into account (Videen et al., 1987), since their half-lives are comparable to the duration of the measurement. This is in contrast to tissue autoradiographic measurements of cerebral blood flow and metabolism in laboratory animals with long-lived isotopes such as ^{14}C (Sokoloff et al., 1977; Sakurada et al., 1978). Finally, it must be possible to collect sufficient counts with the tomograph to permit accurate calculation of the physiologic variable of interest from the model. Factors limiting the number of counts collected include the amount of radiation exposure to the subject, the count rate at which the tomograph can operate accurately, the amount of radiotracer that enters brain tissue, and the scan time available for data collection. Several considerations limit the scan time. These include the physical half-life of the tracer, the requirement for a physiologic steady state during the scan, and patient comfort and cooperation. In addition, the tracer model itself may impose certain limitations on the duration of scanning. For example, at later times radiolabeled metabolites of the tracer not accounted for in the model may appear in the blood, so that it is necessary to complete the scanning period before this occurs.

Tracer kinetic models have been developed to perform a wide variety of physiologic measurements with PET. To date, most studies of brain physiology and pathophysiology with PET have used methods to measure cerebral blood volume, blood flow, and metabolism. These techniques will be described below.

5. Cerebral Blood Volume

The determination of regional cerebral blood volume (rCBV) was one of the earliest, and conceptually is one of the simplest, measurements made with PET. Trace amounts of carbon monoxide, labeled with either ^{11}C (Grubb et al., 1978; Phelps et al., 1979a) or ^{15}O (Mintun et al., 1984) are administered by inhalation. The labeled carbon monoxide binds to hemoglobin in red blood cells, and is thus confined to the intravascular space. Following equilibration of the labeled carboxyhemoglobin throughout the blood pool of the body, which requires approximately 2 min, the local radioactivity recorded from the brain is directly proportional to the local red cell volume. Conceptually, rCBV can be calculated from the ratio of radioactivity concentration in brain to that in peripheral blood. Since the tracer labels only the red cell volume, however, one must correct for the difference between peripheral, large vessel hematocrit, and the hematocrit in the cerebral vasculature. rCBV, in units of mL/100 g of brain (Fig. 6), is calculated from the local tissue radiotracer concentration, C_t (cps/g), and peripheral blood radiotracer concentration, C_{bl} (cps/mL), as follows:

$$\mathrm{rCBV} = \frac{C_t}{C_{bl}R}\,100 \tag{1}$$

where R is the ratio of the cerebral hematocrit to peripheral, large vessel hematocrit. A value of 0.85 has been used for R, based on an average of values obtained in both animal and human studies (Grubb et al., 1978; Phelps et al., 1979a). Recent tomographic studies in which cerebral hematocrit was calculated from measurements of both plasma volume and red cell volume indicate, however, that the routine use of this single value may be incorrect. This ratio may be less than 0.85, and also it may vary in different

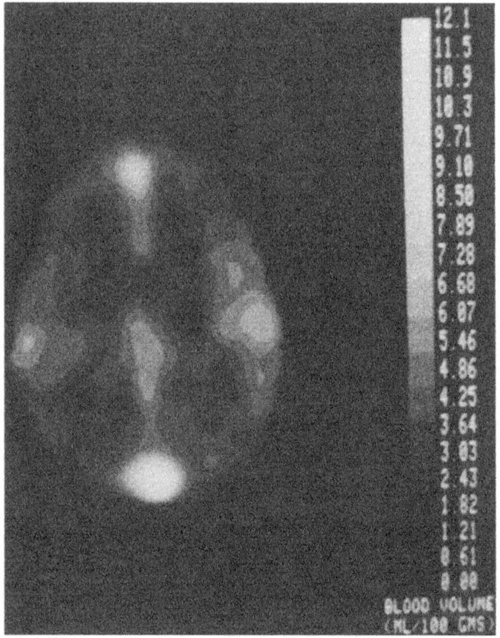
12.1
11.5
10.9
10.3
9.71
9.10
8.50
7.89
7.28
6.68
6.07
5.46
4.86
4.25
3.64
3.03
2.43
1.82
1.21
0.61
0.00
BLOOD VOLUME
(ML/100 GMS)

Fig. 6A

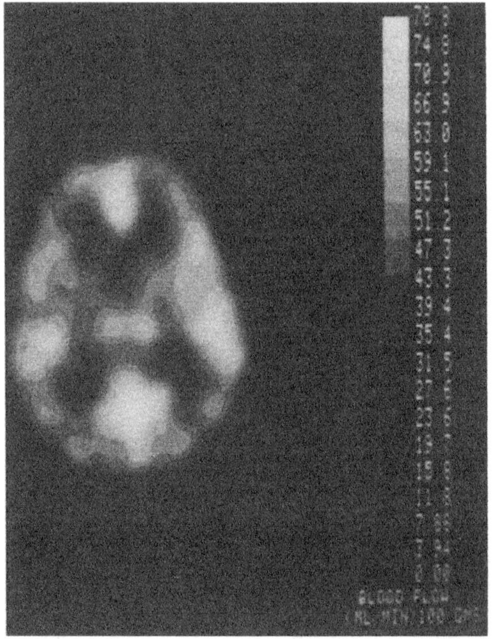

Fig. 6B

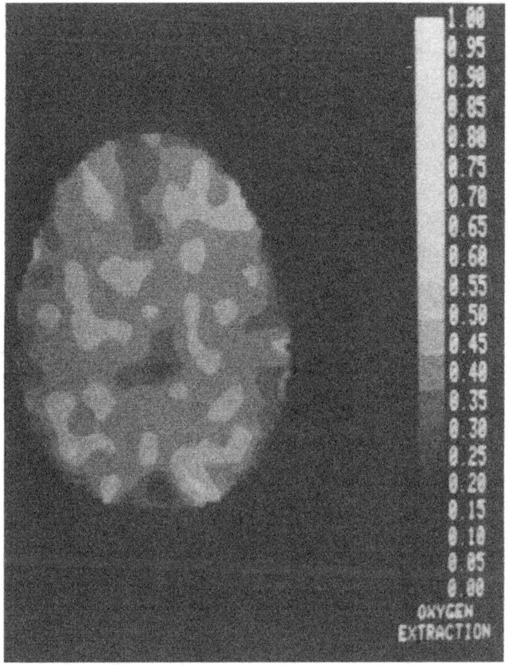

Fig. 6C

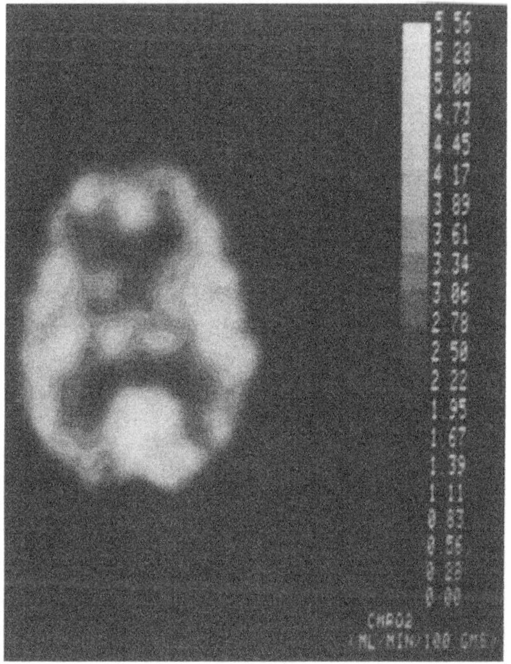

Fig. 6D

Fig. 6. Quantitative PET images from a normal human subject, obtained with ^{15}O-radiotracers and the PETT VI tomograph (Ter-Pogossian et al., 1982). These are horizontal sections at the level of the thalamus, oriented such that anterior is up and left is to the reader's left. (A) Cerebral blood volume measured following the inhalation of $C^{15}O$. The superior saggital sinus is seen anteriorly and posteriorly, and differences in vascular density between gray and white matter are delineated. (B) Quantitative image of cerebral blood flow obtained following the bolus intravenous injection of $H_2{}^{15}O$. (C) Quantitative image of cerebral oxygen extraction fraction calculated from scan data obtained following the brief inhalation of $^{15}O_2$. The image is relatively uniform throughout the brain because oxygen metabolism and blood flow are matched throughout the brain in the resting state. (D) Quantitative image of the cerebral metabolic rate for oxygen.

physiological conditions, such as during CO_2 inhalation (Lammertsma et al., 1984; Sakai et al., 1985).

Although the same tracer model is applied when either ^{11}C- or ^{15}O-labeled carbon monoxide is used, there are certain practical advantages to using ^{15}O. First, the use of the shorter-lived nuclide results in rapid decay of the radioactive background following the rCBV measurement, so that other PET determinations, some of which may require the use of longer-lived nuclides, such as ^{11}C or ^{18}F, can be made. Second, the use of ^{15}O results in a lower radiation exposure to the subject (Kearfott, 1982). Finally, since the rCBV determination is often made in conjunction with measurements of cerebral blood flow and oxygen metabolism that require the use of ^{15}O, the use of ^{15}O for the rCBV measurement as well avoids changing the cyclotron targetry used for radionuclide production during the patient study. ^{11}CO may be preferable, however, when a single slice tomograph is used. Several tomographic slices can be obtained sequentially following a single administration of the longer-lived tracer by repositioning the subject within the tomograph.

The measurement of rCBV is of physiological importance since it may reflect local vasodilation of the cerebral vasculature in response to decreased perfusion pressure (Grubb et al., 1975; Powers et al., 1983). In addition, determination of rCBV is often an important component of other PET radiotracer methods. This is because rCBV data are often required to correct measurements of local radiotracer concentration for radiotracer located in the intravascular space of the brain, so that the amount of radiotracer actually entering brain tissue can be accurately determined (Lammertsma and Jones, 1983; Mintun et al., 1984; Sasaki et al., 1986).

6. Cerebral Blood Flow

6.1. Background

Most methods for measuring regional cerebral blood flow (rCBF) with PET employ ^{15}O-labeled water ($H_2{}^{15}O$) as the flow tracer. $H_2{}^{15}O$ has several desirable characteristics for this application. Water is a biologically inert, naturally occurring, body constituent, and therefore $H_2{}^{15}O$ has no undesirable physiologic or pharmacologic side effects. It is also easily synthesized in large mCi quantities (Welch and Kilbourn, 1985). Because of the short half-life of ^{15}O, relatively large amounts of radioactivity can be administered. This permits the collection of statistically satisfactory images over brief time periods, while keeping the radiation exposure to the subject within acceptable limits. In addition, the short half-life results in rapid decay of the radioactive background once the rCBF determination is completed. Other PET measurements can then be easily performed with little time delay.

Methods for measuring rCBF with PET are based upon a model developed by Seymour Kety and colleagues to describe the in vivo behavior of inert tracers that can diffuse freely between blood and tissue (Kety, 1951, 1960; Landau et al., 1955; Sakurada et al., 1978). Once such a tracer is introduced into the circulation, the rate of change of tracer concentration in a tissue region is equal to the difference in the rate at which the tracer is transported to the tissue in the arterial blood and the rate at which it is washed out from the tissue in its venous drainage. This concept, known as the Fick principle, is expressed mathematically as

$$\frac{dC_t}{dt} = fC_a - fC_v \tag{2}$$

where f is the tissue blood flow (mL/min·g), C_t is the tissue radiotracer concentration (cps/g), and C_a and C_v are the respective tracer concentrations (cps/mL) in the arterial input and venous drainage of the tissue. Since C_v cannot be measured in practice on a regional basis, Kety (1951, 1960) introduced the following substitution: $C_v = C_t/\lambda$. Here, λ is the brain–blood partition coefficient for the tracer. It is the ratio between the tissue and blood radiotracer concentrations at equilibrium. λ can be determined from independent experiments or calculated as the ratio of the solubilities of the tracer in brain and blood (Kety, 1951; Herscovitch and Raichle, 1985a). When there is no limitation to diffusion of the tracer across the blood–brain barrier, local venous radiotracer concentration remains in equilibrium with that of the tissue, so that C_v in Eq. (2) can be replaced by C_t/λ:

$$\frac{dC_t}{dt} = f(C_a - C_t/\lambda) \tag{3}$$

This equation, originally developed to measure rCBF in laboratory animals with tissue autoradiography, forms the basis for tracer models used to measure rCBF with PET.

6.2. Steady-State Method

The steady-state inhalation technique was the earliest-described method for measuring rCBF with ^{15}O and PET (Jones et al., 1976; Subramanyam et al., 1978; Frackowiak et al., 1980). rCBF is measured during the continuous inhalation of trace amounts of ^{15}O-labeled CO_2 ($C^{15}O_2$), delivered at a constant rate. The catalytic action of carbonic anhydrase in the red blood cells in the pulmonary vasculature results in rapid transfer of the ^{15}O label to water. $H_2{}^{15}O$ is constantly generated in the lungs and circulates throughout the body. After approximately 10 min of inhalation, a steady state is reached in which the amount of radioactivity delivered to a brain region equals that leaving the region by the combined processes of radioactive decay and washout into the venous circulation. As a result, the distribution of regional radioactivity in the brain remains constant. Under these circumstances, Eq. (3) may be reformulated as

$$\frac{dC_t}{dt} = f(C_a - C_t/\lambda) - \alpha C_t = 0 \qquad (4)$$

where α is the decay constant of ^{15}O, that is, ln 2/half-life of ^{15}O. Flow, f, can then be expressed in terms of arterial and tissue radiotracer concentrations:

$$f = \frac{\alpha}{C_a/C_t - 1/\lambda} \qquad (5)$$

C_a is determined from arterial blood samples and C_t is measured with PET. λ is calculated from the ratio of the solubility of water in brain and in blood, which is equivalent to the ratio of the water contents of brain and blood. Its value is 0.91 mL/g (Herscovitch and Raichle, 1985a).

The steady-state approach is applicable with all tomographs. It is particularly suited to those that operate accurately only at relatively low count rates, since the count data required to construct the tomographic image can be accumulated over several minutes of $C^{15}O_2$ inhalation. It is also convenient for use with single-ring tomographs, since multiple tomographic slices can be simply obtained by repositioning the patient during a prolonged period of $C^{15}O_2$ inhalation. In addition, only a few samples of arterial blood are required, and the timing of these samples is not critical. This method provides rCBF measurements that vary appropriately in response to variations in arterial pCO_2 in experimental animals (Baron et al., 1981; Rhodes et al., 1981), and rCBF measurements obtained in normal human subjects are comparable to those that have been obtained by other techniques (Frackowiak et al., 1980). CBF measured with this technique in a baboon at three different levels of pCO_2 was similar to that measured in the same animal with a standard microsphere technique (Steinling et al., 1985).

The advantages and limitations of steady-state rCBF method have been extensively analyzed (Huang et al., 1979; Lammertsma et al., 1981, 1982; Steinling and Baron, 1982; Jones et al., 1982; Herscovitch and Raichle, 1983; Meyer and Yamamoto, 1984). Most of its limitations arise from the nonlinear relationship between rCBF and measured tissue radiotracer concentration [Eq. (5)]. At higher flow levels, a large change in rCBF produces a relatively smaller change in brain ^{15}O concentration. Thus, errors in measurement of tissue radioactivity produce proportionately larger errors in flow measurement. The relationships between the statistical accuracy of tomographic radioactivity quantitation, the total amount of radioactivity administered, and the accuracy of rCBF determination have been analyzed in detail. Calculated flow values are also sensitive to errors in the measurement of C_a and to any difference in the value of λ used in Eq. (5) and the actual value of λ, which may vary in pathological conditions (Herscovitch and Raichle, 1985a). Another limitation resulting from the nonlinearity of the model relates to its behavior in heterogeneous tissue regions. This technique, as well as all other PET-CBF methods, assumes that the region of interest in which the measurement is made is uniform in tissue composition and physiologic properties. However, because of the limited spatial resolution of PET, a region will receive tissue count contributions from both gray and white matter (Mazziotta et al., 1981). Because of the nonlinear nature of the steady-state operational equation, the flow measured in such a heterogeneous region underestimates the true weighted regional flow by up to 20%. Deviation from the steady-state requirement of constant arterial radiotracer concentration, caused either by fluctuation in cyclotron delivery of $C^{15}O_2$, or by variation in the patient's respiratory pattern, has been a matter of concern. It has demonstrated, however, that such variations can be accounted for if the average concentration from multiple arterial samples is used.

Jones and colleagues (Jones et al., 1985) have implemented a modification of the steady-state inhalation method by using continuous intravenous administration of $H_2{}^{15}O$. The advantages of this approach include lowered radiation dose to the lungs and trachea, reduced radioactivity in the nasopharynx, which decreases random coincidence counts, and ease of tracer administration in patients unable to maintain a regular respiratory rate. These investigators have also modified the operational equation of the steady-state technique so that equilibrium conditions are not required during the scanning period. Thus, tomographic data collection can begin at the onset of radiotracer administration, and more counts are collected for the same radiation dose to the subject. This method also produces a more linear relationship between local brain radiotracer concentration and flow.

6.3. PET/Autoradiographic Method

An alternative approach to measuring rCBF with $H_2{}^{15}O$ is the adaptation to PET (Herscovitch et al., 1983; Raichle et al., 1983; Kanno et al., 1984) of Kety's tissue autoradiographic method for measuring rCBF in laboratory animals (Landau et al., 1955; Kety, 1960; Sakurada et al., 1978). With Kety's technique, a biologically inert, freely diffusible radiotracer is infused for a brief time period, T, and the animal is then decapitated. The behavior of the radiotracer is described by Eq. (3), which can be integrated and rearranged to give

$$C_t(T) = f \int_0^T C_a(t) \exp[-f/\lambda(T-t)]\,dt$$
$$= fC_a(T) * \exp(-fT/\lambda) \qquad (6)$$

where $*$ denotes the mathematical operation of convolution. One solves this equation for flow, f, using the local brain radiotracer concentration at the end of the infusion [$C_t(T)$], as measured by quantitative tissue autoradiography, and the arterial time-activity curve [$C_a(t)$], obtained by frequent blood sampling. To adapt this method to PET, one must take into account that, because of limited temporal resolution, current tomographs cannot measure the brain radiotracer concentration, $C_t(T)$, instantaneously. A scan must be performed over many seconds, essentially summing the decay events occurring during that time interval. Thus, the operational equation [Eq. (6)] was modified (Herscovitch et al., 1983; Raichle et al., 1983) for PET by an integration of the instantaneous count rate, $C_t(T)$, over the scan time, T_1 to T_2, to correspond to the summing process of tomographic data collection:

$$C = \int_{T_1}^{T_2} C_t(T)dT = f \int_{T_1}^{T_2} C_a(t)^* \exp(-ft/\lambda)dT \qquad (7)$$

Here C is the local number of counts per unit weight of tissue recorded by the tomograph. To implement this approach (Raichle et al., 1983), $H_2{}^{15}O$ is administered by bolus intravenous injection, and a 40-s scan is obtained following arrival of radiotracer in the head (Fig. 6). Blood is sampled every 4–5 s via an indwelling radial artery catheter. Once a value for λ is specified (Herscovitch and Raichle, 1985a), Eq. (7) can be solved numerically for flow. rCBF measurements with the PET/autoradiographic technique have been validated in the baboon by direct comparison to flow measured in the same animals by intracarotid injection of radiotracer and external residue detection (Raichle et al., 1983).

In order to obtain statistically adequate images during the brief tomographic data collection time, a relatively large amount of radioactivity must be administered, typically 50–75 mCi. The scanner must be able to operate accurately at the resulting high count rates (Ter-Pogossian, 1982). In addition, a multislice scanner is preferable, so that a large portion of the brain can be imaged following a single radioisotope injection.

The relationship between tissue counts, C, and rCBF for the PET/autoradiographic approach is almost linear. This has several advantageous consequences (Herscovitch et al., 1983). Errors in measurement of tissue radioactivity result in approximately equivalent errors in rCBF; there is no amplification of error at higher flows, as occurs with the steady-state technique. The method works well in the unavoidable situation of tissue heterogeneity, with the measured flow approximately equal to the true weighted flow in a mixed gray matter-white matter brain region. Any inaccuracy in the value of λ used in the operational equation results in only minimal error in flow. In addition, the tomographic image of tissue counts closely reflects the relative differences in flow in different brain regions. Thus, in functional mapping studies, relative changes in rCBF can be determined from images alone. Arterial sampling to provide absolute flow measurements is not necessarily required (Herscovitch et al., 1983; Fox et al., 1984; Mazziotta et al., 1985). A relative disadvantage of the technique is the need for frequent, accurately timed, arterial blood samples. The peripheral arterial time–activity curve is assumed to be equal to the arterial input to the brain. In fact, the bolus of radioactivity arrives at the radial artery sampling site several seconds later than in the brain. This time difference can be measured and the peripheral arterial curve appropriately shifted in time (Raichle et al., 1983).

6.4. Other rCBF Methods

Huang and colleagues (Huang et al., 1982, 1983) have described a technique for measuring both rCBF and the local value of λ for water using $H_2{}^{15}O$. Following the bolus intravenous administration of $H_2{}^{15}O$, scan data are collected over a 10-min period, and arterial sampling is performed. Two image sets are reconstructed, using decay-corrected and non-decay-corrected scan data, respectively. The operational equations for this method, derived from the basic Kety formulation [Eq. (3)], permit the estimation of both rCBF and λ for $H_2{}^{15}O$. The values for λ obtained in humans by this method were about 15% lower than would be expected based on the known water content of brain. A potential explanation for this discrepancy is that not all water in cerebral tissue is freely exchangeable with water in blood. Further studies are required to clarify this issue. Simulation studies of this approach have shown that the error in rCBF measurement in heterogenous tissue is quite small, and that propagation of error in tissue count measurement is not excessive. Scan data are collected over 10 min, a period over which physiological variations in rCBF may occur. If such changes occur more than 1 min after injection of $H_2{}^{15}O$, however, the error in estimating the CBF at injection time is modest.

Alpert et al. (1984) and Carson et al. (1986) have described extensions of Huang's technique of using scan data in two different forms. These approaches both involve multiplying the terms in the basic Kety equation [Eq. (3) or (7)] by two different, time-dependent weighting functions. The resultant two equations are then numerically solved for both rCBF and λ. The mathematical form of the weighting functions can be selected by use of parameter-estimation techniques to minimize the effect of statistical noise in tomographic radioactivity measurements on the final flow calculations (Carson et al., 1986).

All of these methods for measuring rCBF assume that the tracer is freely diffusible across the blood–brain barrier, so that the amount of tracer entering tissue depends only upon the local flow. However, $H_2{}^{15}O$ does not exhibit this ideal behavior. At higher flow levels, there is a progressive decline in the extraction of $H_2{}^{15}O$ from blood by brain (Eichling et al., 1974). This results in less radiotracer entering tissue at higher flows than would be predicted and leads to an underestimation of rCBF. This has been experimentally demonstrated in baboons with the PET/autoradiographic method (Raichle et al., 1983) and would be expected to occur with the other methods using $H_2{}^{15}O$ as well (Lammertsma et al., 1981; Huang et al., 1983). A recent study (Herscovitch et al., 1987) using the PET/autoradiographic method has compared flow measurements obtained in the same subjects using both $H_2{}^{15}O$ and ^{11}C-butanol, a flow tracer that is not diffusion-limited. In comparison to ^{11}C-butanol, $H_2{}^{15}O$ was found to underestimate rCBF by approximately 15% because of its diffusion limitation.

An alternative radiopharmaceutical for rCBF measurement is ^{18}F-labeled fluoromethane (Holden et al., 1983; Koeppe et al., 1985). This inert gas is administered by inhalation. The Kety model [Eqs. (3) and (7)] is used to describe the behavior of the radiotracer. Scan data are collected in the form of 8–12 sequential 1-min images, an approach originally suggested by Kanno and Lassen (1979). Both local CBF and λ are calculated from scan data and measurements of arterial radioactivity using parameter estimation techniques. Koeppe et al. (1985) have recently described a modification of this technique that uses measurements of expired air and venous blood radioactivity to estimate the arterial time–activity curve. The use of this approach may be limited, however, if arterial puncture is required for other PET measurements, such as oxygen metabolism, made in conjunction with the flow determination. Preliminary data obtained in isolated dog brain indicate that fluoromethane is freely diffusible at average whole brain flows of up to 70 mL/(min·100g) (Holden et al., 1983). Potential disadvantages of the fluoromethane technique include the relatively long time interval required for its clearance (approximately 30 min) before another PET study can be performed and the more complex inhalation mode of administration.

7. Cerebral Oxygen Metabolism

7.1. Background

Two methods have been developed to measure regional cerebral oxygen metabolism ($rCMRO_2$) with PET. One uses continuous inhalation of ^{15}O-labeled oxygen ($^{15}O_2$) and was developed in conjunction with the steady-state technique for measuring rCBF

(Subramanyam et al., 1978; Frackowiak et al., 1980). The other, which uses a brief inhalation of $^{15}O_2$, is a companion method to the PET/autoradiographic approach (Mintun et al., 1984). The principles underlying both methods are the same. A fraction of the oxygen delivered to the brain by its arterial input is extracted and used in the oxidative metabolism of glucose. Both methods measure this fraction, termed the regional oxygen extraction fraction (rOEF). Since there are no stores of oxygen in brain, the rate of oxygen utilization can be determined from the product of rOEF and the rate of oxygen delivery, which equals rCBF multiplied by arterial oxygen content. Strategies using $^{15}O_2$ as a tracer for measuring rOEF and $rCMRO_2$ must adequately describe the fate of the ^{15}O label following $^{15}O_2$ inhalation. $^{15}O_2$ that is extracted from arterial blood by brain tissue is converted to ^{15}O-labeled water of metabolism, which is then washed out of the brain. Labeled water of metabolism, produced by both the brain and the rest of the body, will subsequently recirculate to brain tissue via its arterial input. The tracer kinetic model must take into account the various sources from which the measured ^{15}O activity in brain arises: $^{15}O_2$ in incoming arterial blood; extracted $^{15}O_2$ that is converted to ^{15}O-water of metabolism and washed out of the brain; unextracted $^{15}O_2$ in the capillary and venous circulation of the brain; and recirculating ^{15}O-water of metabolism that washes into and out of brain tissue.

7.2. Steady-State Method

With the steady-state method, rCBF is first determined using $C^{15}O_2$ inhalation. Scanning is then performed during the continuous inhalation of $^{15}O_2$. A two-compartment model is used to compute rOEF from the ratio of tissue counts during the $^{15}O_2$ and $C^{15}O_2$ inhalations and from measurements of blood radioactivity (Subramanyam et al., 1978; Frackowiak et al., 1980). As originally formulated, this model did not include a term to account for intravascular $^{15}O_2$. This lead to an overestimation of both rOEF and $rCMRO_2$ (Lammertsma et al., 1981). The amount of overestimation depends upon the local blood volume and is greater at low levels of rCBF and rOEF. A technique for correcting for intravascular $^{15}O_2$ has been developed (Lammertsma and Jones, 1983; Lammertsma et al., 1987b) and implemented using data from a separate determination of rCBV. The importance of this correction has been demonstrated (Lammertsma et al., 1983).

The steady-state measurement of $rCMRO_2$ has the same practical features as the $C^{15}O_2$ flow technique. It is particularly suited to tomographs requiring low count rates and to single-slice tomographs. Of note, tissue heterogeneity does not affect the accuracy of the OEF calculation (Lammertsma and Jones, 1985; Herscovitch and Raichle, 1985). The accuracy of rOEF and $rCMRO_2$ measurements in relation to tomographic counting statistics has been assessed (Lammertsma et al., 1982; Lammertsma and Jones, 1985). rOEF measurements with the steady-state method have been compared to direct measurements of the cerebral arterial-venous oxygen difference in baboons (Baron et al., 1981). There was a consistant 13% overestimation of rOEF, most likely because of the lack of correction for intravascular tracer in these experiments.

7.3. Brief Inhalation Method

Mintun and colleagues (1984) have described an alternative method for measuring rOEF and $rCMRO_2$ (Fig. 6) that uses scan data obtained following brief inhalation of $^{15}O_2$. The method also involves measurement of rCBF with $H_2{}^{15}O$ and the PET/autoradiographic approach, and of rCBV with $C^{15}O$. A two-compartment model is used to analyze the scan data. It accounts for the production and egress of water of metabolism in the tissue, recirculating water of metabolism, and the arterial, venous, and capillary contents of $^{15}O_2$ in the brain. To implement this technique, a 40-s emission scan is obtained following brief inhalation of approximately 80–100 mCi of $^{15}O_2$, and frequent arterial blood samples are obtained for measurements of blood radioactivity.

As with the PET/autoradiographic flow technique, this method requires a tomograph capable of operating at high count rates. Although the equation for the calculation of rOEF from the measured PET and arterial blood curve data is mathematically complex, an accurate simplification of this equation that facilitates the calculation has been described (Herscovitch et al., 1985b). The accuracy of the brief inhalation technique was demonstrated in baboons by the direct comparison of rOEF measured with PET to OEF measured by intracarotid injection of $^{15}O_2$ in the same animals (Mintun et al., 1984). Simulation studies of this method have demonstrated that measurement errors in rCBV or rCBF cause approximately equivalent percent errors in rOEF and $rCMRO_2$ determinations at high or normal levels of $rCMRO_2$, although errors are amplified at low metabolic rates.

Finally, it should be noted that both methods for measuring cerebral oxygen metabolism with PET, the steady-state technique and the brief inhalation method, require the combination of tomographic data obtained from three separate emission scans. Implicit in these approaches is the assumption of a physiologic steady state with respect to cerebral blood flow, blood volume, and metabolism over the period during which these scans are performed. In addition, it is important that the subject's head be maintained in a constant position throughout the study so that proper registration of the three images will be obtained.

8. Glucose Metabolism

8.1. Deoxyglucose Method

8.1.1. ^{14}C-Deoxyglucose Model

Methods for measuring regional cerebral glucose metabolism (rCMRGlu) with PET are largely based on the 2-deoxyglucose (DG) technique of Sokoloff (Sokoloff et al., 1977; Sokoloff and Smith, 1985). This method, developed to measure rCMRGlu in laboratory animals, uses ^{14}C-labeled DG as the metabolic tracer and autoradiography to determine local tissue radioactivity. DG is an analog of glucose, differing from the native molecule by the substitution of a hydrogen atom for the hydroxyl group on the second carbon atom. DG is transported bidirectionally across the blood–brain barrier between plasma and brain tissue by the same transport mechanism as glucose. In tissue, DG is phosphorylated, as is glucose, by hexokinase, and 2-deoxyglucose-6-phosphate (DG-6-P) is formed. Because of its anomalous structure, DG-6-P is not metabolized further through the glycolytic pathway. Also, because of the low activity of glucose-6-phosphatase in brain, there is little dephosphorylation of DG-6-P back to DG. As a result of this "metabolic trapping" of DG-6-P and its low membrane permeability, there is negligible loss of DG-6-P from cerebral tissue over the time course of the experiment. Because metabolized tracer is retained in the brain, the calculation of rCMRGlu from measurements of local tissue radioactivity is greatly facilitated.

Sokoloff and colleagues developed a three-compartment model to describe the in vivo behavior of DG (Fig. 7). These compartments, in part physical, in part biochemical, consist of (1) DG in the plasma in brain capillaries; (2) DG in tissue; and (3) DG-6-P in

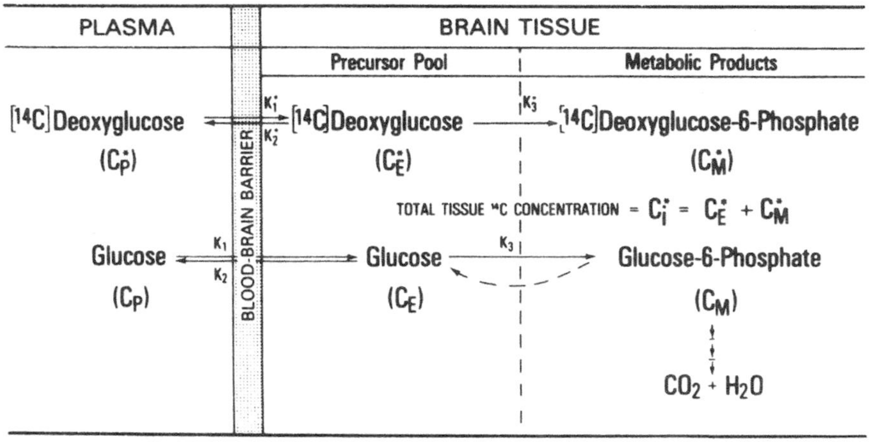

Fig. 7. Diagrammatic representation of the three compartment model used to measure cerebral glucose metabolism with deoxyglucose (DG). The three compartments consist of DG in the plasma in brain capillaries, DG in brain tissue, and DG-6-P in tissue. The lower portion of the figure shows the metabolic fate of glucose, and the upper portion, that of DG. In the adaptation of this model to PET, ^{18}F-labeled DG is used. In addition, a fourth rate constant, k_4*, is added to account for dephosphorylation of DG-6-P back to DG (from Sokoloff et al., 1977, courtesy of Dr. L. Sokoloff and by permission of Raven Press).

tissue. Three first-order rate constants are used to describe the movement of tracer between these compartments. Each rate constant denotes the fraction of the total tracer in a specific compartment that would leave the compartment per unit time and enter another compartment. These rate constants, k_1*, k_2*, and k_3* are, respectively, for the transport of DG from plasma to tissue across the blood–brain barrier, for the transport of DG back from tissue to plasma, and for the phosphorylation of DG to DG-6-P. An operational equation (Fig. 8) was developed to calculate rCMRGlu from a single measurement of total tissue radiotracer concentration (i.e., DG and DG-6-P), as well as arterial plasma concentration of DG and glucose measured as a function of time, the rate constants, and an additional factor, the lumped constant (LC). The LC is a multiplicative term in the operational equation that accounts for the differences between glucose and DG in transport across the blood–brain barrier and in phosphorylation.

Neither the rate constants nor the LC are determined in each experimental animal. Calculation of the rate constants requires

$$\mathrm{CMRGlu} = \frac{C_p}{LC} \cdot \frac{C(T) - k_1^* \exp[-(k_2^*+k_3^*)T] \int_0^T C_p^*(t) \exp[(k_2^*+k_3^*)t]\, dt}{\int_0^T C_p^*(t)\,dt - \exp[-(k_2^*+k_3^*)T] \int_0^T C_p^*(t) \exp[(k_2^*+k_3^*)t]\, dt}$$

$$= \frac{C_p}{LC} \cdot \frac{C(T) - \text{term A}}{\text{term B}}$$

Fig. 8. Operational equation of the deoxyglucose (DG) method of Sokoloff et al. (1977) used to measure the cerebral metabolic rate for glucose. The simplified form of the equation (lower equation) aids in understanding the model. $C(T)$ is the local radioactivity concentration in tissue measured at time T. Term A is the concentration of free DG in tissue at time T, calculated from the plasma concentration of DG over time [$C_p^*(t)$] and the rate constants. The difference between these two values equals the local concentration of DG-6-P that has been formed. Term B essentially represents the total amount of DG delivered to tissue. Therefore, the ratio on the right-hand side of the equation represents the fractional rate of phosphorylation of DG. Multiplying this by the plasma glucose concentration (C_p) would give the rate of glucose phosphorylation if DG and glucose had the same behavior. This is not the case, and to account for this difference, an additional term, the lumped constant (LC), is included. In the adaptation of this model to PET, a fourth rate constant, k_4^*, is included to account for dephosphorylation of DG-6-P, and the resultant operational equation is more complex (Huang et al., 1980).

multiple autoradiographic measurements of tissue radioactivity as a function of time following intravenous DG injection. Therefore, they were determined in a separate group of animals. The rate constant values were found to be uniform throughout gray matter structures and also throughout white matter structures. Thus, mean values for gray and white matter are used in the operational equation. A strategy was developed to minimize the error associated with the use of these mean values. This was based on the observation that the terms containing the rate constants in the operational equation approach zero with increasing time. A 45-min time interval between DG injection and animal sacrifice was chosen to minimize error associated with uncertainty in the rate constants. A longer period was avoided to minimize potential loss of DG-6-P from tissue.

The measurement of the LC in animals on a regional basis would be very difficult. Theoretical arguments were developed, however, to demonstrate that the LC should be relatively uniform throughout brain. Therefore, an average whole brain value was measured. This measurement was based on the mathematical demonstration that the LC is equivalent to the ratio of the steady-state arterial-venous extraction fraction by brain of glucose to that of DG.

8.1.2. Adaptation of the DG Method to PET

The DG method was subsequently adapted for use with PET (Reivich et al., 1979; Phelps et al., 1979b; Huang et al., 1980). The ^{14}C-label used for tissue autoradiography was replaced with the positron emitter, ^{18}F, to produce ^{18}F-2-deoxyglucose (FDG). FDG behaves similarly to DG in the brain, being phosphorylated to FDG-6-P. In order to obtain sufficient counts in the PET image, it is necessary to continue scanning beyond the 45-min time interval that is used between tracer administration and animal sacrifice in tissue autoradiography. Dephosphorylation of FDG-6-P, although small, becomes important. To account for this, a fourth rate constant, k_4^*, was added to the model. The four rate constants were determined in a group of normal subjects using regional radioactivity measurements from serial PET scans obtained over several hours following FDG administration. Average gray and white matter values were computed for use in the operational equation. Originally, the value for the LC in humans was not explicitly measured. Rather, its value was selected so that the calculated average value for global CMRGlu in normal subjects would equal that determined by earlier investigators using the more invasive Kety-Schmidt technique (Phelps et al., 1979b).

The standard FDG method is implemented as follows. About 45 min after the intravenous injection of 5–10 mCi of FDG, scanning is begun. Blood samples are obtained from the time of injection to the end of scanning to measure the concentration of glucose and FDG in plasma as a function of time. These are obtained by sampling peripheral arterial blood or by venous sampling from a hand heated to 44°C to "arterialize" venous blood by the production of arterial-venous shunts. If a single rather than a multiple slice scanner is used, multiple brain slices can be imaged following a single FDG administration by repositioning the subject in the tomograph. With a multislice tomograph, the subject can be repositioned in the axial direction by a fraction of the slice thickness (e.g., one-half) to further increase axial sampling. Scan duration is adjusted to account for ^{18}F decay and loss from tissue so that each image contains sufficient counts for accurate reconstruction. Local CMRGlu is calculated from the measured data by means of the operational equation, with the standard average values for the rate constants and LC, as discussed above.

The tracer kinetic model assumes that carbohydrate metabolism in the brain is in a steady state from the time of FDG administration to the time of scanning. This might not always be the case, for example with an epileptic seizure. Huang et al. (1981) calculated the effect of a change in rCMRGlu during the experiment. The measured rCMRGlu is approximately a weighted average of the metabolic values during the experiment, with the weightings proportional to the plasma FDG concentration at the corresponding times. It is not possible, however, to determine the rCMRGlu values for each of the different physiologic states during tracer uptake.

It has recently been observed that some of the methods used to synthesize FDG result in the production of varying amounts of ^{18}F-labeled deoxymannose as well (Reivich et al., 1985). This labeled contaminant may behave differently in vivo from FDG. Therefore, its presence may lead to errors in the calculation of rCMRGlu from scan data. The magnitude of these errors would depend upon the amount of contamination present. This problem highlights the importance of ensuring that the radiopharmaceuticals used in PET are pure.

Reivich and colleagues (1982) applied the DG method with PET using ^{11}C rather than ^{18}F as the label. The shorter half-life of

^{11}C (20 min) in comparison to ^{18}F (110 min) results in more rapid decay of the radioactive background following scanning. A repeat scan can therefore be performed in the same subject approximately 2 h later. This facilitates experimental designs in which a subject is used as his or her own control. Because of its shorter half life, approximately three times the amount of ^{11}C, in comparison to ^{18}F, must be administered to achieve comparable counts in the tomographic images. Therefore, the radiation dose per study is approximately the same for both radionuclides.

The accuracy of the DG method has been the subject of considerable analysis and discussion (Cunningham and Cremer, 1985). The major issue is the appropriateness of the use of standard values for the rate constants and LC in the operational equation across different brain regions and in pathological conditions. The originators of the DG method stressed that the value of these "constants" may be altered in abnormal conditions (Sokoloff et al., 1977; Reivich et al., 1979; Sokoloff and Smith, 1985). Recent work has included studies of the impact of using incorrect rate constant values, the development of more accurate methods for determining the rate constants, and the measurement of the LC in humans on both a global and regional basis.

The mathematical structure of the operational equation lessens the influence of the rate constant values on calculated rCMRGlu when the interval between FDG administration and scanning is relatively long, ie., 45 min or greater. When rCMRGlu deviates considerably from normal, however, such as in ischemia, the use of standard rate constants results in substantial error (Huang et al., 1981; Hawkins et al., 1981; Wienhard et al., 1985). Several investigators have suggested rearrangement or alternative formulations of the operational equation to decrease its sensitivity to differences between the actual and standard rate constant values (Brooks, 1982; Hutchins et al., 1984; Wienhard et al., 1985; Lammertsma et al., 1987a). In addition, methods used to calculate the rate constants from sequential tomographic images have been refined and applied in both normal and pathological conditions (Hawkins et al., 1986; Sasaki et al., 1986). For accurate measurement of rate constants, it is necessary to include a term for the local fraction of FDG that is present in the vascular space of the brain rather than in brain tissue. Since this fraction is very high initially, neglecting its contribution to measured radioactivity leads to large errors in the measurement of the rate constants and to overestimations of rCMRGlu.

The value for the LC for whole brain in normal humans has been measured from determinations of steady-state arterial and internal jugular venous concentrations of FDG and glucose (Reivich et al., 1985). The measured value of the LC, 0.52, was substantially higher than the previously calculated value of 0.42 (Phelps et al., 1979b). Data obtained in animal studies and more recently in humans indicate that the routine use of the normal value for the LC is in error in abnormal states, such as hyperglycemia and hypoglycemia (Sokoloff and Smith, 1985), and in conditions with low tissue glucose content, such as ischemia and seizures (Pardridge et al., 1982). In acute cerebral ischemia in the cat, the LC increased by a factor of 2.4 (Ginsberg and Reivich, 1979). Since the calculated rCMRGlu is inversely proportional to the value of the LC (see Fig. 8), the use of a control value for the LC in this case would lead to an overestimation of rCMRGlu by the same factor. Gjedde et al. (1985) developed a method to measure the LC regionally in humans with PET using FDG and 3-*O*-^{11}C-methylglucose. The latter tracer is transported bidirectionally across the blood–brain barrier, but is not metabolized. Its distribution in brain is used to calculate the local glucose content and thence the value of the LC. The LC was found to be uniform throughout normal brain, but in some regions of cerebral infarction its value was increased 3–6-fold. Thus, in pathological conditions, it is necessary to redetermine the values of both the LC and the rate constants to avoid errors in the calculation of rCMRGlu.

8.2. Use of ^{11}C-Glucose

An alternative approach to measuring rCMRGlu with PET involves the use of glucose labeled with ^{11}C (Raichle et al., 1975; Mintun et al., 1985; Blomqvist et al., 1985). ^{11}C-Glucose is biologically indistinguishable from glucose and is transported and metabolized in the same fashion and at the same rate as natural glucose. One does not require a correction factor, e.g., the lumped constant, to account for differences between the radiotracer and glucose. A disadvantage of ^{11}C-glucose in comparison with FDG is that the labeled metabolites of glucose, such as $^{11}C\text{-}CO_2$, are not all trapped within the tissue. Thus, tracer kinetic models must account for the rapid formation and loss of labeled metabolites.

A model consisting of four compartments has been used to describe the behavior of ^{11}C-glucose (Mintun et al., 1985; Blomqvist et al., 1985). The compartments consist of tracer in plasma, unmetabolized tracer in tissue, ^{11}C-labeled metabolites in tissue, and ^{11}C-labeled metabolites such as ^{11}C-CO_2 that can leave the tissue and enter blood. Rate constants are used to characterize the movement of tracer and metabolites between these compartments. Numerous brief, sequential scans are obtained following the intravenous administration of tracer to determine local tissue radioactivity as a function of time. Measurements of the concentration of natural glucose, ^{11}C-glucose, and ^{11}C-CO_2 in peripheral arterial blood are also obtained. The measured data are used to solve the differential equations of the tracer kinetic model for the rate constants and thence to calculate rCMRGlu. This approach also provides the values of physiologic parameters other than rCMRGlu, such as the concentration of free glucose in the brain and the fluxes of glucose across the blood–brain barrier. Because the extraction of glucose by brain is relatively low, a large portion of local radioactivity in brain is in the intravascular space, especially at early scan times. Measurement of rCBV is therefore required so that the amount of radiotracer actually entering tissue can be determined.

Methods for measuring rCMRGlu using ^{11}C-glucose are at an earlier stage of development than those using DG. Improved techniques to account for local loss of labeled metabolites are required. With further work, the ^{11}C-glucose approach may become more widely used for measuring rCMRGlu, especially in pathological conditions.

9. Analysis of PET Data

After a PET study has been performed, a computer is used to generate quantitative images representing the physiological variable being measured. This is done by application of the appropriate tracer kinetic model to the tomographic images of tissue radioactivity on a point-by-point basis. In the resultant images, the image intensity is directly proportional to the local value of the physiologic variable. A bar scale is used to display the physiologic value that corresponds to each level of image intensity (*see* Fig. 6). The use of a tracer model to convert PET measurements of radioactivity to physiological measurements is a necessary step in the implementation of PET. Tissue radioactivity measurements alone are insufficient. In some applications, such as rCBV measurement, the local radioactivity measurement is directly proportional to the value of the physiologic variable [*see* Eq. (1)]. Thus, information can be obtained about the *relative* value of the variable in different brain regions in a given subject. Comparisons cannot be made between subjects, however, and the information available from absolute, quantitative measurements is lost. For many PET tracer-kinetic models, the relationship between the local amount of radioactivity and the physiologic variable is nonlinear. This occurs with the steady-state rCBF method and with both techniques for measuring $rCMRO_2$. In these cases, images of radioactivity cannot be used even to determine relative regional differences.

The quantitative images obtained from a typical PET study contain a large amount of data. For example, a study of cerebral hemodynamics and oxygen metabolism with ^{15}O-labeled tracers and a tomograph with seven slices (e.g., Powers et al., 1986) results in a total of 28 images per patient (seven each of regional CBF, CBV, OEF, and $CMRO_2$). Although simple visual inspection of the images may show gross abnormalities in certain cases, one requires data analysis techniques that are systematic, quantitative, and statistically correct to obtain scientifically valid information from a PET study.

Analysis of the information contained in PET images is greatly facilitated by the use of specially designed, interactive computer programs. Both global and regional data can be obtained from the images with such programs. Global measurements are obtained by averaging over multiple contiguous PET slices (Herscovitch et al., 1986). This provides whole brain flow and metabolic data similar to those obtained with the Kety-Schmidt technique (Kety and Schmidt, 1948). More importantly, one can obtain quantitative, *regional* data from the PET image. Small regions of interest of arbitrary size and shape can be placed in selected areas of the image, and the average physiologic variable then computed for the region. The location of regions specified on one image, e.g. CBF, can be stored and used with other physiologic images, e.g., $CMRO_2$, obtained in the same patient. It is possible to compare two sets of images obtained in a subject under different physiological conditions, for example, at rest and during a physiologic stimulus (Fox and Raichle, 1984), or before and after administration of a drug (Perlmutter and Raichle, 1985). This is accomplished by a computer subtraction or division of the two images to obtain a third image representing the percent or absolute change in local flow or metabolism that occurred between the two scan states.

Data obtained from physiologic images representing local cerebral blood flow and metabolism must be related to the corresponding anatomic brain regions. It is not possible to perform such anatomical localization simply by visual inspection of the tomographic images. This approach is subjective and liable to observer bias. The relatively poor resolution of current tomographs limits the amount of detail in the image. More importantly, PET images based on *physiology* may not necessarily delineate the underlying anatomy, especially in cases of local pathology. An example of this occurs in early Huntington's disease (Kuhl et al., 1982), in which the caudate nucleus is difficult to visualize on the PET image because of its decreased metabolism, although it is still structurally intact. Several approaches to the problem of anatomical localization have been proposed (Mazziotta and Koslow, 1987). One method is to obtain the PET images in standard tomographic planes relative to external landmarks, such as the canthomeatal line. The PET images are then visually compared to anatomic brain sections in equivalent planes from a brain atlas (Duara et al., 1983). It may not always be possible to obtain coplanar tomographic and atlas brain slices, however. In addition, region placement on the PET images is subjective. A modification of this method is to store the brain atlas in digital form as part of the computer image analysis program and use it to overlay the tomographic images (Adair et al., 1981).

An alternative approach uses a stereotactic method of anatomical localization for PET regions of interest (Fox et al., 1985). With this technique, a correspondence is established between anatomical regions in a stereotactic brain atlas and specific regions on the PET images. At the time of the PET study, a plastic plate with embedded vertical wires corresponding to the tomographic slices is attached to the head rest of the scanner. A lateral skull X-ray is obtained to record the position of the PET slices, indicated by the radio-opaque wires, in relation to the bony landmarks of the skull. This information is used to set up a transformation between the coordinate system of a stereotactic brain atlas and the coordinate system defining the location of regions of interest in the PET slices. With this technique, regions of interest corresponding to specific brain structures can be placed on the PET images, and conversely, the anatomic location of a region of interest selected on the PET image can be determined. This method provides objective and reproducible region-of-interest localization when brain anatomy is normal. Other approaches are required, however, when there are structural abnormalities in the brain such as cerebral atrophy, or when there is a focal brain lesion for which regional PET measurements are desired. In these cases, one can obtain anatomical brain images in the same planes as the PET slices using either X-ray computed tomography (CT) or magnetic resonance imaging. If these anatomic images can be accessed by the computer image analysis program, regions of interest can be transferred between them and the PET images. Coplanar PET and CT images can be obtained by use of a special head holder that fits onto both the PET and CT couches (Bergstrom et al., 1981; Mazziotta et al., 1982). Alternatively, certain CT scanners can provide a lateral view of the skull in the CT gantry. If a lateral skull X-ray is taken at the time of PET to record the location of the PET slices in relation to the skull, as described above, these two lateral skull views can be used to angulate the CT gantry to obtain coplanar PET and CT slices (Herscovitch et al., 1986).

Analysis of regional PET data is complicated by the fact that there are often relatively wide variations in measured values of global cerebral blood flow and metabolism among subjects in a study (e.g., Duara et al., 1984). These variations may reflect true physiologic differences or may result from methodological errors, such as incorrect scanner calibration. As a result, the absolute measurement in a given region, e.g., the visual cortex, may vary widely among subjects, although the relative difference measured between two brain regions, e.g., visual cortex and white matter, may be similar in all subjects. Alternatively, during a sensory stimulus, a specific brain region may demonstrate increased flow in relation to the rest of the brain, but the absolute flow value may be lower than that seen in the same region in the resting state in the same or another subject. Therefore, approaches have been developed to facilitate the detection and quantification of regional changes. Methods to account for the effect of global variations on regional data include dividing each regional value by the global value (Duara et al., 1984; Fox and Raichle, 1984) or expressing data as a ratio of the values for homologous right and left regions of the brain (Perlmutter and Raichle, 1985). Such techniques may result in a loss of information contained in the absolute values of regional measurements, however, especially if more widespread changes accompany a focal change. For example, the lowered levels of CBF and $CMRO_2$ observed in cerebral tumors is accompanied by a depression of flow and metabolism in the contralateral cerebral hemisphere (Beaney, 1984). This change might not be detected without the use of absolute values.

One requires appropriate control data to define a regional abnormality in a group of subjects. This is true for relative as well as absolute measurements. For example, it cannot be assumed that regional measurements obtained from homologous regions on the right and left side of the brain are normally equal. Significant asymmetries in CBF and $CMRO_2$ in the sensorimotor, occipital, and superior temporal regions have been demonstrated to occur in normal resting subjects (Perlmutter et al., 1987). In addition to the variations in PET measurements that occur among different subjects, the variation in repeat measurements made in the same subject at different times must be considered. Again, these differences may be either physiological or methodological in origin. The normal range for a specific measurement must be known before attributing changes in that measurement in a given subject to a specific therapeutic or physiological intervention.

10. Applications of PET

Most PET studies to date have used either the ^{15}O techniques described above to measure CBF, CBV, and $CMRO_2$ or have used FDG to measure CMRGlu. A detailed discussion of the applications of PET is beyond the scope of this chapter, and the reader is directed to recent reviews (Phelps et al., 1982; Leenders et al., 1984; Beaney, 1984; Powers and Raichle, 1985; Baron, 1985; Phelps and Mazziotta, 1985). We will, however, provide a selective overview of the use of PET to measure regional cerebral hemodynamics and metabolism. This will demonstrate its widespread applicability to the study of cerebral physiology and pathophysiology.

^{15}O-Labeled radiotracers have been widely applied to study cerebrovascular disease (Powers and Raichle, 1985; Baron, 1985). These methods are useful because they provide measurements not only of rCBF, but also of rCBV, $rCMRO_2$, and rOEF, the latter indicating the balance between local oxygen supply and demand. In acute cerebral infarction, the nature and time course of the alterations in rCBF and $rCMRO_2$ have been studied. Areas of decreased rCBF and $rCMRO_2$ have been found in brain regions distant to the site of cerebral infarction, probably resulting from functional depression of local neuronal activity caused by interruption of afferent or efferent pathways associated with these regions (Fig. 9). The response of ischemic, noninfarcted brain to decreased perfusion pressure caused by occlusive vascular disease has been studied (Gibbs et al., 1984; Powers et al., 1984). Increased oxygen extraction and dilatation of intraparenchymal blood vessels, both serving to maintain oxygen metabolism, have been observed. Thresholds of rCBF and $rCMRO_2$ for normal neuronal function and for irreversible cerebral infarction have been examined (Powers et al., 1986), suggesting that it may be possible to differentiate reversible ischemia from irreversible infarction. The hemodynamic and metabolic response of the brain to therapeutic intervention, such as carotid endarterectomy, is a subject of ongoing investigation (Gibbs et al., 1985). The measurement of rCBF alone in cerebrovascular disease is insufficient to characterize tissue integrity. For example, blood flow may be normal in acutely infarcted tissue because of luxury perfusion, although metabolism is markedly decreased (Powers and Raichle, 1985). Similarly, rCBF may be decreased in noninfarcted tissue distal to an occluded internal carotid artery, although metabolism is relatively maintained. Thus, measurements of metabolism are required as well. Although FDG is widely used to measure regional cerebral glucose metabolism, the accuracy of this method in ischemic or infarcted tissue is open to question (Ginsberg and Reivich, 1979; Choki et al., 1983;), and the ^{15}O techniques to measure $rCMRO_2$ are preferable.

PET measurements have been performed in a wide variety of neurologic and psychiatric diseases that are not caused by disturbances of cerebral blood flow or metabolism. Such measurements are used as indirect indicators of the level of local neuronal activity, however. Therefore, they may demonstrate the site and degree of local abnormalities of neuronal function and the response of these abnormalities to therapeutic interventions. A few specific examples will be quoted. Abnormalities of rCBF have been demonstrated in basal ganglia and mesocortical regions in hemiparkinsonism, and the effects of L-dopa on rCBF and $rCMRO_2$ have been investigated (Perlmutter and Raichle, 1985; Leenders et al., 1985). In senile dementia of the Alzheimer type, declines in rCBF, $rCMRO_2$, and rCMRGlu have been observed (Frackowiak et al., 1981), presumably paralleling the loss of functioning neurons. In contrast, rCMRGlu has been reported not to decrease in associa-

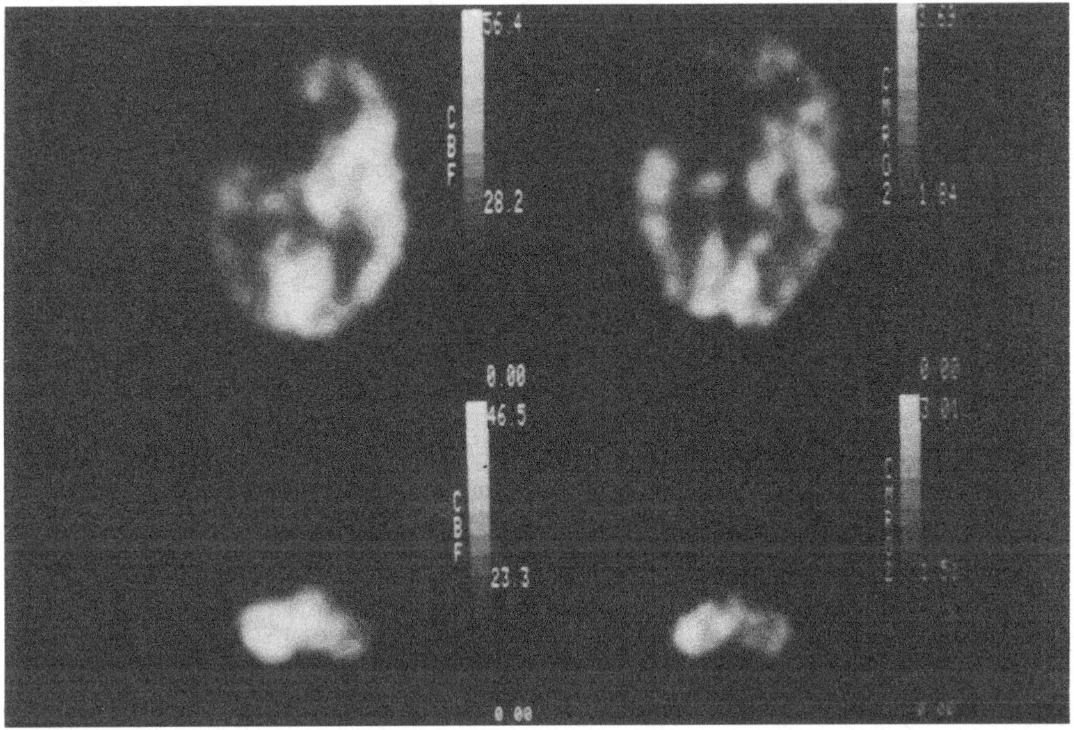

Fig. 9. Example of crossed cerebellar diaschisis. Upper scans show a marked decrease in both CBF and $CMRO_2$ in an area of infarction in the left frontal lobe. In the lower scans of cerebellar blood flow (left) and cerebral oxygen metabolism (right), both flow and metabolism are decreased in the right cerebellar hemisphere, contralateral to the cerebral infarction.

tion with normal aging (Duara et al., 1984). rCBF has been measured in patients with panic disorder (Reiman et al., 1986), which is characterized by recurrent anxiety attacks in the absence of a frightening stimulus. Analysis of rCBF in brain regions thought to mediate symptoms of panic and anxiety demonstrated an abnormal asymmetry of flow in a region of the parahippocampal gyrus. In patients with complex partial seizures, hypometabolic regions in the temporal lobe have frequently been demonstrated in interictal scans with FDG. The physiologic basis of this finding is an area of ongoing investigation (Mazziotta and Engel, 1984).

PET has been widely applied to study the response of the brain to a variety of sensory, motor, and cognitive tasks (Phelps and Mazziotta, 1985). Most of these studies have used FDG to measure local changes in glucose utilization. This approach has certain limitations, however (Mazziotta et al., 1985). The tracer strategy requires a period of approximately 45 min between FDG administration and onset of scanning. It may be difficult to maintain a physiological steady state during the neurobehavorial task over the period of radiotracer uptake by the brain. Because of the relatively long half-life of ^{18}F (110 min), repeat studies must be performed on consecutive days, and only one or two such studies can be obtained in a subject. These factors introduce the difficulty of repositioning subjects for follow-up studies and limit the range of experimental conditions that can be studied in a given subject. Some of these difficulties may be partially overcome by the use of the shorter-lived radiotracer, ^{11}C-deoxyglucose (Reivich et al., 1982). Chang et al. (1987) have recently described an approach in which two FDG studies are performed in the same scan session. Data used to compute rCMRGlu during the second scan are corrected for the residual ^{18}F radioactivity remaining from the first procedure. An alternative approach to functional activation studies involves the measurement of local CBF responses using bolus intravenous $H_2{}^{15}O$ and the PET/autoradiographic method (Fig. 10) (Raichle et al., 1983; Fox et al., 1984). Each flow determination is performed in less than 1 min. Because of the short half-life of ^{15}O, up to eight measurements of rCBF can be performed in the same subject in a 2-h period, allowing for great flexibility in experimental design.

These examples have served to demonstrate the widespread applicability of PET measurements of cerebral blood flow and metabolism to the study of cerebral physiology and pathophysiology.

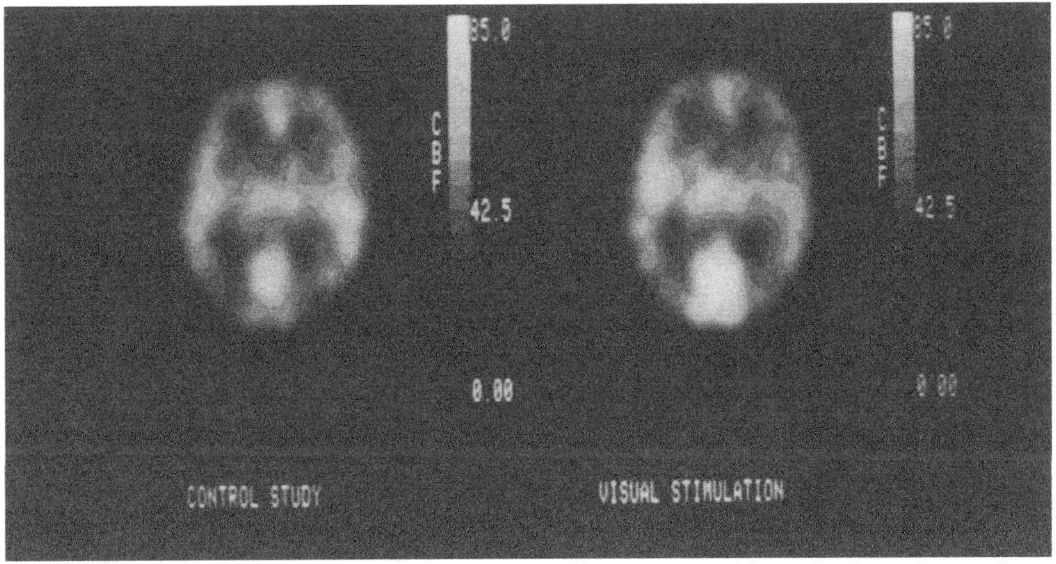

Fig. 10. Local blood flow response to visual stimulation in a normal subject. The control study was obtained with the subject's eyes patched. During pattern-flash photic stimulation, there is an increase in flow in the visual cortex (lower portion of image). Note that the same flow scale is used in both images.

References

Adair T., Karp P., Stein A., Bajesy R., and Reivich M. (1981) Computer assisted analysis of tomographic images of the brain. *J. Comput. Assist. Tomogr.* **5,** 929–932.

Alpert N. M., Eriksson L., Chang J. Y., Bergstrom M., Litton J. E., Correia J. A., Bohm C., Ackerman R. H., and Taveras J. M. (1984) Strategy for the measurement of regional cerebral blood flow using short-lived tracers and emission tomography. *J. Cereb. Blood Flow Metab.* **4,** 28–34.

Baron J. C. (1985) Positron tomography in cerebral ischemia. *Neuroradiology* **27,** 509–516.

Baron J. C., Steinling M., Tanaka T., Cavalheiro E., Soussaline F., and Collard P. (1981) Quantitative measurement of CBF, oxygen extraction fraction (OEF) and $CMRO_2$ with the ^{15}O continuous inhalation technique and positron emission tomography (PET): Experimental evidence and normal values in man. *J. Cereb. Blood Flow Metab.* **1** (suppl. 1), S5–S6.

Beaney R. P. (1984) Positron emission tomography in the study of human tumors. *Sem. Nucl. Med.* **14,** 324–341.

Bergstrom M., Boethius J., Eriksson L., Greitz T., Ribbe T., and Widen L. (1981) Head fixation device for reproducible position alignment in transmission CT and positron emission tomography. *J. Comput. Assist. Tomogr.* **5,** 136–141.

Blomqvist G., Bergstrom K., Bergstrom M., Ehrin E., Eriksson L., Garmelius B., Lindberg B., Lilja A., Litton J.-E., Lundmark L., Lundqvist H., Malmborg P., Mostrom U., Nilsson L., Stone-Elander S., and Widen L. (1985) Models for ^{11}C-Glucose, in *The Metabolism of the Human Brain Studied with Positron Emission Tomography* (Greitz T., Ingvar D. H., and Widen L., eds.) Raven, New York.

Brooks R. A. (1982) Alternative formula for glucose utilization using labeled deoxyglucose. *J. Nucl. Med.* **23,** 538–539.

Brooks R. A. and DiChiro G. (1976) Principles of computer assisted tomography (CAT) in radiographic and radioisotopic imaging. *Phys. Med. Biol.* **21,** 689–732.

Brooks R. A., Sank V. J., Friauf W. S., Leighton S. B., Cascio H. E., and DiChiro G. (1981) Design considerations for positron emission tomography. *IEEE Trans. Biomed. Eng. BME* **28,** 158–177.

Budinger T. F., Derenzo S. E., Greenberg W. L., Gullberg G. T., and Huesman R. H. (1978) Quantitative potentials of dynamic emission computed tomography. *J. Nucl. Med.* **19,** 309–315.

Carson R. E., Huang S.-C., and Green M. V. (1986) Weighted integration method for local cerebral blood flow measurements with positron emission tomography. *J. Cereb. Blood Flow. Metab.* **6,** 245–258.

Chang J. Y., Duara R., Barker W., Apicella A., and Finn R. (1987) Two behavioral states studied in a single PET/FDG procedure: theory, method, and preliminary results. *J. Nucl. Med.* **28,** 852–860.

Choki J., Greenberg J., and Reivich M. (1983) Regional cerebral glucose metabolism during and after bilateral cerebral ischemia in the gerbil. *Stroke* **14,** 568–574.

Comar D., Berridge M., Maziere B., and Crouzel C. (1982) Radiopharmaceuticals Labelled with Positron-Emitting Radioisotopes, in *Computed Emission Tomography* (Ell P. A. and Holman B. L., eds.) Oxford University Press, New York.

Cunningham V. and Cremer J. E. (1985) Current assumptions behind the use of PET scanning for measuring glucose utilization in brain. *Trends Neurosci.* **8,** 96–99.

Duara R., Margolin R. A., Robertson-Tchabo E. A., London E. D., Schwartz M., Renfrew J. W., Koziarz B. J., Sundaram M., Grady C., Moore A. M., Ingvar D. H., Sokoloff L., Weingartner H., Kessler R. M., Manning R. G., Channing M. A., Cutler N. R., and Rapoport S. I., (1983) Cerebral glucose utilization, as measured with positron emission tomography in 21 resting healthy men between the ages of 21 and 83 years. *Brain* **106,** 761–775.

Duara R., Grady C., Haxby J., Ingvar D., Sokoloff L., Margolin R. A., Manning R. G., Cutler N. R., and Rapoport S. I. (1984) Human brain glucose utilization and cognitive function in relation to age. *Ann. Neurol.* **16,** 702–713.

Eichling L. O., Raichle M. E., Grubb R. J., Jr, and Ter-Pogossian M. M. (1974) Evidence of the limitations of water as a freely diffusible tracer in the brain of the rhesus monkey. *Circ. Res.* **35,** 358–64.

Eichling L. O., Raichle M. E., Grubb R. J., Jr, Larson K. B., and Ter-Pogossian M. M. (1975) In vivo determination of cerebral blood volume with radioactive oxygen-15 in the monkey. *Circ. Res.* **37,** 707–714.

Eichling J. O., Higgins C. S., and Ter-Pogossian M. M. (1977) Determination of radionuclide concentration with positron CT scanning. *J. Nucl. Med.* **18,** 845–847.

Fox P. T. and Raichle M. E. (1984) Stimulus rate dependence of regional cerebral blood flow in human striate cortex, demonstrated by positron emission tomography. *J. Neurophysiol.* **51,** 1109–1120.

Fox P. T., Mintun M. A., Raichle M. E., and Herscovitch P. (1984) A noninvasive approach to quantitative functional brain mapping with $H_2{}^{15}O$ and positron emission tomography. *J. Cereb. Blood Flow Metab.* **4,** 329–333.

Fox P. T., Perlmutter J. S., and Raichle M. E. (1985) A stereotactic method of anatomical localization for positron emission tomography. *J. Comput. Assist. Tomogr.* **9,** 141–153.

Frackowiak R. S. J., Lenzi G.-L., Jones T., and Heather J. D. (1980) Quantitative measurement of regional cerebral blood flow and oxygen metabolism in man using ^{15}O and positron emission tomography: Theory, procedure and normal values. *J. Comput. Assist. Tomogr.* **4,** 727–736.

Frackowiak R. S. J., Pozzilli C., Legg N. J., DuBoulay G. H., Marshall J., Lenzi G. L., and Jones T. (1981) Regional cerebral oxygen supply and utilization in dementia: A clinical and physiological study with oxygen-15 and positron tomography. *Brain* **104,** 753–778.

Gibbs J. M., Wise R. J. S., Leenders K. L., and Jones T. (1984) Evaluation of cerebral perfusion reserve in patients with carotid-artery occlusion. *Lancet* **1,** 310–314.

Gibbs J. M., Wise R. J. S., Mansfield A. O., Ross Russell R. W., Thomas D. J., and Jones T. (1985) Cerebral circulatory reserve before and after surgery for occlusive carotid artery disease. *J. Cereb. Blood Flow Metab.* **5** (suppl. 1), S19–S20.

Ginsberg M. D. and Reivich M. (1979) Use of the 2-deoxyglucose method of local cerebral glucose utilization in the abnormal brain: Evaluation of the lumped constant during ischemia. *Acta Neurol. Scand.* **60** (suppl. 72), 226–227.

Gjedde A., Wienhard K., Heiss W.-D., Kloster G., Diemer N. H., Herholz K., and Pawlik G. (1985) Comparative regional analysis of 2-fluorodeoxyglucose and methylglucose uptake in brain of four stroke patients. With special reference to the regional estimation of the lumped constant. *J. Cereb. Blood Flow Metab.* **5,** 163–178.

Graham M. M., Bassingthwaighte J. B., and Chan J. (1985) Validation of compartmental models of deoxyglucose kinetics using data from a distributed model. *J. Cereb. Blood Flow Metab.* **5** (suppl. 1), S573–S574.

Grubb R. L., Jr, Raichle M. E., Phelps M. E., and Ratcheson R. A. (1975) Effects of increased intracranial pressure on cerebral blood volume, blood flow and oxygen utilization in monkeys. *J. Neurosurg.* **43,** 385–398.

Grubb R. L., Jr, Raichle M. E., Higgins C. S., and Eichling J. O. (1978) Measurement of regional cerebral blood volume by emission tomography. *Ann. Neurol.* **4,** 322–328.

Hawkins R. A., Phelps M. E., Huang S.-C., and Kuhl D. E. (1981) Effect of ischemia on quantification of local cerebral glucose metabolic rate in man. *J. Cereb. Blood Flow Metab.* **1,** 37–51.

Hawkins R. A., Phelps M. E., and Huang S.-C. (1986) Effects of temporal sampling, glucose metabolic rates, and disruptions of the blood-brain barrier on the FDG model with and without a vascular compartment: Studies in human brain tumors with PET. *J. Cereb. Blood Flow Metab.* **6,** 170–183.

Herscovitch P. and Raichle M. E. (1983) Effect of tissue heterogeneity on the measurement of cerebral blood flow with the equilibrium $C^{15}O_2$ inhalation technique. *J. Cereb. Blood Flow Metab.* **3**, 407–415.

Herscovitch P. and Raichle M. E. (1985a) What is the correct value for the brain-blood partition coefficient for water? *J. Cereb. Blood Flow Metab.* **5**, 65–69.

Herscovitch P. and Raichle M. E. (1985b) Effect of tissue heterogeneity on the measurement of regional cerebral oxygen extraction and metabolic rate with positron emission tomography. *J. Cereb. Blood Flow Metab.* **5** (suppl. 1), S671–S672.

Herscovitch P., Markham J., and Raichle M. E. (1983) Brain blood flow measured with intravenous $H_2{}^{15}O$. I. Theory and error analysis. *J. Nucl. Med.* **24**, 782–789.

Herscovitch P., Raichle M. E., Kilbourn M. R., and Welch M. J. (1987) Positron emission tomographic measurement of cerebral blood flow and permeability-surface area product of water using ^{15}O-water and ^{11}C-butanol. *J. Cereb. Blood Flow Metab.*, in press.

Herscovitch P., Mintun M. A., and Raichle M. E. (1985) Brain oxygen utilization measured with oxygen-15 radiotracers and positron emission tomography: Generation of metabolic images. *J. Nucl. Med.* **26**, 416–417.

Herscovitch P., Auchus A. P., Gado M., Chi D., and Raichle M. E. (1986) Correction of positron emission tomography data for cerebral atrophy. *J. Cereb. Blood Flow Metab.* **6**, 120–124.

Hoffman E. J. (1982) Instrumentation for Quantitative Tomographic Determination of Concentrations of Positron-Emitting, Receptor Binding Radiotracers, in *Receptor-Binding Radiotracers* vol. II, (Eckelman W. C., ed.) CRC, Boca Raton.

Hoffman E. J., Huang S.-C., and Phelps M. E. (1979) Quantitation in positron emission computed tomography. 1. Effect of object size. *J. Comput. Assist. Tomogr.* **3**, 299–308.

Hoffman E. J., Huang S.-C., Phelps M. E., and Kuhl D. E. (1981) Quantitation in positron emission computed tomography. 4. Effect of accidental coincidences. *J. Comput. Assist. Tomogr.* **5**, 391–400.

Holden J. E., Gatley S. J., Nickles R. J., Koeppe R. A., Celesia G. G., and Polcyn R. E. (1983) Regional Cerebral Blood Flow Measurement with Fluoromethane and Positron Emission Tomography, in *Positron Emission Tomography of the Brain* (Heiss W.-D. and Phelps M. E., eds.) Springer-Verlag, Berlin.

Huang S.-C., Phelps M. E., Hoffman E. J., and Kuhl D. E. (1979) A theoretical study of quantitative flow measurements with constant infusion of short-lived isotopes. *Phys. Med. Biol.* **24**, 1151–1161.

Huang S.-C., Phelps M. E., Hoffman E. J., Sideris K., Selin C. J., and Kuhl D. E. (1980) Non-invasive determination of local cerebral metabolic rate of glucose in man. *Am. J. Physiol.* **238,** E69–E82.

Huang S.-C., Phelps M. E., Hoffman E. J., and Kuhl D. E. (1981) Error sensitivity of fluorodeoxyglucose method for measurement of cerebral metabolic rate of glucose. *J. Cereb. Blood Flow Metab.* **1,** 391–401.

Huang S.-C., Carson R. E., and Phelps M. E. (1982) Measurement of local blood flow and distribution volume with short-lived isotopes: A general input technique. *J. Cereb. Blood Flow Metab.* **2,** 99–108.

Huang S.-C., Carson R. E., Hoffman E. J., Carson J., MacDonald N., Barrio J. R., and Phelps M. E. (1983) Quantitative measurement of local cerebral blood flow in humans by positron computed tomography and ^{15}O-water. *J. Cereb. Blood Flow Metab.* **3,** 141–153.

Hutchins G. D., Holden J. E., Koeppe R. A., Halama J. R., Gatley S. J., and Nickles R. J. (1984) Alternative approach to single-scan estimation of cerebral glucose metabolic rate using glucose analogs, with particular application to ischemia. *J. Cereb. Blood Flow Metab.* **4,** 35–40.

Jones T., Chesler D. A., and Ter-Pogossian M. M. (1976) The continuous inhalation of oxygen-15 for assessing regional oxygen extraction in the brain of man. *Br. J. Radiol.* **49,** 339–343.

Jones S. C., Greenberg J. H., and Reivich M. (1982) Error analysis for the determination of cerebral blood flow with the continuous inhalation of ^{15}O-labeled carbon dioxide and positron emission tomography. *J. Comput. Assist. Tomogr.* **6,** 116–124.

Jones S. C., Greenberg J. H., Dann R., Robinson, G. D., Jr., Kushner M., Alavi A., and Reivich M. (1985) Cerebral blood flow with the continuous infusion of oxygen-15 labeled water. *J. Cereb. Blood Flow Metab.* **5,** 566–575.

Kanno I. and Lassen N. A. (1979) Two methods for calculating regional cerebral blood flow from emission computed tomography of inert gas concentrations. *J. Comp. Assist. Tomogr.* **3,** 71–76.

Kanno I., Lammertsma A. A., Heather J. D., Gibbs J. M., Rhodes C. G., Clark J. C., and Jones T. (1984) Measurement of cerebral blood flow using bolus inhalation of $C^{15}O_2$ and positron emission tomography: Description of the method and its comparison with the $C^{15}O_2$ continuous inhalation method. *J. Cereb. Blood Flow Metab.* **4,** 224–234.

Kearfott K. J. (1982) Absorbed dose estimates for positron emission tomography (PET): $C^{15}O$, ^{11}CO, and $CO^{15}O$. *J. Nucl. Med.* **23,** 1031–1037.

Kety S. S. (1951) The theory and applications of the exchange of inert gas at the lungs and tissues. *Pharmacol. Rev.* **3,** 1–41.

Kety S. S. (1960) Measurement of local blood flow by the exchange of an inert diffusible substance. *Meth. Med. Res.* **8,** 228–236.

Kety S. S. and Schmidt C. F. (1948) The nitrous oxide method for the quantitative determination of cerebral blood flow in man: Theory, procedure, and normal values. *J. Clin. Invest.* **27,** 476–483.

Koeppe R. A., Holden J. E., Polcyn R. E., Nickles R. J., Hutchins G. D., and Weese J. L. (1985) Quantitation of local cerebral blood flow and partition coefficient without arterial sampling: Theory and validation. *J. Cereb. Blood Flow Metab.* **5,** 214–223.

Kuhl D. E., Phelps M. E., Markham C. H., Metter E. J., Riege W. H., and Winter J. (1982) Cerebral metabolism and atrophy in Huntington's disease determined by ^{18}FDG and computed tomographic scan. *Ann. Neurol.* **12,** 425–434.

Lammertsma A. A. and Jones T. (1983) Correction for the presence of intravascular oxygen-15 in the steady state technique for measuring regional oxygen extraction ratio in the brain. 1. Description of the method. *J. Cereb. Blood Flow Metab.* **13,** 416–424.

Lammertsma A. A., Jones T., Frackowiak R. S. J., and Lenzi G.-L. (1981) A theoretical study of the steady-state model for measuring regional cerebral blood flow and oxygen utilisation using oxygen-15. *J. Comput. Assist. Tomogr.* **5,** 544–550.

Lammertsma A. A., Heather J. D., Jones T., Frackowiak R. S. J., and Lenzi G.-L. (1982) A statistical study of the steady state technique for measuring regional cerebral blood flow and oxygen utilisation using ^{15}O. *J. Comput. Assist. Tomogr.* **6,** 566–573.

Lammertsma A. A., Wise R. J. S., Heather J. D., Gibbs J. M., Leenders K. L., Frackowiak R. S. J., Rhodes C. G., and Jones T. (1983) Correction for the presence of intravascular oxygen-15 in the steady-state technique for measuring regional oxygen extraction ratio in the brain. 2. Results in normal subjects and brain tumour and stroke patients. *J. Cereb. Blood Flow Metab.* **3,** 425–431.

Lammertsma A. A., Brooks D. J., Beaney R. P., Turton D. R., Kensett M. J., Heather J. D., Marshall J., and Jones T. (1984) In vivo measurement of regional cerebral haematocrit using positron emission tomography. *J. Cereb. Blood Flow Metab.* **4,** 317–322.

Lammertsma A. A., Brooks D. J., Frackowiak R. S. J., Beany R. P., Herold S., Heather J. D., Palmer A. J., and Jones T. (1987a) Measurement of glucose utilisation with [^{18}F]2-fluoro-2-deoxy-D-glucose: A comparison of different analytical methods. *J. Cereb. Blood Flow Metab.* **7,** 161–172.

Lammertsma A. A., Baron J.-C., and Jones T. (1987b) Correction for intravascular activity in the oxygen-15 steady-state technique is independent of the regional hematocrit. *J. Cereb. Blood Flow Metab.* **7,** 372–374.

Landau W. M., Freygang W. H., Jr, Rowland L. P., Sokoloff L., and Kety S. (1955) The local circulation of the living brain; values in the unanesthetized and anesthetized cat. *Trans. Am. Neurol. Assoc.* **80,** 125–129.

Larson K. B., Markham J., Herscovitch P., and Raichle M. E. (1987) A distributed-parameter tracer-kinetic model for regional CBF measurement with PET. *J. Cereb. Blood Flow Metab.* **7,** S575.

Lassen N. A. and Ingvar D. H. (1972) Radioisotopic assessment of regional cerebral blood flow. *Progr. Nucl. Med.* **1,** 376–409.

Leenders K. L., Gibbs J. M., Frackowiak R. S. J., Lammertsma A. A., and Jones T. (1984) Positron emission tomography of the brain: New possibilities for the investigation of human cerebral pathophysiology. *Prog. Neurobiol.* **23,** 1–38.

Leenders K. L., Wolfson L., Gibbs J. M., Wise R. J. S., Causon R., Jones T., and Legg N. J. (1985) The effects of L-Dopa on regional cerebral blood flow and oxygen metabolism in patients with Parkinson's disease. *Brain* **108,** 171–191.

Mazziotta J. C. and Engel J., Jr. (1984) The use and impact of positron computed tomography scanning in epilepsy. *Epilepsia* **25** (suppl. 2), S86–S104.

Mazziotta J. C. and Koslow S. H. (1987) Assessment of goals and obstacles in data acquisition and analysis from emission tomography: Report of a series of international workshops. *J. Cereb. Blood Flow Metab.* **7,** S1–S31.

Mazziotta J. C., Phelps M. E., Plummer D., and Kuhl D. E. (1981) Quantitation in positron computed tomography. 5. Physical-anatomical effects. *J. Comput. Assist. Tomogr.* **5,** 734–743.

Mazziotta J. C., Phelps M. E., Meadors A. K., Ricci A., Winter J., and Bentson J. R. (1982) Anatomical localization schemes for use in positron computed tomography using a specially designed headholder. *J. Comput. Assist. Tomogr.* **6,** 848–853.

Mazziotta J. C., Huang S.-C., Phelps M. E., Carson R. E., MacDonald N. S., and Mahoney K. (1985) A noninvasive positron computed tomography technique using oxygen-15-labeled water for the evaluation of neurobehavioral task batteries. *J. Cereb. Blood Flow Metab.* **5,** 70–78.

Meyer E. and Yamamoto Y. L. (1984) The requirement for constant arterial radioactivity in the $C^{15}O_2$ steady-state blood-flow model. *J. Nucl. Med.* **25,** 455–460.

Mintun M. A., Raichle M. E., Martin W. R. W., and Herscovitch P. (1984) Brain oxygen utilization measured with 0–15 radiotracers and positron emission tomography. *J. Nucl. Med.* **25,** 177–187.

Mintun M. A., Raichle M. E., Welch M. J., and Kilbourn M. R. (1985) Brain glucose metabolism measured with PET and U-^{11}C-glucose. *J. Cereb. Blood Flow Metab.* **5** (suppl. 1), S623–S624.

Muehllehner G. and Karp J. S. (1986) Positron emission tomography imaging—technical considerations. *Sem. Nucl. Med.* **16,** 35–50.

Obrist W. D., Thompson H. K., King C. H., and Wang H. S. (1967) Determination of regional cerebral blood flow by inhalation of xenon-133. *Circ. Res.* **20,** 124–135.

Pardridge W. M., Crane P. D., Mietus L. J., and Oldendorf W. H. (1982) Kinetics of regional blood-brain barrier transport and brain phosphorylation of glucose and 2-deoxyglucose in the barbituate-anesthetized rat. *J. Neurochem.* **38,** 560–568.

Perlmutter J. S. and Raichle M. E. (1985) Regional blood flow in hemiparkinsonism. *Neurology* **35,** 1127–1134.

Perlmutter J. S., Powers W. J., Herscovitch P., Fox P. T., and Raichle M. E. (1987) Regional asymmetries of cerebral blood flow, blood volume, oxygen utilization and extraction in normal subjects. *J. Cereb. Blood Flow Metab.* **7,** 64–67.

Phelps M. E. and Mazziotta J. C. (1985) Positron emission tomography: Human brain function and biochemistry. *Science* **228,** 799–809.

Phelps M. E., Hoffman E. J., Huang S.-C., and Ter-Pogossian M. M. (1975) Effect of positron range on spatial resolution. *J. Nucl. Med.* **16,** 649–652.

Phelps M. E., Huang S. C., Hoffman E. J., and Kuhl D. E. (1979a) Validation of tomographic measurement of cerebral blood volume with C-11-labeled carboxyhemoglobin. *J. Nucl. Med.* **20,** 328–334.

Phelps M. E., Huang S. C., Hoffman E. J., Selin C., Sokoloff L., and Kuhl D. E. (1979b) Tomographic measurement of local cerebral glucose metabolic rate in humans with (F-18) 2-fluoro-2-deoxy-D-glucose: Validation of method. *Ann. Neurol.* **6,** 371–388.

Phelps M. E., Mazziotta J. C., and Huang S.-C. (1982) Study of cerebral function with positron computed tomography. *J. Cereb. Blood Flow Metab.* **2,** 113–162.

Powers W. J. and Raichle M. E. (1985) Positron emission tomography and its application to the study of cerebrovascular disease in man. *Stroke* **16,** 361–376.

Powers W. J., Martin W., Herscovitch P., Grubb R. L., Jr., and Raichle M. E. (1983) The value of regional blood volume measurements in the diagnosis of cerebral ischemia. *J. Cereb. Blood Flow Metab.* **3** (suppl. 1), S598–S599.

Powers W. J., Grubb R. L., Jr., and Raichle M. E. (1984) Physiologic responses to focal cerebral ischemia in humans. *Ann. Neurol.* **16,** 546–552.

Powers W. J., Grubb R. L., Jr., Darriet D., and Raichle M. E. (1986) CBF and $CMRO_2$ requirements for cerebral function and viability in humans. *J. Cereb. Blood Flow Metab.* **5,** 600–608.

Raichle M. E., Larson K. B., Phelps M. E., Grubb R. L., Jr., Welch M. J., and Ter-Pogossian M. M. (1975) In vivo measurement of brain glucose transport and metabolism employing glucose-^{11}C. *Am. J. Physiol.* **228,** 1936–1948.

Raichle M. E., Martin W. R. W., Herscovitch P., Mintun M. A., and Markham J. (1983) Brain blood flow measured with intravenous $H_2{}^{15}O$. II. Implementation and validation. *J. Nucl. Med.* **24,** 790–798.

Reiman E. M., Raichle M. E., Robins E., Butler F. K., Herscovitch P., Fox P., and Perlmutter J. (1986) The application of positron emission tomography to the study of panic disorder. *Am. J. Psychiatry* **143,** 469–477.

Reivich M., Kuhl D., Wolf A., Greenberg J., Phelps M., Ido T., Casella V., Fowler J., Hoffman E., Alavi A., Som P., and Sokoloff L. (1979) The (^{18}F)-fluorodeoxy-glucose method for the measurement of local cerebral glucose utilization in man. *Circ. Res.* **44,** 127–137.

Reivich M., Alavi A., Wolf A., Greenberg J. H., Fowler J., Christman D., MacGregor R., Jones S. C., London J., Shiue C., and Yonekura Y. (1982) Use of 2-deoxy-D[1-^{11}C]glucose for the determination of local cerebral glucose metabolism in humans: Variation within and between subjects. *J. Cereb. Blood Flow Metab.* **2,** 307–319.

Reivich M., Alavi A., Wolf A., Fowler J., Russell J., Arnett C., MacGregor R. R., Shiue C. Y., Atkins H., Anand A., Dann R., and Greenberg J. H. (1985) Glucose metabolic rate kinetic model parameter determination in humans: The lumped constants and rate constants for [^{18}F]fluorodeoxyglucose and [^{11}C]deoxyglucose. *J. Cereb. Blood Flow Metab.* **5,** 179–192.

Rhodes C. G., Lenzi G. L., Frackowiak R. S. J., Jones T., and Pozzilli C. (1981) Measurement of CBF and $CMRO_2$ using continuous inhalation of $C^{15}O_2$ and $^{15}O_2$: Experimental validation using CO_2 reactivity in the anaesthetised dog. *J. Neurol. Sci.* **50,** 381–389.

Sakai F., Nakazawa K., Tazaki Y., Ishii K., Hino H., Igarushi H., and Kanda T. (1985) Regional cerebral blood volume and hematocrit measured in normal human volunteers by single-photon emission computed tomography. *J. Cereb. Blood Flow Metab.* **5,** 207–213.

Sakurada O., Kennedy C., Jehle J., Brown J. D., Carbon G. L., and Sokoloff L. (1978) Measurement of local cerebral blood flow with iodo[^{14}C]antipyrine. *Am. J. Physiol.* **234,** H59–H66.

Sasaki H., Kanno I., Murakami M., Shishido F., and Uemera K. (1986) Tomographic mapping of kinetic rate constants in the fluorodeoxyglucose model using dynamic positron emission tomography. *J. Cereb. Blood Flow Metab.* **6,** 447–454.

Sokoloff L. and Smith C. B. (1985) Basic Principles Underlying Radioisotopic Methods for Assay of Biochemical Processes In Vivo, in *The Metabolism of the Human Brain Studied with Positron Emission*

Tomography (Greitz T., Ingvar D. H., and Widen L., eds.) Raven, New York.

Sokoloff L., Reivich M., Kennedy C., Des Rosiers M. H., Patlak C. S., Pettigrew K. D., Sakurada O., and Shinohara M. (1977) The [^{14}C]deoxyglucose method for the measurement of local cerebral glucose utilization: Theory, procedure, and normal values in the conscious and anesthetized albino rat. *J. Neurochem.* **28,** 897–916.

Steinling M. and Baron J. C. (1982) Mesure du debit sanguin cerebral local par inhalation continue de $C^{15}O_2$ et tomographie d'emission: etude des limites du modele. *J. Biophys. Med. Nucl.* **6,** 89–95.

Steinling M., Baron J. C., Maziere B., Lasjaunias P., Loc'h C., Cabanis E. A., and Guillon B. (1985) Tomographic measurement of cerebral blood flow by the ^{68}Ga-labelled-microsphere and continuous-$C^{15}O_2$-inhalation methods. *Eur. J. Nucl. Med.* **11,** 29–32

Subramanyam R., Alpert N. M., Hoop B., Jr., Brownell G. L., and Taveras J. M. (1978) A model for regional cerebral oxygen distribution during continuous inhalation of $^{15}O_2$, $C^{15}O$, and $C^{15}O_2$. *J. Nucl. Med.* **19,** 48–53.

Ter-Pogossian M. M. (1981) Special characteristics and potential for dynamic function studies with PET. *Sem. Nucl. Med.* **11,** 13–23.

Ter-Pogossian M. M., Eichling J. O., Davis D. O., Welch M. J., and Metzger J. M. (1969) The determination of regional cerebral blood flow by means of water labeled with radioactive oxygen 15. *Radiology* **93,** 31–40.

Ter-Pogossian M. M., Eichling J. O., Davis D. O., and Welch M. J. (1970) The measure in vivo of regional cerebral oxygen utilization by means of oxyhemoglobin labeled with radioactive oxygen-15. *J. Clin. Invest.* **49,** 381–391.

Ter-Pogossian M. M., Phelps M. E., Hoffman E. J., and Mullani N. A. (1975) A positron-emission transaxial tomograph for nuclear imaging (PETT). *Radiology* **114,** 89–98.

Ter-Pogossian M. M., Ficke D. C., Hood J. T., Sr., Yamamoto M., and Mullani N. A. (1982) PETT VI: A positron emission tomograph utilizing cesium fluoride scintillation detectors. *J. Comput. Assist. Tomogr.* **6,** 125–133.

Videen T. O., Perlmutter J. S., Herscovitch P., and Raichle M. E. (1987) Brain blood volume, flow, and oxygen utilization measured with 0–15 radiotracers and positron emission tomography: Revised metabolic computations. *J. Cereb. Blood Flow Metab.*, **7,** 513–516.

Welch M. J. and Kilbourn M. R. (1984) Positron Emitters for Imaging, in *Freeman and Johnson's Clinical Radionuclide Imaging* (Freeman L. M., ed.) Grune & Stratton, Orlando, Florida.

Welch M. J. and Kilbourn M. R. (1985) A remote system for the routine production of oxygen-15 radiopharmaceuticals. *J. Labeled Cmpds.* **22,** 1193–1200.

Wienhard K., Pawlik G., Herholz K., Wagner R., and Heiss W.-D. (1985) Estimation of local cerebral glucose utilization by positron emission tomography of [^{18}F]2-fluoro-2-deoxy-D-glucose: A critical appraisal of optimization procedures. *J. Cereb. Blood. Flow Metab.* **5,** 115–125.

Wolf A. (1981) Special characteristics and potential for radiopharmaceuticals for positron emission tomography. *Sem. Nucl. Med.* **11,** 2–12.

NMR Spectroscopy of Brain Metabolism In Vivo

James W. Prichard and Robert G. Shulman

1. Introduction

Nuclear magnetic resonance (NMR) methods applicable to the study of functioning brain *in situ* are new in neuroscience, and many neuroscientists may not be familiar with them. Although we have provided some explanatory material in this review, many readers may wish to consult other works (Gadian, 1982; Moore, 1984) which provide access to a wider range of basic NMR literature, as well as introductions to NMR theory for scientists in other disciplines. For an excellent discussion of in vivo NMR studies of muscle, which are not covered here, the reader is referred to a recent review (Radda et al., 1984).

NMR spectroscopy is possible because some atomic nuclei act like tiny bar magnets when placed in a magnetic field. They line up with or against the field and can be excited in a controlled way by irradiation with radio frequency energy. During relaxation from the excitation, they emit radio frequency signals, which contain a great deal of information about the molecules they are in. The process is practical with samples ranging from crystalline solids to living people, though only compounds in or near the m*M* range can be detected in vivo. It is thought to be harmless for the great majority of human subjects (Budinger and Cullander, 1983; NRPB, 1983; Saunders and Smith, 1984). Instruments implementing it are comparable in cost to electron microscopes and modern X-ray diagnostic equipment.

The NMR phenomenon was discovered independently in two laboratories in 1946 (Bloch et al., 1946; Purcell et al., 1946). It was of such clearly far-reaching importance that Felix Bloch and Edward Purcell received the Nobel prize for physics in 1952. NMR methods for the study of test tube-size samples have become steadily more productive over the last four decades. By the late 1970s, improvements in magnet technology made similar methods applicable to much larger samples, including the human body. A short history of these developments is available (Andrew, 1984).

Current NMR research on living animals and humans has two branches which are related but must not be confused. NMR *imaging* uses spatial resolution of some strong signal, usually that of the hydrogen nuclei in water molecules, to make pictures of anatomical structure (Bydder, 1984). The images are more detailed than those obtained by X-ray methods, including computed tomography (CT) scanning, and making them involves no ionizing radiation. Imaging methods based on ^{23}Na (Maudsley and Hilal, 1984) and ^{31}P (Maudsley et al., 1984) are being developed. NMR *spectroscopy*—the subject of this review—uses much weaker signals from phosphorus, carbon-13, nonwater hydrogen, and some other nuclei to obtain quantitative information about specific compounds present in the tissue, with relatively crude spatial resolution. Imaging can be done with weaker and less homogeneous magnetic fields than those needed for spectroscopy. Because it is somewhat less demanding technically and is analagous to CT scanning, its development for in vivo use has been faster. In the most visible example of this, imaging is already in widespread use for routine medical diagnosis; spectroscopy is unlikely to reach that stage for several more years. The most advanced instruments now available can obtain both images and spectra, but each modality has its own optimum requirements, and the means of combining localization of signal with spectral information are still being developed (Aue et al., 1984; Bendall and Gordon, 1983; Bottomley et al., 1985).

There are two principal reasons to do NMR spectroscopy in vivo. The first is purely scientific—one wishes to study properties of tissue which are sensitive to tissue disruption. Metabolic studies free of agonal artifact on organs in their natural hormonal, neural, and hemodynamic environments can be expected to produce new understanding of how complex organisms function. The other reason is largely medical—one wishes to study otherwise inaccessible human tissues, both to investigate pathophysiological processes which may be unique to humans and, eventually, to provide information useful in the management of individual patients. Phosphate energy stores, lactate concentrations, and intracellular pH in the human brain are good examples of NMR-measurable variables which are important for both purposes and cannot be studied easily by any other technique.

2. Phosphorus (^{31}P) Studies

The first NMR measurements possibly including signals from living brain were ^{31}P spectra (Chance et al., 1978). These were obtained from an anesthetized mouse inserted into an 18 mm NMR tube and studied in a conventional spectrometer with a radio frequency coil that surrounded the entire head; muscle probably contributed much of the signal under these circumstances. Brain spectra reliably uncontaminated by signals from other tissues first became possible when surface coils were developed (Ackerman et al., 1980), and nearly all of the results mentioned later in this review were obtained with them. However, NMR spectra can be obtained from more sharply localized regions of the body by techniques that rely on manipulation of radio frequency fields (Bendall and Gordon, 1983) or magnetic field gradients (Aue et al., 1984; Bottomley et al., 1985).

2.1. Animals

The first NMR observations of living brain with a surface coil (Ackerman et al., 1980) were made on an anesthetized rat, and demonstrated that ^{31}P spectra from that organ resemble the one in Fig. 1, which is from rabbit. At least eight resonances were present; five could be assigned with confidence to the particular compounds indicated by the labeling in Fig. 1. The resonances labeled "sugar phosphates" and "phosphodiesters" were later shown to be more variable with species and age (see below) than the others; they both probably contain contributions from several compounds, but these have not been firmly identified. An eighth, very broad resonance was attributed to relatively immobile phosphates in bone. It interferes with quantitation of the other resonances, which are superimposed on it. An effective method for eliminating it is presaturation (Ackerman et al., 1984b).

Ackerman and colleagues (1980) noted that the concentration ratio of phosphocreatine (PCr) to adenosine triphosphate (ATP) derived from their spectrum was 1.93, which is higher than the highest values obtained by chemical assay of freeze-clamped brain samples. They also estimated that in vivo adenosine diphosphate (ADP) and inorganic phosphate (Pi) concentrations were probably much lower than those measured by destructive analytical techniques. If these things are true, the phosphorylation state of brain

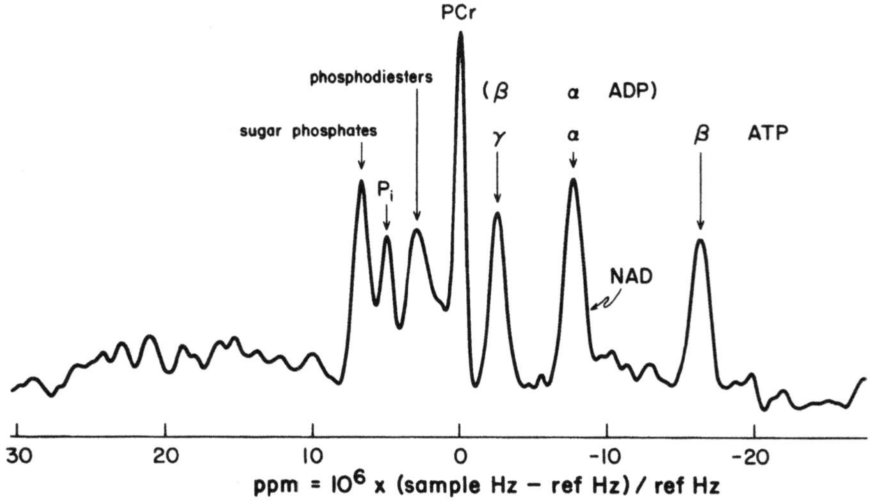

Fig. 1. Phosphorus spectrum of rabbit brain. The seven labeled resonances are present in brain spectra from other species, including humans. The "ATP" resonances contain contributions from other nucleotide di- and triphosphates. Labels in parentheses indicate that ADP phosphates resonate at nearly the same frequencies as the ATP alpha and gamma phosphates; their contribution to the spectrum is undetectable due to the low concentration of ADP. Resonances from nicotinamide adenine dinucleotide (NAD) and the alpha phosphate of ATP are too close to be resolved. The "sugar phosphates" and "phosphodiesters" resonances are probably composite signals from phosphomono- and phosphodiester configurations in several compounds; the largest contribution to the former may be from phosphoethanolamine (see text). The formula on the axis explains the calculation of "chemical shift"—a measure of small but characteristic differences in the resonant frequencies of nuclei of the same species in different molecular environments; these small shifts are the basis of the chemical selectivity of NMR spectroscopy. "ref Hz" is the frequency of some resonance chosen as the reference point, in this case that of PCr; "sample Hz" is the frequency of any other resonance in the spectrum. Chemical shifts in parts per million are independent of magnetic field strength and therefore facilitate comparison of data from different spectrometers. By convention, chemical shift numbers are plotted increasing to the left, which is the "downfield" direction of decreasing magnetic field strength. The variations between 10 and 30 ppm are noise. The spectrum was made from 512 scans obtained in 5 min from a paralyzed, pump-ventilated rabbit under nitrous oxide analgesia, using a 4-cm surface coil in an Oxford TMR 32/200 spectrometer operating at a magnetic field of 1.9 T (reprinted with permission from *Neurology*).

tissue reflected by the expression (ATP)/[(ADP)(HPO_4^{2-})] must be well above the generally accepted value of 3400/*M*. More recently, workers using perfused heart (Matthews et al., 1982) and rabbit brain in vivo (Prichard et al., 1983) assumed that the creatine kinase reaction was at equilibrium and calculated the concentration of ADP participating in it from quantities measurable in ^{31}P spectra; in both cases the calculated values were 20–30 μM. There is also indirect biochemical evidence that chemical assay gives values higher than are present in vivo (Veech et al., 1979). Contamination of chemical measurements by agonal change is a plausible explanation for the discrepancy; because the reactions involved are fast, it could occur even during very rapid freezing of tissue. However, it is conceivable that both the NMR and chemical estimates are right, as they would be if much ADP present in vivo is bound to protein or otherwise sequestered from the pool exposed to creatine kinase.

Once the practical utility of surface coils was established, the way was open for development of animal models in which some aspects of pathological brain function could be studied in vivo. Several groups (Ackerman et al., 1984a; Bottomley et al., 1982; Decorps et al., 1984; Delpy et al., 1982; Hilberman et al., 1984; Naruse et al., 1984, 1983; Prichard et al., 1983; Thulborn et al., 1982) validated ^{31}P spectroscopy for this purpose by showing that it could detect changes in phosphate energy stores, Pi, and intracellular pH (pHi) known to occur during a variety of metabolic stresses. The area of a resonance is proportional to the concentration of the compound producing it; changes in area reflect changes in concentration directly. The resonant frequency of Pi is sensitive to pH, and since the intracellular fluid volume of brain is large relative to other compartments containing Pi, it can be used to measure pHi of the brain.

Thulborn et al. (1982) observed a simultaneous decline in PCr and ATP, a rise in Pi, and tissue acidification in Mongolian gerbil brain ipsilateral to carotid occlusion. With the surface coil over the contralateral hemisphere or cerebellum, the changes were much less pronounced. The NMR changes correlated well with cerebral edema estimated from the specific gravity of gray matter in brains removed 1 h after the occlusion and with histological signs of cell damage. This study was the first systematic demonstration that data from in vivo NMR methods and conventional measurements can supplement each other in analysis of a specific pathophysiological problem.

Prichard et al. (1983) studied the behavior of cerebral phosphate energy stores and pHi in paralyzed, pump-ventilated rabbits during hypoglycemia, hypoxia, and status epilepticus, simultaneously with measurements of conventional physiological variables. Insulin shock caused PCr and ATP to fall and Pi and pHi to rise within minutes of the disappearance of the electroencephalogram. All of these changes reversed when glucose was given. Blood pressure, electrocardiogram, and arterial pO_2, pCO_2, and pH remained within normal limits throughout the experiment. Similarly, hypoxic hypoxia- and bicuculline-induced seizures caused the expected changes in the same range of NMR and physiological variables. This study showed that the physiological measurements were routinely feasible with the animal in the spectrometer and that the NMR measurements were unperturbed by such experimental arrangements.

Decorps et al. (1984) showed that radio frequency surface coils permanently fixed to the skulls of rats remained usable for as long as a month, enabling repeated observations from exactly the same region of brain to be made on different days. They also observed the time course of reversible cerebral acidosis and loss of phosphate energy stores following intraperitoneal injection of sublethal doses of potassium cyanide, and found that it correlated well with the behavioral abnormalities of freely moving rats given the same doses.

Hilberman et al. (1984) demonstrated the usefulness of line-fitting routines for quantitation of individual resonances in a group of overlapping ones. This is an especially important problem for NMR studies done in vivo, which must be carried out at relatively low [approximately 2 Tesla (T)] magnetic field strengths and always yield spectra with broad lines. In experiments on hypoxia in dogs, they found that the phosphodiester resonance accounted for nearly 40% of the total ^{31}P signal. It was relatively more intense than in published spectra from adult rodents and their own unpublished spectra from newborn puppies. They reported preliminary observations on chloroform–methanol–HCl extracts of dog brain which suggest that the signal in the region around 2 ppm in the in vivo spectrum is predominately from phosphodiester-containing phospholipids.

Assignments in the phosphodiester region may be important for future NMR studies of cerebral development and pathology. The few published spectra from brains of adult humans have an intense resonance there, whereas those from neonates do not (*see* below). Perchloric acid extracts of guinea pig brain identified glycerol 3-phosphorylcholine, glycerol 3-phosphorylethanolamine, and certain of their metabolites as principal sources of the phosphodiester signal (Glonek et al., 1982). It now appears that species, age, and method of extraction must all be considered if correct assignments are to be made. Moreover, NMR signals obtained in vivo may originate from mobile portions of large polymers, as well as from small, rapidly tumbling molecules. Glycogen is 100% detectable by ^{13}C spectroscopy (Sillerud and Shulman, 1983); the presence in the ^{31}P spectrum of signals from relatively mobile phospholipid components of membranes would not be surprising.

A study of bicuculline-induced status epilepticus in rabbits (Petroff et al., 1984) confirmed and extended preliminary observations by the same group (Prichard et al., 1983). The PCr/Pi ratio fell 50% and pHi fell to 6.7–6.9 from control values near 7.1 during the first hour of status. These remained depressed for up to 3 h, despite virtual disappearance of intense seizure discharge after 1 h. In all animals, ATP remained in the normal range throughout the experiments. Repeat doses of bicuculline demonstrated that the brain retained its capacity to mount massive electrical seizures under these circumstances. Calculations based on the assumption of equilibrium in the creatine kinase reaction indicated that the cerebral acidosis was responsible for most of the PCr decline. A later ^{1}H study documented a persistent rise in brain lactate caused by the seizure discharge (Petroff et al., 1986).

Measurement of cerebral pHi in vivo is a special capability of NMR spectroscopy. Invasive methods are available for use with experimental animals [*see* Petroff et al., (1985) for references], and positron emission tomographic methods are under development for use in humans (Brooks et al., 1984; Rottenberg et al., 1984; Syrota et al., 1983). In the foreseeable future, however, the accuracy, safety, and relative simplicity of the ^{31}P NMR method ensure that it will be widely employed for both experimental and clinical research. Since existing NMR titration data on Pi were not fully appropriate for work on brain, a study was done to establish suitable constants for conversion of the Pi-PCr chemical shift difference (ΔPi) to pH in the Henderson-Hasselbach equation (Petroff et al., 1985). From new titration data, the relation

$$pH = 6.77 + \log[(\Delta Pi - 3.29)/(5.68 - \Delta Pi) \quad (1)$$

was obtained. Mg affected the constants appreciably only in concentrations above 2.5 m*M*, which is well above estimates of free Mg concentration in brain. PCr was a satisfactory internal chemical shift reference down to approximately pH 6.5, below which titration of PCr introduced a progressively larger error. Both for this reason and because PCr is usually depleted in metabolic states that cause severe tissue acidosis, some external reference is necessary for pHi measurements in such states.

Cerebral pHi values calculated using the new formula were 7.14 ± 0.04 (SD) and 7.13 ± 0.03, respectively, for paralyzed, mechanically ventilated rabbits and rats under nitrous oxide analgesia. These values are toward the alkaline end of the rather wide range reported by workers using destructive analytical methods.

2.2. Magnetization Transfer Experiments

It is possible to "label" nuclei having the same resonant frequency by irradiating them selectively with radio frequency energy that briefly changes their magnetization state. If such nuclei are transferred enzymatically from the molecules they were in at the time of irradiation to other molecules at a rate comparable to the lifetime of the changed magnetic state, their presence in the second molecular population may be detectable at its (different) resonant frequency. Under favorable conditions, the rate constants of an enzyme-catalyzed reaction and the fluxes through it can be calculated from this measurement in living systems. The principles of such experiments and results obtained in various preparations have been reviewed recently (Alger and Shulman, 1984).

Only two studies using magnetization transfer techniques in living brain have been published, both done on rats (Balaban et al., 1983; Shoubridge et al., 1982). The results, which are somewhat different, raise interesting issues for future work. Shoubridge et al. (1982) found a unidirectional ATP synthetase rate of 0.33 μmol/g wet wt/s. This is only 10–15% of the rate in yeast and perfused heart. These workers also reported unidirectional fluxes through creatine kinase of 1.64 and 0.68 μmol/g wet wt/s, respectively, for the forward (PCr hydrolysis) and reverse reactions. Since PCr participates in no other known reaction, the fluxes must actually be equal during any period of stable PCr concentration. The results obtained therefore imply that some ATP is in a metabolic compartment which does not contain creatine kinase, or that ATP participates in other reactions with total fluxes amounting to a substantial fraction of the flux through creatine kinase. The latter explanation has recently been shown to be the correct one in perfused heart (Ugurbil et al., 1984). Proof of either explanation in brain would advance understanding of cerebral biochemistry. The problem was complicated by the report of Balaban et al, (1983), who found equal forward and reverse fluxes of about 2 μmol/g wet wt/s. They used a different magnetization transfer technique, a different anesthetic, and a coil which included the entire head of the rat, so that the signals may have come in part from muscle.

Magnetization transfer experiments are difficult to do, but they provide important, unique kinetic information about the function of some enzymes in their natural cellular environment. Beyond the resolvable discrepancies in presently available data lies the further question of how the measurable fluxes behave in different functional states of the brain; results from other systems show that magnetization transfer techniques open a window on some aspects of enzyme regulation in vivo (Alger and Shulman, 1984).

2.3. Humans

The first ^{31}P spectra from the living human brain were obtained in a study of newborn infants, most of whom suffered from some degree of perinatal brain damage (Cady et al., 1983). The same group later published additional observations on normal as well as abnormal infants (Hope et al., 1984). A practical result of this work is that the PCr/Pi ratio appears to be of use both for assessment of tissue damage and for prediction of outcome after metabolic stress. The ratio was 1.35 ± 0.22 (SD) in six normal infants. Lower ratios found in some infants after severe asphyxia returned toward normal as the infants' clinical conditions improved. In others, falling ratios over the several days after birth asphyxia heralded death or neurological impairment. An infant who had persistently low ratios for the first 26 d of life developed multiple porencephalic cysts during the same period. Low ratios rose in some infants after mannitol infusion, which presumably reduced cerebral edema. A trend toward an inverse relation between PCr/Pi and pHi noted in the first study by interested readers (Petroff and Prichard, 1983) was reinforced in the second; the most likely explanation for it is an increase in the proportion of extra- to intracellular fluid, as would be expected in developing porencephaly and some kinds of cerebral edema. A prominent phosphomonoester resonance was a consistent feature of all but the most abnormal spectra; data from several sources suggest that it is more prominent in infant than adult brains of both humans and rats and may therefore reflect metabolic conditions in the rapidly growing brain. The compounds responsible for it have not been identified with certainty. Extract work suggested ribose-5-phosphate as a possible source (Glonek et al., 1982), but Hope et al., (1984), referring to unpublished data, expressed the view that phosphoethanolamine is a more likely assignment.

Another group confirmed the presence of an intense phosphomonoester resonance in newborn human infants and marshalled reasons for its assignment principally to the phosphoryl esters of choline and ethanolamine (Younkin et al., 1984). They did not find a correlation of PCr/Pi ratio or other features of the ^{31}P spectrum with age or clinical condition, possibly because their series was small.

By the end of 1984, two groups had published ^{31}P spectra from brains of adult humans (Bottomley et al., 1984; Radda et al., 1984). All seven peaks in Fig. 1 were identifiable in these spectra, but the phosphodiester resonance was relatively more intense. As noted earlier, this difference between adult and infant brains within the same species appears to exist in humans and dogs, but not in rodents, whereas the phosphomonoester resonance may be more prominent in the infant in some species. When these resonances are firmly assigned and correlated with species and age, they may well be a rich source of new information about brain development under normal and pathological conditions.

3. Carbon (^{13}C) Studies

Due to inherent properties of ^{13}C, the chemical shift range of resonances in carbon spectra of organic compounds is very wide, being some 200 ppm, compared to about 10 and 30 ppm, respectively, in ^{1}H and ^{31}P spectra. The difference is reflected in the chemical shift axes of Figs. 1–3. Its practical significance is that ^{13}C spectra from organic samples are likely to contain resonances which are well resolved (separated) from each other, so that precise measurements on the compounds or chemical groups they represent are possible. However, ^{13}C signals are weaker than ^{31}P signals for three reasons: First, the inherent signal strength of ^{13}C is only about one quarter that of ^{31}P. Next, whereas ^{31}P is nearly 100% abundant in nature, only 1.1% of carbon nuclei are the magnetic isotope ^{13}C; the rest are ^{12}C, which are not magnetic and give no

NMR signal. Finally, magnetic interaction of ^{13}C with nearby ^{1}H nuclei reduces the detectable signal from ^{13}C by splitting its resonances into multiple smaller ones; the interaction can be removed by selective radio frequency irradiation of the ^{1}H nuclei, but this involves the potential risk of tissue heating. The total effect of these factors is an obstacle to ^{13}C spectroscopy of living systems, but not an insurmountable one. A sizable body of literature has accumulated on ^{13}C metabolic studies on cell suspensions, perfused organs, and, more recently, organs *in situ;* this work has been reviewed (Alger and Shulman, 1984).

In our laboratory, ^{13}C spectroscopy has been done on rat and rabbit brain in vivo (Behar et al., 1986). Figure 2 illustrates an experiment in which rabbit brain was observed before and after intravenous infusion of 1-^{13}C-glucose combined with hypoxia. In Fig. 2A, the natural abundance spectrum from the brain contains resonances from ^{13}C atoms in carboxyl, olefinic, and methylenic configurations, mostly in lipid molecules. In Fig. 2B, additional resonances from the alpha and beta anomers of 1-^{13}C-glucose in brain are evident. The spectrum (Fig. 2C) created by subtracting spectrum 2A from spectrum 2B reveals resonances from the methyl carbon of lactate and carbons 2, 3, and 4 of glutamate and glutamine. Data from the whole experiment showed that the ^{13}C-lactate signal rose, fell, and rose again in the course of two descents into hypoxia, whereas the signals from the amino acids rose steadily. These results showed that detection of brain metabolites enriched with ^{13}C is practical in vivo. That being true, the low natural abundance of ^{13}C is a fortunate circumstance in that it allows turnover studies of concentrated metabolites to be done. The experiment illustrated in Fig. 2 was done at 1.9 T —a magnetic field strength which is available for research on humans. Metabolic studies similar to the illustrated experiment could be done in humans, but still more recent spectroscopic developments suggest that combined ^{13}C and ^{1}H methods will be more effective, for reasons given below in the section on editing techniques.

4. Hydrogen (^{1}H) Studies

The NMR signal from ^{1}H is inherently much stronger than that from any other nucleus, and nearly all concentrated metabolites contain ^{1}H nuclei, which in principle could be used to identify

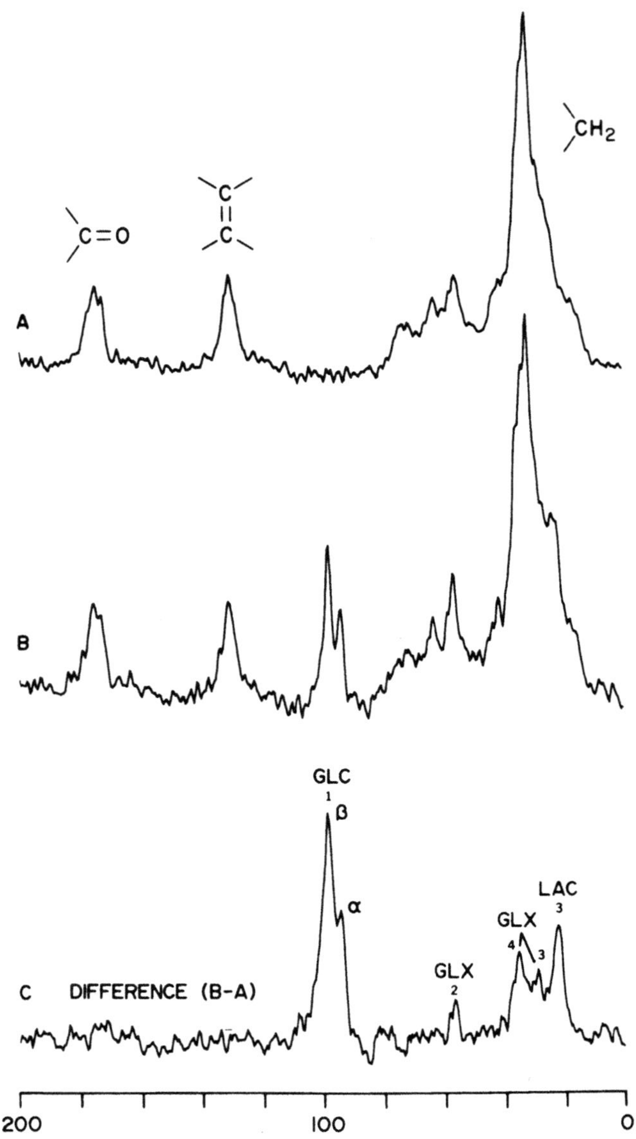

Fig. 2. Carbon 13 spectra of rabbit brain. Naturally abundant ^{13}C (1.1% of total carbon) principally in carboxyl, olefinic, and methylenic bond configurations is responsible for the three most prominent resonances in spectrum A. After intravenous infusion of 1-^{13}C-glucose and

them in ^{1}H spectra. Until recently, two problems prevented this from being done in living systems. First, the signal from ^{1}H nuclei in water is so strong that the weaker signals from most other compounds cannot be selectively detected in its presence. In studies of non-living systems by ^{1}H spectroscopy, the problem is solved by dehydrating the sample and resuspending it a ^{1}H-free solvent to achieve molecular mobility. Since 1983, techniques for eliminating the water resonance have been successfully applied to living systems in our laboratory, with the results discussed below. Second, the chemical shift range of ^{1}H is narrow, so that resonances from many different compounds overlap each other, which prevents useful measurement of most of them. We and our colleagues have developed procedures for editing spectra at the time of acquisition so that several metabolites can be selectively and quantitatively detected; these procedures are discussed in the next section.

Figure 3A is an unmodified ^{1}H spectrum of living rat brain, entirely dominated by the resonance from water protons (this signal is the basis of NMR imaging; variations in its intensity and rates of relaxation in different parts of the body provide a high degree of contrast.) Figure 3B shows the result of applying the simplest technique for eliminating the water resonance. Selective irradiation of water protons at their resonant frequency temporarily destroyed their orderly alignment with the magnetic field, which is a necessary condition for generating an NMR signal. Immediately after this "saturating" radiation, before most of the water protons had become realigned with the magnetic field, spectrum 3B was acquired. The intensity of the water resonance is

←

induction of hypoxia, resonances from the alpha and beta anomers of the infused glucose appeared (B). Subtraction of spectrum A from spectrum B revealed other new resonances (C) caused by flow of ^{13}C into the methyl carbon of lactate (LAC) and carbons 2, 3, and 4 of glutamate and glutamine, labeled GLX to indicate that signals from the two compounds are not resolved from each other. The lower case delta beneath the axis is commonly used to mean "chemical shift." The rabbit was maintained as described in the Fig. 1 legend and studied in the same spectrometer, using a 2.3-cm surface coil and proton decoupling; data collection time for spectra A and B was 20 min (reprinted with permission from *Magn. Res. Med.*).

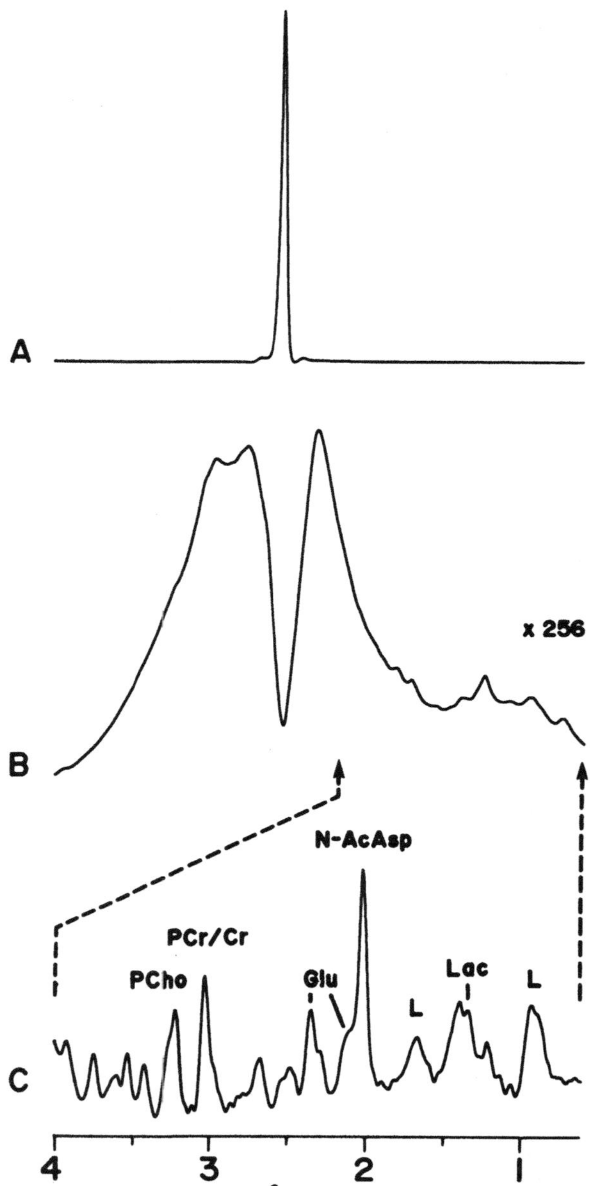

Fig. 3. Proton spectra of rat brain. The large resonance from water protons (A) was greatly reduced by presaturation (B), and further processing revealed several much smaller resonances from other compounds

reduced several hundred times compared to 3A; consequently, resonances from other metabolites are visible on its upfield (viewer's right) side. That region of the same spectrum was further processed to straighten its baseline and is displayed at greater amplification in Fig. 3C. Virtually all of the resonances that can be seen in Fig. 3C are from specific concentrated metabolites or groups of metabolites having ^{1}H nuclei in similar chemical configurations; signal-to-noise ratios in ^{1}H spectra are so much greater than those in ^{31}P and ^{13}C spectra that the baseline noise evident in Figs. 1 and 2 is nearly invisible. Resonances in Fig. 3C that have been assigned are labeled.

Assignment of resonances to particular compounds or chemical groups depends on procedures illustrated in Fig. 4. Spectrum A is from a dehydrated perchloric acid extract of freeze-clamped rat brain resuspended in D_2O; sharp resonance lines from numerous water-soluble metabolites are present. These can be identified in such extracts in three principal ways: (1) The extract spectrum is compared to spectra from a known pure compound, PCr for instance, in the same solvent. If resonances from the known compound have the same relative intensities and chemical shifts from some common reference point as resonances which can be found in the extract spectrum, there is a high likelihood that the compound is present in the extract. In a complex spectrum, however, this test may not be conclusive, and additional ones are necessary. (2) A pure compound is added to the extract. Precise augmentation by this procedure of resonances already present in the extract spectrum is strong evidence that the compound was present in the

←

on the upfield (right) side of it (C). The chemical shift scale pertains only to spectrum (C). Resonances have so far been assigned to a pool of choline-containing compounds (PCho), PCr and creatine together (PCr/Cr), glutamate (Glu), *N*-acetylaspartate (*N*-AcAsp), lactate (Lac), and unidentified lipids (L). However, nearly all of the peaks, including the unlabeled ones, are metabolite resonances; noise variations are comparable in amplitude to the tic marks on the chemical shift axis and are much smaller in proportion to the metabolite signals than in the ^{31}P and ^{13}C spectra of Figs. 1 and 2. These spectra were accumulated in 2.5 min and, like all in vivo spectra in later figures, were obtained from paralyzed, pump-ventilated rats maintained under nitrous oxide analgesia and studied with surface coils in a Bruker WH360 spectrometer.

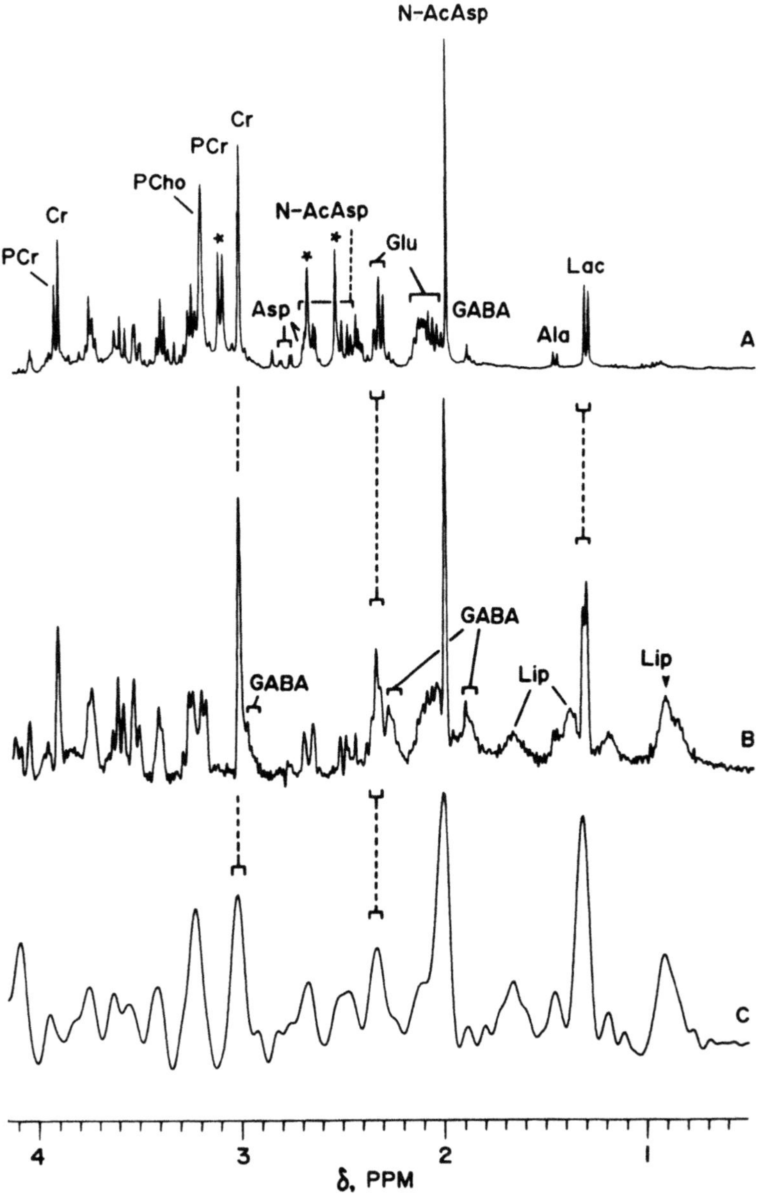

Fig. 4. Proton spectra of perchloric acid extract of rat brain resuspended in D_2O (A), excised but not extracted rat brain (B), and rat brain in vivo (C). Assigned resonances are from the same compounds labeled in Fig. 3, as well as creatine (Cr), aspartate (Asp), gamma-aminobutyric acid

original material. Finally, (3) selective removal of particular sets of resonances from the extract spectrum by exposure of the original material to a specific enzyme provides evidence that is independent of and complementary to that from procedure (2). The assignments in Fig. 4A are based on a large body of previous work of this kind. Many of the less intense resonances in the spectrum have not yet been assigned.

Figure 4B is a spectrum from excised but not extracted rat brain tissue in D_2O; the water proton resonance was suppressed by presaturation. The four most prominent resonances correspond to the ones assigned to PCr/Cr, *N*-acetylaspartate (NAA), and lactate in spectrum 4A. They are broader because the metabolites are less mobile in the excised tissue than in the extract. The lactate resonance is more intense because the excised tissue was not frozen quickly *in situ* to minimize agonal change. Lipid resonances are present because lipids were not extracted.

The spectrum of Fig. 4C is from the brain of a living, hypoxic rat. The major resonance lines are again identifiable by comparison with spectra 4A and 4B, though all resonances are much broader due to inhomogenities of the magnetic field caused by the animal's body. In every respect except the hypoxia, the spectra in Figs. 4C and 3C were obtained under the same experimental conditions; the most prominent difference between them is the greater intensity of the lactate resonance in Fig. 4C, attributable to hypoxia. This kind of manipulation of a resonance by a well-understood metabolic stress is the final step in validation of a spectroscopic technique for in vivo use. Studies from our laboratory (Behar et al., 1983, 1984) have shown that reversible elevations of the lactate resonance by hypoxia are detectable at both 8.4 T (the magnetic field strength used in the studies of Fig. 3 and 4) and 1.9 T, which is available in instruments physically large enough for human work.

Bottomley et al. (1985) showed that ^1H spectra could be

(GABA), alanine (Ala), unidentified lipids (Lip), and extraction reagents (*). Because of the narrow line widths characteristic of extract spectra, the complex line structure caused by signals from ^1H in different parts of the same molecule can be appreciated. *See* text for explanation of the differences among the three spectra. Animal preparation for spectrum (C) was as described in Fig. 3 legend, and all spectra were made in the Bruker WH360 spectrometer (reprinted with permission from *Proc. Natl. Acad. Sci. USA*).

obtained from the living human brain in an instrument operating at 1.5 T when the water proton resonance was removed by presaturation, as in Fig. 3. Their most important result was the demonstration that the anatomical source of the spectrum can be controlled by a pulsed magnetic field gradient and surface coil detection method. Resonances from NAA and pools of compounds containing choline and creatine are clearly evident in their published spectrum. Lactate would surely be detectable in such spectra if it were elevated, even with the reduced spectral resolution of the 1.5 T magnetic field. However, selective detection of amino acids other than NAA and accurate quantitation of anything at the field strengths which must be used for human work will require the editing techniques described in the next section.

Combined ^{1}H and ^{31}P observation in vivo is possible with a surface coil double-tuned for use in both frequency ranges. Figure 5 depicts the course of such a combined study of hypoglycemic encephalopathy in the rat (Behar et al., 1985). The ^{31}P spectra (Fig. 5E–H) show virtually complete loss of phosphate energy stores and a corresponding increase in Pi caused by insulin, followed by recovery after glucose infusion. Proton spectra (Fig. 5A–D) documented the fall of glutamate and rise of aspartate which are known to occur during profound hypoglycemia as glutamate metabolized via the aspartate amino transferase reaction becomes a principal source of carbon for the tricarboxylic acid cycle (Siesjo and Agardh, 1983).

5. Editing Techniques

Any measurement method dependant on detection of specific compounds will be confounded by samples containing many compounds which yield signals that are close together. NMR spectroscopy is no exception. The problem is especially great in vivo, since the usual biochemical strategy of simplifying the signal source by physical or chemical fractionation of the sample is not available. Several remarkably successful methods for simplifying in vivo spectra have been developed quite recently.

5.1. Homonuclear Decoupling

The magnetic interaction, or coupling, between nearby nuclei can be used to distinguish the signal of a single compound from

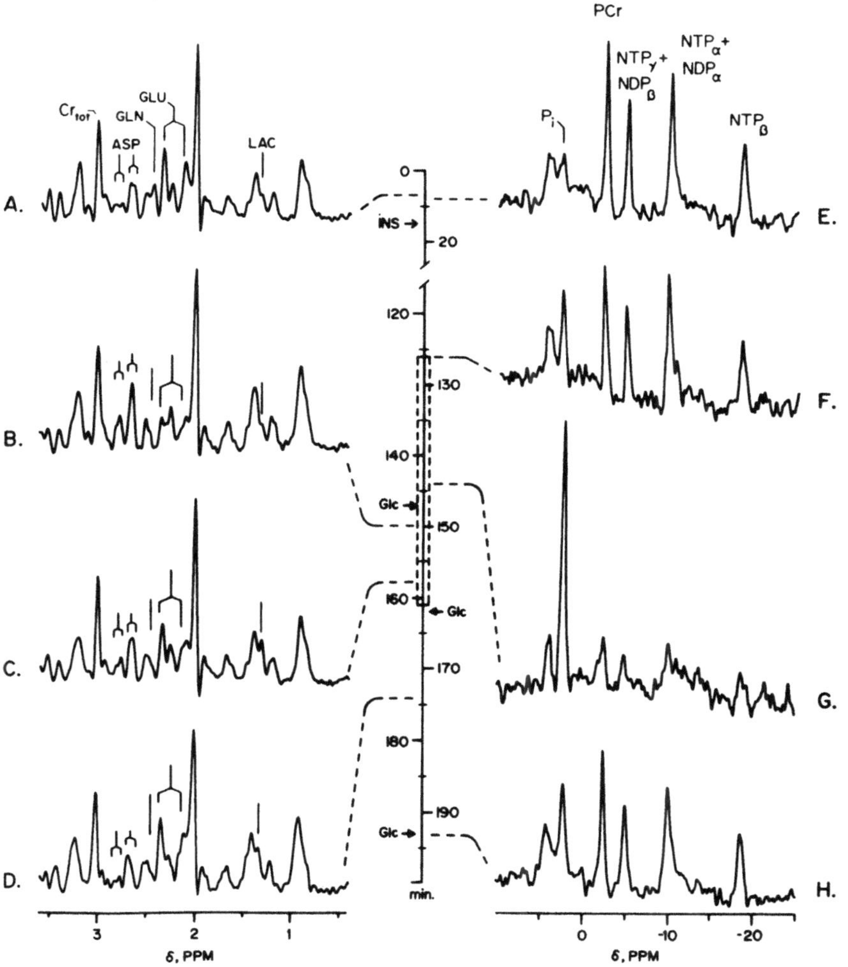

Fig. 5. Proton (A–D) and phosphorus (E–H) spectra from rat brain before, during, and after profound insulin-induced hypoglycemia. The temporal relation of spectra is shown on the time line in the center; electroencephalographic activity was absent during the period indicated by dotted lines flanking the time line; glucose was given at times indicated by arrows labeled "Glc." Labeling of resonances is as in earlier figures, except for glutamine (GLN), the composite resonance from PCr and creatine (Cr_{tot}), and designation of nucleotide di- and triphosphate resonances as "NDP" and "NTP" to indicate some contribution from bases other than adenine. *See* text for description of the metabolic changes. Animal preparation and spectrometer were the same as for Fig. 3 (reprinted with permission from *J. Neurochem.*).

many resonances in the same spectral region. The principle is subtraction of an unmodified spectrum from one acquired under conditions that decouple some particular interaction between neighboring nuclei. The only signals in the resulting difference spectrum will be from the compound in which the decoupled interaction is normally present. The first case we consider is that of homonuclear decoupling, in which the interacting nuclei are both protons. Figure 6 illustrates the mechanism and result of this procedure, as applied to selective detection of glutamate in the ^{1}H spectrum of a rat brain in vivo. Two spectra were built up from sets of scans taken with and without irradiation of the protons bonded to the beta carbon of glutamate; the effect of the irradiation was to induce a transient change in the magnetization state of the protons bonded to the alpha and gamma carbons by uncoupling them from the beta-carbon protons. The difference spectrum (Fig. 6A) contains resonances from the alpha- and gamma-carbon protons, whereas no glutamate signal could be detected in either the unmodified spectrum (Fig. 6B) or the irradiated one (not shown). So far, resonances from alanine, taurine, gamma-aminobutyric acid, and lactate, as well as glutamate, have been isolated this way (Rothman et al., 1984). Quantitative studies of these compounds in vivo are therefore possible.

5.2. Heteronuclear Decoupling

Compounds in which magnetic coupling exists between nearby nuclei of different species can be selectively detected by an editing procedure analagous to selective homonuclear decoupling. The heteronuclear procedure enables an entirely different kind of measurement in which the observed nucleus is ^{1}H and the irradiated one is ^{13}C. With proton observe-carbon decouple editing, the sensitivity of the ^{1}H resonance is retained, and the ^{13}C label is followed. This makes possible a far more powerful form of ^{13}C enrichment experiment than the one illustrated in Fig. 2. The principle is presented schematically in Fig. 7. Three populations of lactate molecules in a magnetic field are represented by the three formulas: In one (center), the methyl carbon is ^{12}C, which exerts no magnetic influence on the protons bonded to it; in the other two populations (left and right), the methyl carbons are ^{13}C, half of them aligned spin up and half spin down with respect the magnetic field. If 50% of total lactate methyl carbons were ^{13}C, the portion of the ^{1}H spectrum where the signals from lactate methyl protons

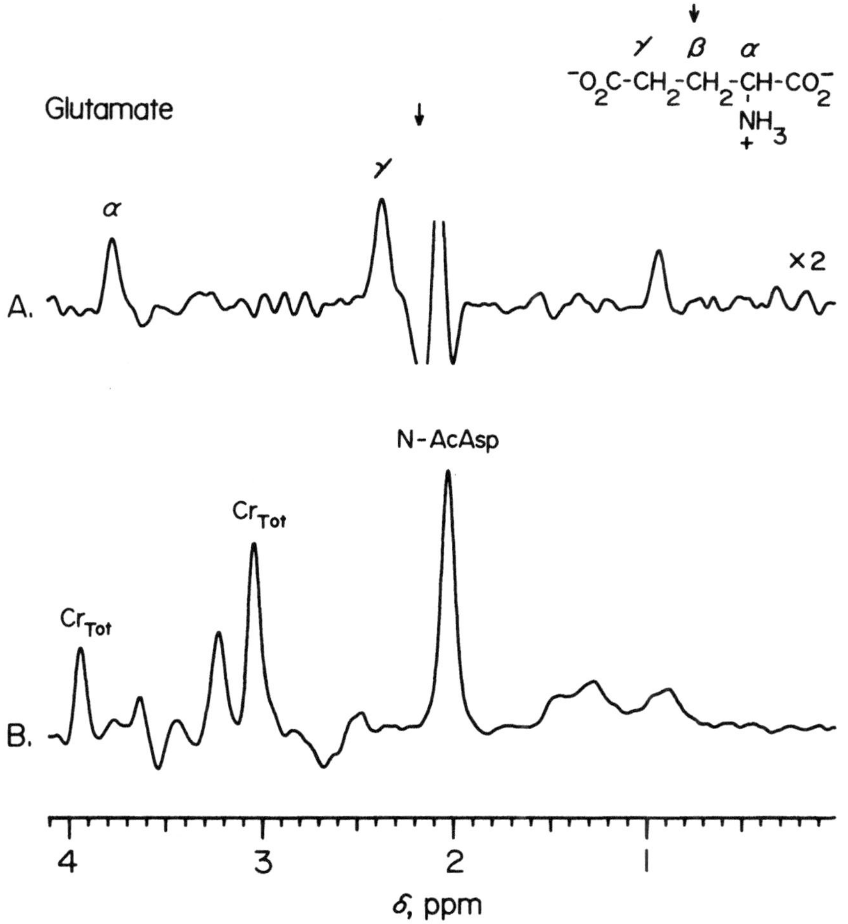

Fig. 6. Homonuclear editing of rat brain proton spectrum to yield resonances from protons bonded to the alpha and gamma carbons of glutamate (A). These resonances are not detectable in the raw ^1H spectrum (B). Single-frequency irradiation of the beta-carbon protons (inset) was given at the frequency indicated by the arrow above spectrum (A); *see* text for explanation. Animal preparation and spectrometer were the same as for Fig. 3 (reprinted with permission from *Proc. Natl. Acad. Sci. USA*).

resonate would look like the mock spectrum labeled "coupled." A central resonance from protons bonded to ^{12}C is flanked by two resonances that are shifted away from the central one by the magnetic effect of the up and down spins of ^{13}C on the protons bonded to them. If the sample were irradiated at the resonant frequency of ^{13}C, the side resonances would collapse into the central one as shown in the "^{13}C-decoupled" mock spectrum, because the effect of the magnetic carbons on their attached protons would be removed. The areas of the single resonance in the decoupled spectrum and the central resonance in the coupled one would be measures, respectively, of total lactate in the sample and the proportion of it with ^{12}C in the methyl position; subtraction would give the fraction of the lactate pool labeled at that position with ^{13}C. In principle the method can be used in living systems to follow the flow of ^{13}C through any detectable metabolite pool that can be enriched with it. Direct observation of ^{13}C as in Fig. 2 can detect only the labeled fraction of a compound. Indirect measurement of ^{13}C labeling by the method of Fig. 7 not only provides full

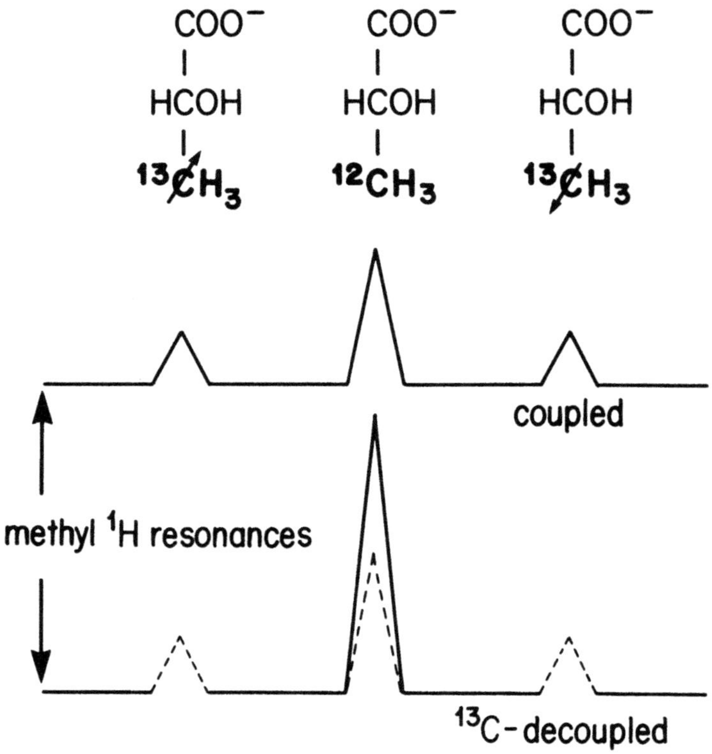

Fig. 7. Schematic illustration of ^{1}H observe-^{13}C decouple method of measuring fractional labeling of lactate methyl carbon with ^{13}C. *See* text for explanation.

labeling information about the compound, it does so by detecting the much stronger ^1H signal and hence is much more sensitive.

The first use of the proton observe-carbon decouple method on a living system was in a study of yeast metabolism (Sillerud et al., 1981). It has recently been shown to be effective for measurement of fractional labeling of lactate and glutamate in living rat brain (Rothman et al., 1985). Figure 8, taken from that study, shows the equivalent of a decoupled spectrum (Fig. 8C), the difference spectrum (Fig. 8B) obtained by subtracting the coupled spectrum from the decoupled one, and the time course of ^{13}C labeling of glutmate at the C4 position (Fig. 8A).

5.3. Selective Excitation

A method much more effective than presaturation for eliminating the water proton resonance from ^1H spectra was developed by spectroscopists working with test tube samples (Hore, 1983), and it has recently been adapted for observations on whole animals with surface coils (Hetherington et al., 1985). It works by minimizing excitation of the water protons, rather than reducing the response of the water proton signal to excitation, as presaturation does. A pulse sequence called "1331" (from the relative durations of some of its components) can be delivered in such a way as to provide nearly zero excitation at the frequency of the water protons, whereas excitation maxima occur on either side of it at frequencies that can be optimized for specific resonances of interest. The technique can be combined with other editing methods.

Hetherington et al., (1985) also showed that another powerful and versatile selective excitation technique can be used for surface coil measurements on intact animals. Its developers (Morris and Freeman, 1978) named it the DANTE (delays alternating with nutations for tailored excitation) sequence after their quite soberly described recognition of the resemblance between the magnetization trajectory it induces and the route of Dante Aligheri's tour of Purgatory in the company of Virgil. The method works by substituting a series of short excitation pulses for one long pulse, so that excitation occurs a little at a time, rather than all at once. The resonant frequency of a nucleus is the frequency at which it precesses around the axis of the magnetic field. Continued precession in the intervals between the short pulses causes loss of phase coherence among the excited nuclei, the dephasing being greater the farther the resonant frequency of a nuclear species is from the

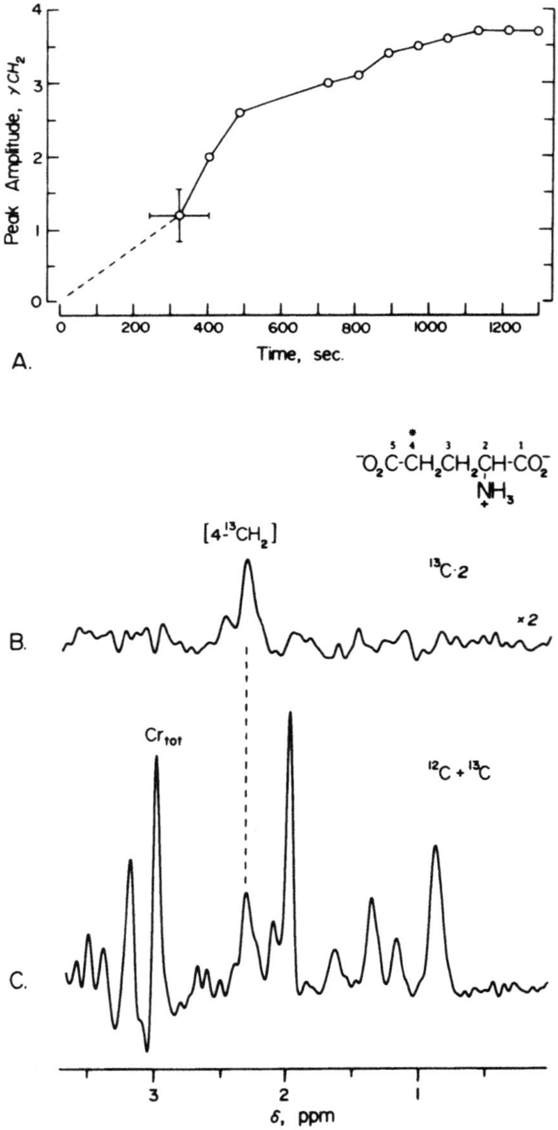

Fig. 8. Glutamate labeling with ^{13}C in rat brain. (A). The amplitude (arbitrary units) of the proton resonance of [4-$^{13}CH_2$] glutamate is plotted as a function of time after the start of a [1-^{13}C] glucose intravenous infusion that began at time 0 and continued throughout the experiment. ^1H observe-^{13}C decouple spectra were acquired with single frequency ^{13}C decoupling centered at the ^{13}C resonance of [4-$^{13}CH_2$] glutamate. Each

transmitter frequency. Because of this, only nuclei precessing at a rate very near the transmitter frequency experience a cumulative effect of successive pulses in the series. The result is excitation highly selective for a narrow portion of the spectrum. The technique has a variety of uses and is compatible with other editing methods.

6. Prospects

Selective NMR spectroscopy of the living brain began in 1980 with the introduction of surface coils. Most of the studies mentioned in this review belong to an early development and validation period in the history of an observational method which has a great deal of potential for generating major advances in the study of normal and deranged brain metabolism. As a whole, the work reviewed here shows that in vivo NMR measurements agree well with measurements made by other techniques; major unexpected artifacts or distortions have not appeared. The capability of in vivo NMR to get information unobtainable any other way can therefore be exploited with cautious confidence.

Spectrometers operating in the range of 2–5 T are becoming available for in vivo work on human and animal brain. At such field strengths it is possible to measure ATP, PCr, Pi, and pHi in the ^{31}P spectrum, which may also yield important information about some

←

data point represents the sum of two ^{13}C difference spectra accumulated over a 162-s time interval (horizontal bar). Vertical bars represent the amplitude uncertainty due to root mean square noise in the spectrum. The total concentrations of glutamate determined from the [4-$^{12+13}CH_2$] glutamate resonance amplitude in the ($^{12}C+^{13}C$) subspectrum remained constant during the infusion. (B) ^1H observe-^{13}C decouple difference spectrum of [4-$^{13}CH_2$] glutamate. The spectrum is the sum of two difference spectra centered at 22 min after the start of the [1-^{13}C] glucose infusion. With the assumption that the concentration of the creatine + PCr (Cr_{tot}) represented by the resonance at 3.03 ppm in spectrum C was 10.5 μmol/g wet wt, the concentration of [4-$^{13}CH^2$] glutamate was calculated to be about 1.5 m*M*. (C) The sum of the two ($^{12}C+^{13}C$) subspectra, which are components of the two ^{13}C difference spectra whose sum is spectrum (B). The resonance of [4-$^{12+13}CH_2$] glutamate is at 2.35 ppm (dashed line) (reprinted with permission from *Proc. Natl. Acad. Sci. USA*).

phosphomono- and phosphodiesters. In the ^1H spectrum, NAA and elevated lactate can be measured in humans and these plus glutamate, glutamine, taurine, alanine, and gamma-aminobutyric acid, in animals. There are excellent prospects for extending these capabilities through editing techniques. The proton observe-carbon decouple technique for measuring flow of ^{13}C through pools of lactate, glutamate, and some other metabolites can provide abundant new data about the rates at which these turn over under normal and pathological conditions. From such data glycolytic and respiratory rates can be estimated in some cases. Most of these measurements require only a few minutes of acquisition time, and there is no known barrier to repeating them as often as desired in the same subject.

The greatest problem presently facing users of in vivo NMR methods is localization of the anatomical source of the signals. With a single exception (Bottomley et al., 1985), all of the studies reviewed here relied entirely on coil placement to achieve anatomical selectivity. Although much useful information can be obtained that way, vigorous current research is likely to provide improved localization methods in the near future. An important biological fact favoring such progress is the nearly exclusive localization of NAA to brain, where it is more concentrated in gray than white matter. Since this compound is the source of the most intense resonance in the ^1H spectrum, it can be used to evaluate the success of localization procedures.

Acknowledgments

Studies from our own laboratory were supported by grants from the United States Public Health Service (AM 27121 NS 21708 and GM 30287), the National Science Foundation (PCM 8402670), and the Esther A. and Joseph Klingenstein Fund. Reproduced, with permission, from the *Annual Review of Neuroscience,* vol. 9, copyright 1986 by Annual Reviews, Inc. Minor changes were made to fit the style of Neuromethods.

References

Ackerman J. J. H., Grove T. H., Wong G. G., Gadian D. G., and Radda G. K. (1980) Mapping of metabolites in whole animals by 31P NMR using surface coils. *Nature* (Lond.) **283,** 167–170.

Ackerman J. J. H., Berkowitz B. A., and Deuel R. K. (1984a) Phosphorus-31 NMR of rat brain in vivo with bloodless perfluorocarbon perfused rat. *Biochem. Biophys. Res. Commun.* **119,** 913–919.

Ackerman J. J. H., Evelhoch J. L., Berkowitz B. A., Kichura G. M., Deuel R. K., and Lown K. S. (1984b) Selective suppression of the cranial bone resonance from 31P NMR experiments with rat brain in vivo. *J. Magn. Reson.* **56,** 318–322.

Alger J. R. and Shulman R. G. (1984) NMR studies of enzymatic rates in vitro and in vivo by magnetization transfer. *Quart. Rev. Biophys.* **17,** 83–124.

Andrew E. R. (1984) A historical review of NMR and its clinical applications. *Br. Med. Bull.* **40,** 115–119.

Aue W. P., Muller S., Cross T. A., and Seelig J. (1984) Volume-selective excitation. A novel approach to topical NMR. *J. Magn. Reson.* **56,** 350–354.

Balaban R. S., Kantor H. L., and Ferretti J. A. (1983) In vivo flux between phosphocreatine and adenosine triphosphate determined by two-dimensional phosphorus NMR. *J. Biol. Chem.* **258,** 12787–12789.

Behar K. L., den Hollander J. A., Petroff O. A. C., Hetherington H., Prichard J. W., and Shulman R. G. (1985) The effect of hypoglycemic encephalopathy upon amino acids, high energy phosphates, and pHi in the rat brain in vivo: Detection by sequential ^{1}H and ^{31}P NMR spectroscopy. *J. Neurochem.* **44,** 1045–1055.

Behar K. L., den Hollander J. A., Stromski M. E., Ogino T., Shulman R. G., Petroff O. A. C., and Prichard J. W. (1983) High-resolution ^{1}H nuclear magnetic resonance study of cerebral hypoxia in vivo. *Proc. Natl. Acad. Sci. USA* **80,** 4945–4948.

Behar K. L., Petroff O. A. C., Prichard J. W., Alger J. R., and Shulman R. G. (1986) Detection of metabolites in rabbit brain by ^{13}C-NMR spectroscopy following administration of [1–^{13}C] glucose. *Magn. Res. Med.* **3,** 911–920.

Behar K. L., Rothman D. L., Shulman R. G., Petroff O. A. C., and Prichard J. W. (1984) Detection of cerebral lactate in vivo during hypoxemia by ^{1}H NMR at relatively low field strengths (1.9 Tesla). *Proc. Natl. Acad. Sci. USA* **81,** 2517–2519.

Bendall M. R. and Gordon R. E. (1983) Depth and refocusing pulses designed for multipulse NMR with surface coils. *J. Magn. Reson.* **53,** 365–385.

Bloch F., Hansen W. W., and Packard M. E. (1946) Nuclear induction. *Phys. Rev.* **69,** 127.

Bottomley P. A., Edelstein W. A., Foster T. H., and Adams W. A. (1985) In vivo solvent suppressed localized hydrogen nuclear magnetic resonance (NMR): A new window to metabolism?. *Proc. Natl. Acad. Sci. USA* **82,** 2148–2152.

Bottomley P. A., Hart H. R., Edelstein W. A., Schenk J. F., Smith L. S., Leue W. M., Mueller O. M., and Reddington R. W. (1984) Anatomy and metabolism of the normal human brain studied by magnetic resonance at 1.5 Tesla. *Radiology* **150,** 441–446.

Bottomley P. A., Kogure K., Namon R., and Alonso O. F. (1982) Cerebral energy metabolism in rats studied by phosphorus nuclear magnetic resonance using surface coils. *Magn. Reson. Imag.* **1,** 81–85.

Brooks D. J., Lammertsma A. A., Beaney R. P., Leenders K. L., Buckingham P. D., Marshall J., and Jones T. (1984) Measurement of regional cerebral pH in human subjects using continuous inhalation of 11CO2 and positron emission tomography. *J. Cereb. Blood Flow Metab.* **4,** 458–465.

Budinger T. F. and Cullander C. (1983) Biophysical Phenomena and Health Hazards of In Vivo Magnetic Resonance, in *Clinical Magnetic Resonance Imaging* (Margulis A. R., Higgins C. B., Kaufman L., and Crooks L. B., eds.) Radiol. Res. Educ. Foundation, San Francisco.

Bydder G. M. (1984) Nuclear magnetic resonance imaging of the brain. *Br. Med. Bull.* **40,** 170–174.

Cady E. B., Costello A. M., Dawson M. J., Delpy D. T., Hope P. L., Reynolds E. O. R., Tofts P. S., and Wilkie D. R. (1983) Noninvasive investigation of cerebral metabolism in newborn infants by phosphorus nuclear magnetic resonance spectroscopy. *Lancet* **ii,** 1059–1062.

Chance B., Nakase Y., Bond M., Leigh J. S., and McDonald G. (1978) Detection of 31P nuclear magnetic resonance signals in brain by in vivo and freeze-trapped assays. *Proc. Natl. Acad. Sci. USA* **75,** 4925–4929.

Decorps M., Lebas J. L., Leviel J. L., Confort S., Remy C., and Benabid A. L. 1984. Analysis of brain metabolism changes induced by acute potassium cyanide intoxication by 31P NMR in vivo using chronically implanted surface coils. *FEBS Lett.* **168,** 1–6.

Delpy D. T., Gordon R. E., Hope P. L., Parker D., Reynolds E. O. R., Shaw D., and Whitehead M. D. (1982) Noninvasive investigation of cerebral ischemia by phosphorus nuclear magnetic resonance. *Pediatrics* **70,** 310–313.

Gadian D. G. (1982) *Nuclear Magnetic Resonance and its Applications to Living Systems.* Clarendon Press, Oxford.

Glonek T., Kopp S. J., Kot E., Pettegrew J. W., Harrison W. H., and Cohen M. M. (1982) P-31 nuclear magnetic resonance analysis of brain: The perchloric acid extract spectrum. *J. Neurochem.* **39,** 1210–1219.

Hetherington H. P., Avison M. J., and Shulman R. G. (1985) ^{1}H homonuclear editing of rat brain using semi-selective pulses. *Proc. Natl. Acad. Sci. USA* **82,** 3115–3118.

Hilberman M., Subramanian V. H., Haselgrove J., Cone J. B., Egan J. W., Gyulai L., and Chance B. (1984) In vivo time-resolved brain phosphorus nuclear magnetic resonance. *J. Cereb. Blood Flow Metab.* **4,** 334–342.

Hope P. L., Cady E. B., Tofts P. S., Hamilton P. A., Costello A. M., Delpy D. T., Chu A., and Reynolds E. O. R. (1984) Cerebral energy metabolism studied with phosphorus NMR spectroscopy in normal and birth-asphyxiated infants. *Lancet* **ii,** 366–370.

Hore P. J. (1983) Solvent suppression in fourier transform nuclear magnetic resonance. *J. Magn. Reson.* **55,** 283–300.

Matthews P. M., Bland J. L., Gadian D. G., and Radda G. K. (1982) A 31P-NMR saturation transfer study of the regulation of creatine kinase in the rat heart. *Biochim. Biophys. Acta.* **721,** 312–320.

Maudsley A. A. and Hilal S. K. (1984) Biological aspects of sodium-23 imaging. *Br. Med. Bull.* **40,** 165–166.

Maudsley A. A., Hilal S. K., Simon H. E., and Wittekoek S. (1984) In vivo MR spectroscopic imaging with P-31. *Radiology* **153,** 745–750.

Moore W. S. (1984) Basic physics and relaxation mechanisms. *Br. Med. Bull.* **40,** 120–124.

Morris G. A. and Freeman R. (1978) Selective excitation in Fourier transform nuclear magnetic resonance. *J. Magn. Reson.* **29,** 433–462.

Naruse S., Horikawa Y., Tanaka C., Hirakawa K., Nishikawa H., and Watari H. (1984) In vivo measurement of energy metabolism and the concomitant monitoring of encephalogram in experimental cerebral ischemia. *Brain Res.* **206,** 370–372.

Naruse S., Takada S., Koizuka I., and Watari H. (1983) In vivo ^{31}P NMR studies on experimental cerebral infarction. *Japan J. Physiol.* **33,** 19–28.

NRPB (1983) Revised guidance on acceptable limits of exposure during nuclear magnetic resonance clinical imaging. (Statement by National Radiological Protection Board). *Br. J. Radiol.* **56,** 974–977.

Petroff O. A. C. and Prichard J. W. (1983) Cerebral pH by NMR. *Lancet* **ii,** 105–106.

Petroff O. A. C., Prichard J. W., Behar K. L., Alger J. R., den Hollander J. A., and Shulman R. G. (1985) Cerebral intracellular pH by ^{31}P nuclear magnetic resonance spectroscopy. *Neurology* **35,** 781–788.

Petroff O. A. C., Prichard J. W., Behar K. L., Alger J. R., and Shulman R. G. (1984) In vivo phosphorus nuclear magnetic resonance spectroscopy in status epilepticus. *Ann. Neurol.* **16,** 169–177.

Petroff O. A. C., Prichard, J. W., Ogina T., Avison M., Alger J. R., and Shulman R. G. (1986)Combined ^{1}H and ^{31}P magnetic resonance spectroscopy studies of bicuculline-induced seizure in vivo. *Ann. Neurol.* **20,** 185–193.

Prichard J. W., Alger J. R., Behar K. L., Petroff O. A. C., and Shulman R. G. (1983) Cerebral metabolic studies in vivo by ^{31}P NMR. *Proc. Natl. Acad. Sci. USA* **80,** 2748–2751.

Purcell E. M., Torrey H. C., and Pound R. V. (1946) Resonance absorption by nuclear magnetic moments in a solid. *Phys. Rev.* **69,** 37–38.

Radda G. K., Bore P. J., and Rajagopalan B. (1984) Clinical aspects of ^{31}P NMR spectroscopy. *Br. Med. Bull.* **40,** 155–159.

Rothman D. L., Behar K. L., Hetherington H. P., and Shulman R. G. (1984) Homonuclear ^{1}H double resonance difference spectroscopy of the rat brain in vivo. *Proc. Natl. Acad Sci. USA* **81,** 6330–6334.

Rothman D. L., Behar K. L., Hetherington H. P., den Hollander J. A., Bendall M. R., Petroff O. A. C., and Shulman R. G. (1985) ^{1}H observed ^{13}C decoupled spectroscopic measurements of lactate and glutamate in the rat brain in vivo. *Proc. Natl. Acad Sci. USA* **82,** 1633–1637.

Rottenberg D. A., Ginos J. Z., Kearfort K. J., Junck L., and Bigner D. D. (1984) In vivo measurement of regional brain tissue pH using positron emission tomography. *Ann. Neurol.* **15** (suppl.), S98–S102.

Saunders R. D. and Smith H. (1984) Safety aspects of NMR clinical imaging. *Br. Med. Bull.* **40,** 148–154.

Shoubridge E. A., Briggs R. W., and Radda G. K. (1982) ^{31}P NMR saturation transfer measurements of the steady state rates of creatine kinase and ATP synthetase in the rat brain. *FEBS Lett.* **140,** 288–292.

Siesjo B. K. and Agardh C. D. (1983) Hypoglycemia, in *Handbook of Neurochemistry,* 2nd Ed. (Lajtha A., ed.) Plenum, New York.

Sillerud L. O. and Shulman R. G. (1983) High-resolution ^{13}C nuclear magnetic resonance studies of glucose metabolism in *Escherichia coli. Biochemistry* **22,** 1087–1094.

Sillerud L. O., Alger J. R., and Shulman R. G. (1981) High-resolution proton NMR studies of intracellular metabolites in yeast using ^{13}C decoupling. *J. Magn. Reson.* **45,** 142–150.

Syrota A., Castaing M., Rougement D., Berridge M., Baron J. C., Bousser M. G., and Pocidalo J. J. (1983) Tissue acid-base balance and oxygen metabolism in human cerebral infarction studied with positron emission tomography. *Ann. Neurol.* **14,** 419–428.

Thulborn K. R., duBoulay G. H., Duchen L. W., and Radda G. (1982) A ^{31}P nuclear magnetic resonance in vivo study of cerebral ischemia in the gerbil. *J. Cereb. Blood Flow Metab.* **2,** 299–306.

Ugurbil K., Maidan R. R., Petein M., Michurski S. P., Cohn J. N., and From A. H. L. (1984) NMR measurements of myocardial CK rates by multiple saturation transfer. *Circulation* **70** (suppl.), II–84.

Veech R. L., Lawson J. W. R., Cornell N. W., and Krebs H. A. (1979) Cytosolic phosphorylation potential. *J. Biol. Chem.* **254,** 6538–6547.

Younkin D. P., Delivoria-Papadopoulos M., Leonard J. C., Subramanian V. H., Eleff S., Leigh J. S., and Chance B. (1984) Unique aspects of human newborn cerebral metabolism evaluated with phosphorus nuclear magnetic resonance spectroscopy. *Ann. Neurol.* **16,** 581–586.

NMR Imaging of the Central Nervous System

Peter S. Allen

1. Introduction

To locate and identify pathology of the central nervous system (CNS) noninvasively and with a high degree of specificity is an objective pursued by means of several imaging modalities and with varying degrees of success. Some, for example X-ray, computed tomography (CT), and ultrasonography, provide purely structural information, whereas others, namely positron emission tomography (PET) and single photon emission computed tomography (SPECT), give rise largely to functional data. The principal advantage of the structural technique of CT is one of high spatial resolution, whereas the primary advantage of low resolution PET is one of high specificity. Ultimately, nuclear magnetic resonance (NMR) promises to provide both high-resolution structural images and lower-resolution functional data, but at the moment only the imaging application is well enough developed for widespread clinical use. It is the imaging application of NMR that is the subject of this article, whereas the highly promising, functional-data-providing spectroscopy will be dealt with in a companion article by J. W. Pritchard.

Diagnostic imaging is one of the more recent applications of the physical technique of NMR, following many years of its highly successful application to the physics, chemistry, and biochemistry of condensed and gaseous matter. The term NMR imaging will be used in this article in preference to the term magnetic resonance imaging (MRI) in order to emphasize that the magnetic nuclei are the resonant centers, not electrons, paramagnetic ions, nor even mesons, which make up the general body of magnetic particles that can resonate under the generic heading of magnetic resonance. In fact, in order to obtain images that have a high spatial resolution and, in addition, exhibit a substantial contrast resolution between tissue types, the restriction on resonant particles is even more severe and applies only to the nucleus of the hydrogen atom, namely the proton.

Our ability to manipulate proton magnetic resonance signals to form images is dependent in no small measure on the mobility of the protons at a molecular level. In simple terms, the signal from highly mobile protons persists long enough for it to be encoded with spatial information in order to generate an image, whereas the signal from protons in structures that are rigid molecular environments dies away very quickly, before such spatial encoding can take place. In consequence, NMR images represent to a large extent the water distribution in the body, detected by means of the protons residing in the water molecules. Protons residing in cortical bone or in large macromolecules are not usually detected in medical images.

In addition to the easier manipulation of NMR signals from mobile protons, certain characteristics of these signals (the relaxation times that are discussed in section 5) are very sensitive to the degree of mobility of those water molecules. This, in turn, means a high sensitivity to the water concentration itself. It is through this high sensitivity to water concentration that NMR provides such good discrimination between soft tissue types. It must be borne in mind, however, that, notwithstanding this excellent contrast resolution, proton images have not yet fulfilled their early promise to provide specificity as well as resolution. It may well be left to localized in vivo NMR spectroscopy to fulfil this role.

The purpose of the following sections of this article is two-fold. First, it is to present an outline of the physical mechanisms that lead to our ability to generate high-quality images using NMR and to our ability to differentiate so well soft tissues within those images. Second, it is to illustrate, by means of examples, some of the more successful applications of NMR imaging of the CNS in comparison with other imaging modalities.

2. Some Basic Principles of Nuclear Magnetic Resonance

Not all nuclear species are admitting of NMR. This is because not all species exhibit a nuclear magnetic moment. However, many nuclear species do possess an intrinsic nuclear magnetic moment, μ, and these species are those whose nuclear spin quantum number I is equal to or greater than one-half. The magnitude and direction of the magnetic moment is related to the spin angular momentum, and is given by

$$\mu = \hbar\gamma I \tag{1}$$

where $\hbar$ is Planck's constant divided by 2π, and γ is a constant called the nuclear gyromagnetic ratio. γ is governed by the nuclear structure and is unique for each nuclear species. The value of the spin, I, governs the available number of energy states in NMR, and although its range includes all integral and half-integral values from $I = 1/2$ to $I = 8$, we are fortunate in that the nucleus that is important from the point of view of medical imaging, namely ^1H, possesses a spin $I = 1/2$. As a result, it conforms to the simplest description of nuclear resonance and I shall, therefore, confine the description here to the case of $I = 1/2$.

When a collection of nuclei of spin $I = 1/2$ is left to come to thermal equilibrium in a uniform static magnetic field B_0, one can only find their magnetic moments taking up one of two well-defined orientations with respect to the direction of B_0, namely parallel and antiparallel alignment, somewhat like a collection of compass needles and anticompass needles. These two alignments give rise to two discrete energy states for the magnetic moments, E_1 and E_2, as illustrated in Fig. 1, and whose energy is given by

$$E = -\mu B \tag{2}$$

It requires work to rotate a compass needle from its parallel to an antiparallel alignment, and thus the antiparallel orientation has a greater energy. In taking up their respective alignments, the nuclear magnetic moments distribute themselves among the two energy states according to a Maxwell-Boltzman population distribution, which is also illustrated in Fig. 1. This means that in thermal equilibrium, an excess of nuclei resides in the lower-energy state. If all the nuclear magnetic moments are now summed to give the net total nuclear magnetic moment, M, the excess population in the lower-energy state ensures that, in thermal equilibrium, M is parallel to B_0, and by convention this direction is defined as the z direction of a laboratory frame of coordinates. In general, the behavior of the whole nuclear spin system may be analyzed in terms of the behavior of its resultant magnetic moment vector, M, which, although parallel to B_0 in thermal equilibrium, undergoes orientation and magnitude changes as a result of the influence of various perturbations. It is through these changes that the nuclear spin system communicates information to the outside world.

In order to stimulate the occurrence of any useful behavior of M, it is necessary to perturb it from its equilibrium state with electromagnetic irradiation at a frequency ω_0, which is called the

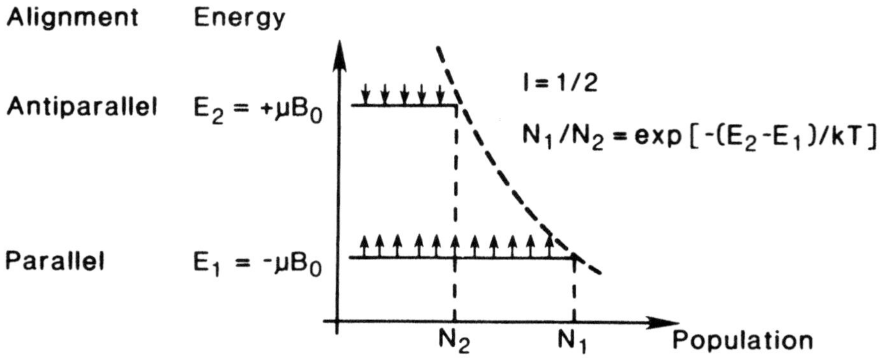

Fig. 1. A diagram illustrating schematically the two energy states E_1 and E_2 of nuclei of spin $I = 1/2$ and magnetic moment μ. Populations of these two energy states are represented by N_1 and N_2, under thermal equilibrium conditions at temperature T. k is Boltzman's constant.

Larmor frequency and whose photon energy exactly matches the energy difference between the two energy states of the nuclear magnetic moments, i.e.,

$$\omega_0 = (E_2 - E_1) = 2\mu B_0 \tag{3}$$

By means of Eq. (1), the Larmor frequency can be written, therefore, as

$$\omega_0 = \gamma B_0 \tag{4}$$

This direct proportionality between the Larmor frequency and the static magnetic field is of crucial importance for the purpose of spatial encoding in NMR images.

Typical values for B_0 in NMR imaging fall in the $0.1T$–$0.3T$ range for resistive magnets, and in the $0.3T$–$2T$ range for superconducting magnets. These magnetic field ranges correspond to proton Larmor frequencies between 4.2 and 84 MHz. These frequencies are in the radiofrequency, RF, region of the electromagnetic spectrum, their energy is insufficient to cause ionization, and the irradiation of the nuclei can be brought about very simply by winding an RF antenna coil of wire that fits around each patient.

The interaction of the electromagnetic radiation with the nuclear magnetic moments, which is what causes them to realign the net nuclear magnetic moment, M, relative to the static field B_0, is brought about by the magnetic component, B_1, of that radiation. Its effect can be understood in vectorial terms if one remembers that a magnetic moment, M, experiences a torque $M \wedge B$, when subjected to an arbitrary magnetic field B. This torque will cause M to rotate around B at the frequency of rotation γB.

During excitation of the nuclei, the total magnetic field B to which they are exposed is composed of two parts, a large static part, B_0, and an electromagnetic part, B_1, which is circularly polarized in a plane at right angles to B_0. Under the influence of these two fields, M experiences two torques and undergoes a precessional motion in which it effectively rotates around B_0 at the angular frequency $\omega_0 = \gamma B_0$, and at the same time rotates around B_1 at an angular frequency $\omega_1 = \gamma B_1$. This motion is easiest to visualize in a coordinate frame (x_ρ, y_ρ, z_ρ) that is rotating about the z axis in synchronism with the electromagnetic field B_1 at the frequency ω_0. In this rotating frame, the rotation about B_0 is frozen, and the effective motion takes the form of a slow rotation of M, at the angular frequency ω_1, about the direction of polarization of the electromagnetic field, as shown in Fig. 2. The mathematical analysis of the transformation of the motion of M to a rotating coordinate system can be found in most NMR textbooks (Abragam, 1961; Slichter, 1964; Farrar and Becker, 1971), and some illuminating diagrams of the precessional motion can be found in an article by Pykett (1982).

By controlling the duration of the stimulating electromagnetic irradiation, it is possible to determine the orientation of M immediately after the radiation is switched off. For example, two such realignments of M are commonly employed. These are the 90 and 180° pulses, brought about, respectively, by irradiation periods t_{90} and t_{180} such that the rotation angle around B_1 in the rotating frame of reference is 90 and 180°, viz

$$\omega_1 t_{90} = \pi/2 \text{ and } \omega_1 t_{180} = \pi$$

The effect of these pulses is to deposit M either along the y_ρ axis or in the negative z_ρ direction, respectively. We shall explain the use of a 180° pulse later, but for the purpose of generating an NMR signal, the 90° pulse is the most effective. The 90° pulse leaves the net nuclear magnetic moment vector, M, in the laboratory xy plane and under the influence of the static B_0 field alone. Because M and B_0 are now perpendicular, M will precess around B_0 in the fixed laboratory coordinate system at a constant angular frequency ω_0, because of the torque $M \wedge B_0$ exerted on it by the field B_0. This precessional motion of M, in the absence of any electromagnetic stimulation, provides the means for detecting the nuclear signal.

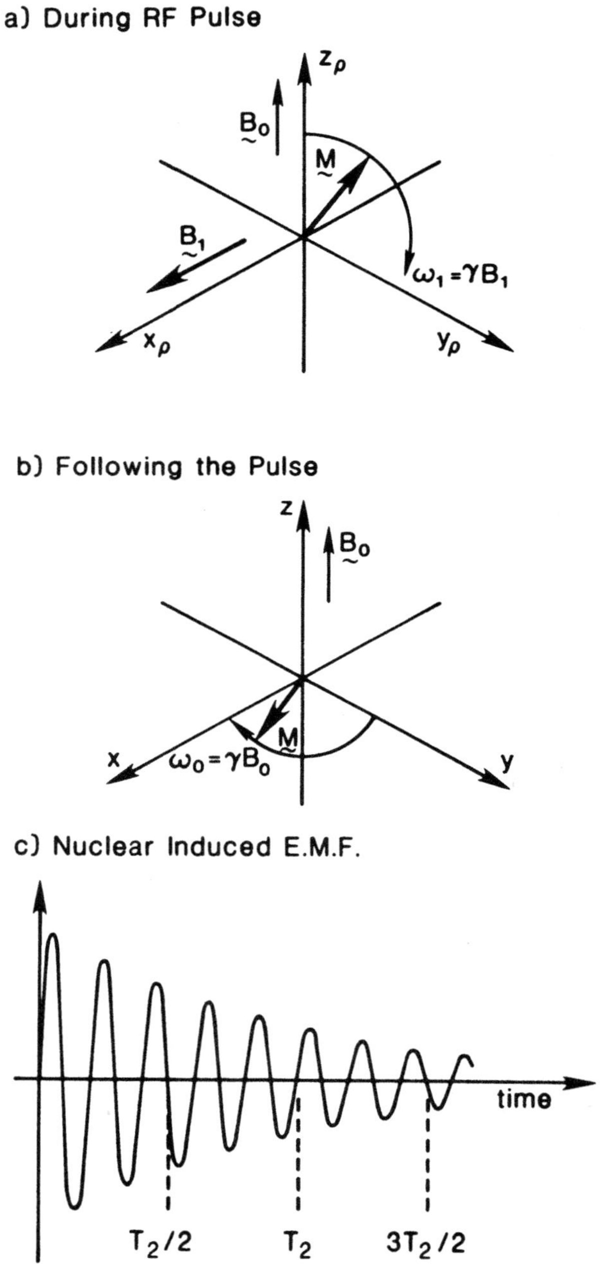
a) During RF Pulse
z_ρ
B_0
M
B_1
$\omega_1 = \gamma B_1$
x_ρ
y_ρ
b) Following the Pulse
z
B_0
M
$\omega_0 = \gamma B_0$
x
y
c) Nuclear Induced E.M.F.
time
$T_2/2$
T_2
$3T_2/2$

For instance, if a pick-up antenna coil were placed around the nuclear system, with its axis in the *xy* plane, the magnetic moment vector rotating at ω_0 would induce an EMF in that coil, oscillating at $\omega_0/2\pi$ and whose magnitude is proportional to the magnitude of the *M*. This oscillating EMF, illustrated in Fig. 2c, constitutes the nuclear signal.

It is important to remember that the frequency ω_0 of the nuclear signal is directly proportional to the static magnetic field B_0 as expressed in Eq. (4). As will be explained in the following section, this proportionality relationship is the crux of our ability to encode spatial information.

3. The Encoding of Spatial Information

The fundamental objective in NMR imaging is to identify, individually, the nuclear signal coming from each location in space. To fulfill this objective, we require a method of encoding spatial information into the nuclear signals. The means by which this is accomplished is the use of a carefully controlled spatial variation in the static magnetic field called a magnetic field gradient.

The term magnetic field gradient describes a spatial variation of only the magnitude of a magnetic field and not a variation in its direction. The most widely used gradients in NMR imaging are linear gradients, and one of these is illustrated in Fig. 3. In the illustration, the gradient is centered at the origin of the coordinate system, and the length of the arrows represents the magnitude of the gradient field to be added to, or subtracted from, the main static homogeneous field, B_0, applied in the *z* direction. This field gradient illustrated endows different locations along the *x* axis with different magnitudes of applied magnetic field and thereby ensures that the resonance condition of Eq. (4) varies in a controlled manner from place to place along the *x* axis.

←

Fig. 2. An illustration of the rotational motion of the net magnetic moment vector *M* brought about first by the radio frequency irradiation field B_1, which is applied at the Larmor frequency ω_o, and second, by the static magnetic field, B_o, after *M* has been tipped in to the *xy* plane of the laboratory coordinate system. This latter rotational motion gives rise to the oscillating EMF illustrated in part (c). The coordinate system (x_p, y_p, z_p) in part (a) is rotating around B_o at ω_o.

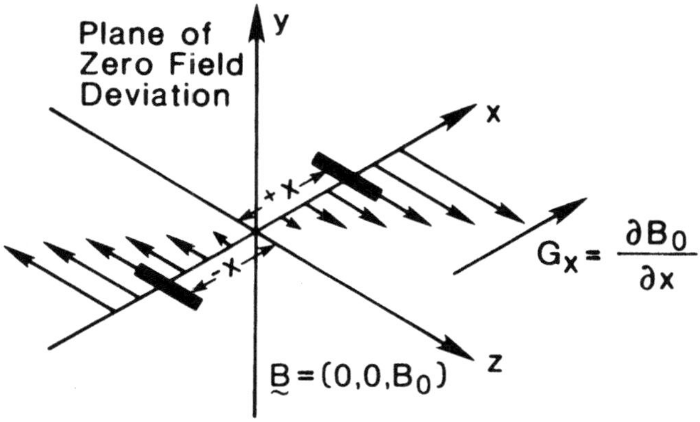

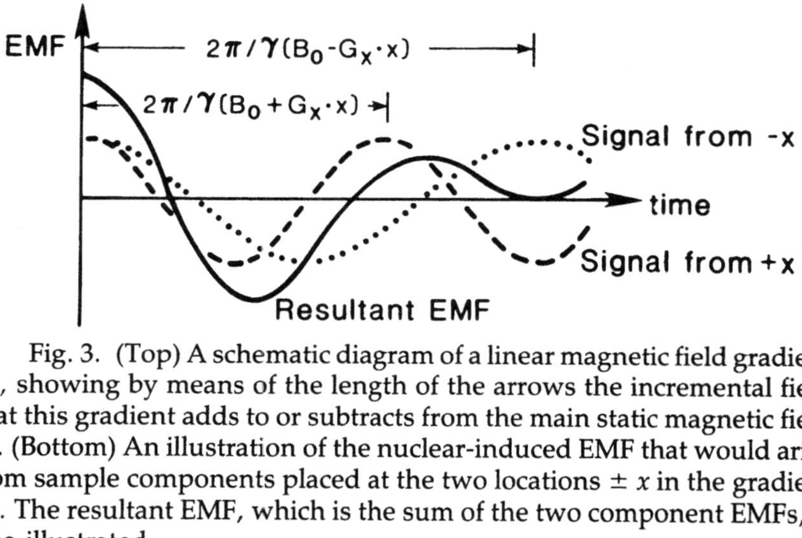

Fig. 3. (Top) A schematic diagram of a linear magnetic field gradient G_x, showing by means of the length of the arrows the incremental field that this gradient adds to or subtracts from the main static magnetic field B_0. (Bottom) An illustration of the nuclear-induced EMF that would arise from sample components placed at the two locations $\pm x$ in the gradient G_x. The resultant EMF, which is the sum of the two component EMFs, is also illustrated.

These gradient field components are typically a thousand times less than the static field B_0, but they are sufficient to cause easily detectable differences in Larmor frequency from regions in a body separated by a fraction of a millimeter.

As a first approximation, the superposition of a linear gradient onto a uniform static field has the effect of dividing the sample into elemental slices perpendicular to the gradient direction, and any particular slice, at a distance x from the origin, will have its own resonance condition given by

$$\omega_{0x} = \gamma \, (B_0 + G_x x)$$

where G_x is the strength of the gradient.

As a result of this, and following the termination of a 90° pulse of RF irradiation, the magnetic moment of each slice now perpendicular to the z direction will precess around z at a frequency ω_{0x} and, therefore, induce an EMF in the receiver antenna coil at its own Larmor frequency. This is illustrated for two elemental and equal samples in an x gradient in Fig. 3 (lower panel). By picking out, from the overall EMF, those components at discrete intervals of frequency and by then analyzing their respective amplitudes, it is possible to obtain a measure of the number of resonating nuclei at each value of x. The process of picking out the individual frequency components and evaluating their amplitudes is carried out by a mathematical procedure called the Fourier transform operation on the time-domain EMF waveform. The resulting variation of signal amplitude along the frequency axis then reflects the projection of the proton density onto the gradient direction.

4. NMR Imaging

The effect of a single magnetic field gradient on the NMR signal has been outlined in the previous section. Three such gradients are used in order to obtain a two-dimensional, slice image of a body, and in doing so a certain strategy is usually employed. The first task of this strategy is usually to define the slice to which the image is to correspond. A typical slice-defining procedure using linear gradients will be outlined below. Once the slice has been defined, the second function of the strategy is to identify the location, within that slice, from which each component of the nuclear signal comes. This spatial encoding within a slice, also using linear magnetic field gradients, will be covered in section 4.2.

4.1. Slice Definition

In order to define a slice, we require that all nuclei within such a slice be selectively and equivalently excited without perturbing the nuclei on either side of that slice. In very simple terms, one can do this if the patient is situated in a field gradient by irradiating the patient selectively at the specific Larmor frequency corresponding exactly to the required slice frequency. To irradiate selectively over a narrow band of frequencies is not a trivial exercise when using pulsed irradiation (Locher, 1980; Mansfield and Morris, 1982) The slice-defining RF pulse shapes, which are applied at the same time as a slice-defining gradient, must be carefully tailored so that their frequency-domain envelope is cut off sharply. The slice-defining procedure of concurrent linear gradient and tailored RF pulse usually prepares the nuclear magnetic moments from within the defined slice in such a way that they are perfectly aligned in the xy plane, just as though they had experienced a perfect 90° pulse. Following such a preparation, encoding within the slice can begin.

4.2. Image Production

Within the slice, two spatial coordinates need to be encoded into the NMR signal for each pixel location. The encoding of one of these is brought about by using a gradient in that particular coordinate direction in order to determine the Larmor frequencies of the individual NMR signal components. This is called frequency encoding, and the spatial information can be decoded by carrying out a Fourier transform operation on the time-domain waveform. The encoding of the other coordinate, called phase encoding, is brought about in a different, but related, manner. For example, we already know that when a field gradient is applied, the magnetic moments from different spatial locations along that gradient precess around the B_0 direction at different frequencies. This means that in a given time interval magnetic moments precessing at different frequencies will have precessed through different angles. Thus, at the end of the given time interval, the phase angle of a magnetic moment component will be determined by its precessional frequency, which is in turn determined by its location in space. To decode this information, one needs to obtain signals at many gradient strengths, but for the same given time interval, and then carry out an additional Fourier transformation (Kumar et al., 1975). A didactic and more detailed explanation of this two-dimensional Fourier transform technique (2-DFT) has been presented by Allen (1983a).

A simplified sequence of RF and gradient pulses that could facilitate the acquisition of a 2-DFT image is illustrated in Fig. 4. Here it is assumed that a transverse slice is defined perpendicular to the z axis, that the x coordinate is phase encoded, and that the y

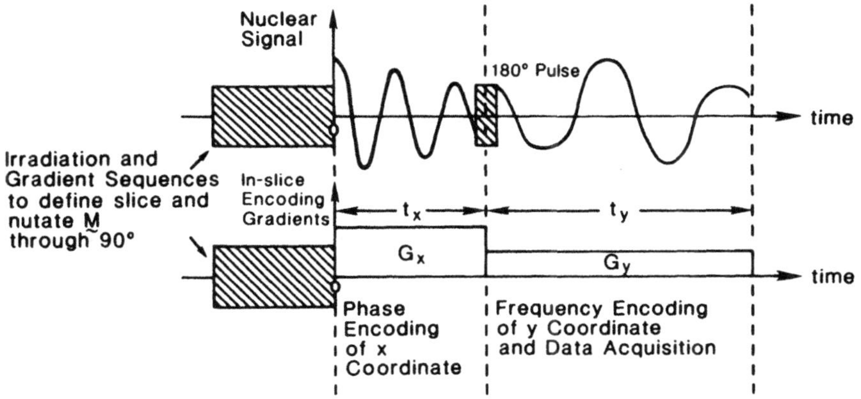

Fig. 4. A diagram illustrating the sequence of in-slice encoding gradients used in two-dimensional Fourier transform imaging. The signal is collected in the period t_y, where its frequency components are governed by G_y, but the starting phase of this signal in the t_y segment is determined by the oscillations in t_x, which are in turn governed by the magnitude of G_x.

coordinate is frequency encoded. The frequency selective 90° pulse and the z gradient waveform give rise to the defined slice. The magnetic moments within that slice will then dephase under the influence of the x gradient, thereby accumulating a phase angle dependent on their x coordinate. The wideband or hard 180° pulse will turn over the disk of dephased magnetic moments in the xy plane and thereby allow them to refocus again. This refocusing signal is acquired under the influence of the y gradient, which causes the precessional frequencies of the magnetic moments to depend on their y coordinate. Typically, such a sequence will be repeated 128 times, with the x gradient incremented before each repeat. The 128 repeated acquisitions will then be subjected to the 2-DFT procedures to generate the image.

5. Nuclear Relaxation and Tissue Discrimination

In section 4, which dealt with the production of an NMR image, no attention was paid to the effects of nuclear-spin relaxation, and, in the interests of clarity, the image was taken to correspond to the nuclear density alone. Such an assumption is not entirely valid, because nuclear relaxation can modify the magnitude of the magnetic moment components and, as a result, affect the intensity distribution in the image.

There are two relaxation processes that can affect the intensity of the NMR signal. The first is called longitudinal, or spin-lattice, relaxation and this concerns the recovery of the nuclear-spin system back to thermal equilibrium after it has been disturbed by RF irradiation. The longitudinal relaxation process is, therefore, concerned with the recovery of the magnitude of the nuclear magnetic moment along the z axis and parallel to the static field B_0. This recovery is normally exponential in time and characterized by a time constant denoted by T_1. The second relaxation process, which is often characterized by an exponential time constant T_2, is called the transverse, or spin-spin, relaxation process. This process again concerns the behavior of the components of the net magnetic moment vector M following the termination of the RF irradiation, but this time it is not the component along the z axis that is of interest, but that within the xy plane of a laboratory frame of coordinates. Transverse relaxation is, therefore, concerned with those components of M whose rotation give rise directly to the nuclear signal.

5.1. Longitudinal Relaxation (T_1)

When, as a result of RF irradiation, the nuclear spin system is disturbed from its thermal equilibrium state, there is a corresponding change in the relative populations of the two nuclear spin energy states (*see* Fig. 1).

Since the amplitude of a nuclear signal depends on the value of M_z at the beginning of the stimulating irradiation sequence, it is clear that by starting one's irradiation sequence before recovery from the previous irradiation sequence is complete, one can reduce the amplitude of that nuclear signal. In section 5.3, we shall describe how this property may be used to encode NMR images with T_1 information.

5.2. Transverse Relaxation (T_2)

Following a 90° pulse of RF irradiation, the net magnetic moment vector M will have been rotated from the z axis into the xy plane of the laboratory coordinate system. The concept of transverse relaxation provides a means with which to describe the behavior of this xy magnetic moment independently of the behavior along the z axis. If all the individual nuclear moments in the

xy plane were to experience identical local environments, then, under the influence of B_0 alone, their frequencies of precession around B_0 would also be identical. In consequence, the vector sum of these precessing moments, i.e., M_{xy}, would also remain constant in magnitude and (neglecting longitudinal relaxation) would precess indefinitely about B_0, with the net result that the nuclear signal would continue forever. In practice, the local magnetic environment is very sensitive to molecular structure and to molecular motion and, as a result, varies from place to place and from time to time on a molecular scale. These variations give rise to a distribution of precessional frequencies for the individual nuclear magnetic moment components and cause them to get out of step with each other as they precess. The net effect of this dephasing is to precipitate a decrease with time in the magnitude of the magnetic moment vector in the xy plane. As this decrease takes place, the reduction in the net effective magnetic moment available for generating a nuclear signal leads to a loss of amplitude of that signal that is called the free induction decay (FID).

5.3. Nuclear Relaxation in Biological Systems

Although the characteristic times T_1 and T_2 reflect the rates of different processes, they are both dependent on the molecular environment, and they are both related to the motion of nuclei at a molecular level. In the human body, the nature of the molecular environments and their internal motions is such that, as far as the protons are concerned, very little other than the water molecules produce nuclear relaxation rates that are admitting of NMR imaging procedures. The more rigid environments give rise to short T_2 values and, therefore, signals that decay too rapidly to be useful. The key factor that influences the relaxation rates in fluid environments is the water concentration itself. Water concentrations that occur in the body are typically between 65 and 80%, and these concentrations show definite changes among the major organs and different types of tissue. It is, moreover, quite fortunate that the changes in relaxation times caused by a change in water concentration exhibit an enhancement over that change in water concentration and thereby provide a very sensitive indicator of water content. An excellent example of this differentiation is that between the gray and the white matter of the brain. Gray matter has about a 15% greater water concentration than white matter, but this difference, nevertheless, shows up as a 50% greater mean T_1 value for the gray matter. This difference in T_1 enables gray and white matter to be differentiated easily by means of their T_1-weighted images as obtained by an inversion recovery technique to be described below.

The principal reason for the dependence of relaxation time on water content follows from its dependence on the relative motional rates of neighboring nuclei (Mansfield and Morris, 1982; Allen, 1983b). In simple terms, water molecules have two types of environment between which they can exchange. One is closely associated with a large macromolecule, which in turn determines the motional rate of the water molecule. In the other, a water molecule is freely mobile, i.e., in an environment much like that of free water. The motion in the former environment is the slower and leads to a more efficient relaxation process. The observed relaxation time, however, is a weighted mean over the two environments sampled by any water molecule. Since the water concentration determines the proportion of water in the free-water-like environment, and hence the likelihood of an individual water molecule exchanging between the two environments, the observed relaxation rate is very dependent on water concentration.

5.4. Some Methods for Encoding Images with Relaxation Information

Usually, in the interest of speed, instead of attempting to make a quantitative measurement of T_1 or T_2 at each pixel location of a two-dimensional image, what is often done is to weight the individual pixel intensities by their corresponding relaxation times of interest. For example, this can be brought about quite easily in the T_1 case by allowing the longitudinal relaxation rate at each pixel location to govern the magnitude of each pixel magnetic moment, instead of allowing that magnetic moment to recover to its full thermal equilibrium value before each successive excitation is carried out. There are a number of variations on this theme, and in the following sections those denoted by the terms saturation recovery, inversion recovery, and spin-echo imaging will be outlined.

Before describing their individual differences, it is useful to recall the general strategy behind all of these methods. Figure 5 represents a single signal acquisition sequence, which, when repeated in order to incorporate successive phase encoding steps, will give rise to a much longer data acquisition period. In Fig. 5 we have subdivided the RF irradiation and the gradient waveforms

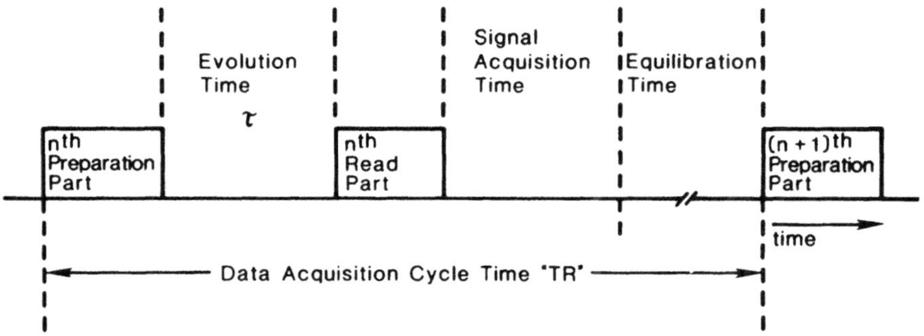

Fig. 5. A generalized schematic diagram of a single data acquisition period. The preparation part together with the evolution time τ determines the proportion of the total net magnetic moment that will be effective during the signal acquisition time. The RF irradiation and gradient waveforms that constitute the read part serve to encode the nuclear signal with spatial information.

into a read part, which stimulates the space-encoded signal for imaging purposes, and a preparation part, which preconditions the magnetic moments and, together with the evolution time τ, controls the magnitude of the net nuclear magnetic moment effective in the read part. If the image were only required to display proton density, then the preparation part would be absent, τ would be zero, and the equilibration time would be sufficient to allow complete recovery to thermal equilibrium before the next acquisition sequence commenced. If, instead, the image is required to display some relaxation information in order to enhance the discrimination, then one or other of the following schemes may be employed.

5.4.1. Saturation Recovery

This technique, often denoted by the initials SR, can be used to discriminate against regions of long T_1, for example cerebrospinal fluid (CSF). Alternatively it can be used to highlight vessels through which fluid is flowing from without to within the slice. To visualize this type of sequence, suppose that the preparation part of Fig. 5 were omitted, that the read part effectively produced a 90° rotation of the z magnetic moment into the xy plane, and that the equilibration time were shortened to prevent full recovery of the z magnetic moment to thermal equilibrium. In effect, the nth read part would now be acting as the $(n + 1)^{\text{th}}$ preparation part, and the sum of the signal acquisition and equilibration times effectively

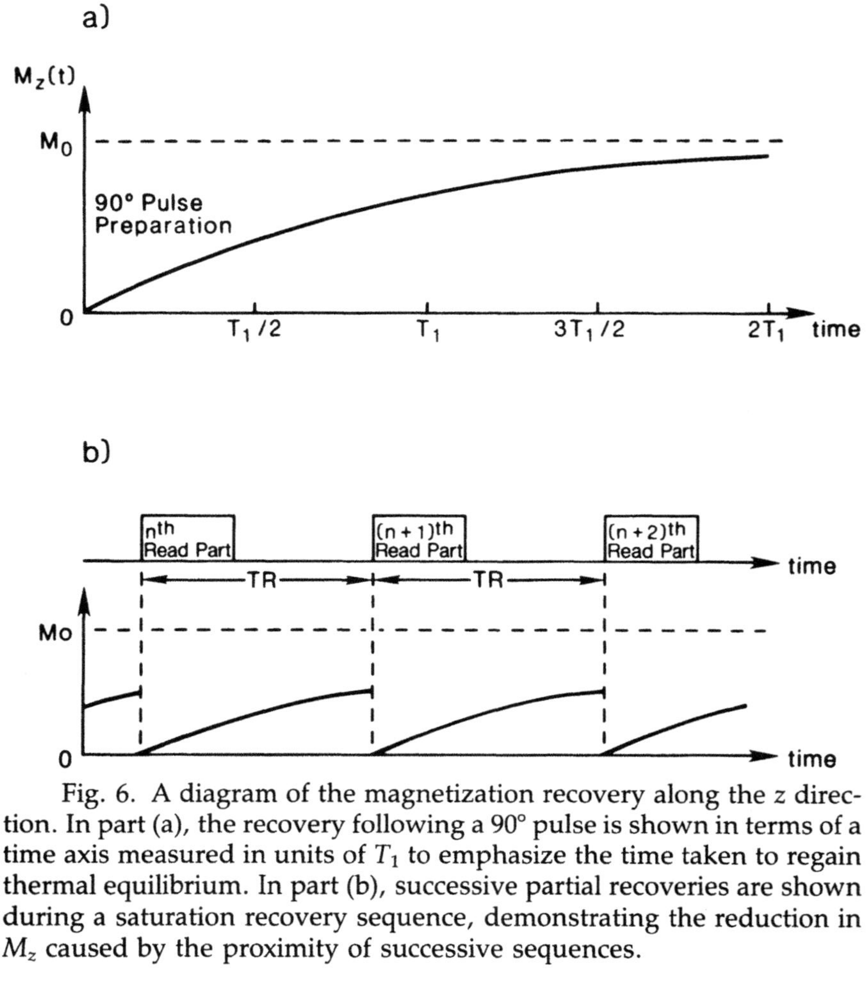

Fig. 6. A diagram of the magnetization recovery along the *z* direction. In part (a), the recovery following a 90° pulse is shown in terms of a time axis measured in units of T_1 to emphasize the time taken to regain thermal equilibrium. In part (b), successive partial recoveries are shown during a saturation recovery sequence, demonstrating the reduction in M_z caused by the proximity of successive sequences.

becomes an evolution time. Following the irradiation of a read part, the net *z* magnetic moment will recover according to the curves in Fig. 6, and the shorter the T_1, the more net magnetic moment will have recovered before the next irradiation. Thus, in a repetitive sequence of many data acquisition intervals, the resultant signal from regions of longer T_1 will be depressed relative to that from shorter T_1 regions, and the short T_1 regions will, therefore, be highlighted.

The ability of the saturation recovery technique to highlight fluid moving into a slice being imaged stems from the movement of that fluid. By making the data acquisition period quite short relative to the T_1 range of biological water, most of the nuclei that reside permanently in the imaged slice will become saturated and give rise to very little signal. In contrast, the fluid that is flowing into the slice will be fresh from an NMR point of view because it will have escaped magnetic moment flipping by irradiation bursts prior to its arrival in the slice. The fluid, therefore, possesses a large net magnetic moment and gives rise to a bright region in the image.

5.4.2. Inversion Recovery

Discrimination against a band of T_1 values can be obtained with this technique and, moreover, the band itself can be adjusted to coincide with either the shorter or the longer end of the T_1 range for biological water. To perform inversion recovery, often referred to as IR sequences, the preparation part of the irradiation pattern needs to cause a 180° inversion of the net z magnetic moment. This is usually brought about by a 180° pulse that can be spatially selective or nonselective. Following the 180° inversion, the z component of net magnetic moment will recover according to the curves in Fig. 7. The read part of the sequence must then rotate the net z magnetic moment, which has recovered during the evolution period τ through 90° and into the xy plane so that it can generate a signal. In addition, it must include the spatial encoding components. For IR sequences, the read part is essentially the whole imaging sequence illustrated in Fig. 4. The crucial point about discrimination by inversion recovery is to choose the evolution time, τ, to be equal to the average t_{null} for the band of T_1 values against which the discrimination is required. Then the regions with that T_1 value will possess zero net z magnetic moment. This is illustrated in Fig. 7 for two tissues with different T_1 values. By adjusting τ, discrimination against different T_1 bands can be brought about. For example, in head images, it is possible to discriminate against regions of short T_1 (T_{1s}) by having $\tau \sim T_{1s} \ln 2$, and in this case white matter will appear darkest, but CSF will be highlighted. Alternatively, by increasing τ to discriminate against regions of long T_1, the CSF can be made to appear darkest in the image, and the white matter can be highlighted. Gray matter will be intermediate in intensity in both cases, but nevertheless closer to the appearance of white matter than to CSF. As a result, by means of the adjustment of τ, it is possible to arrange for one's image to highlight those tissue types that one wishes to highlight.

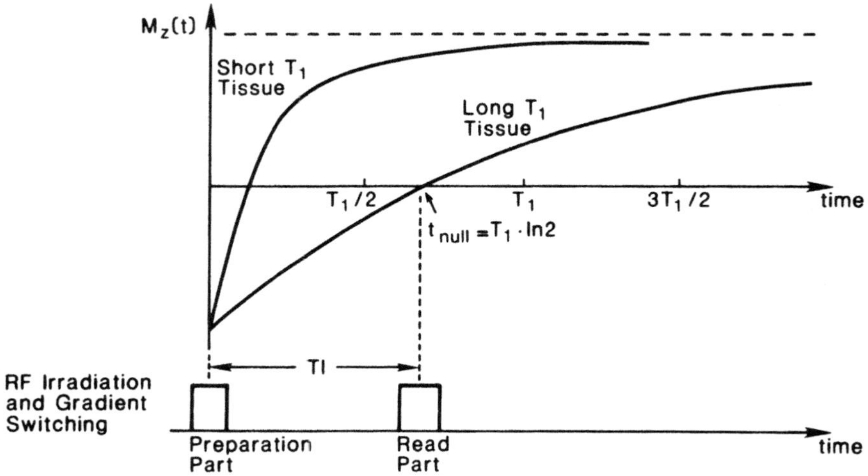

Fig. 7. A diagram showing, for two tissue types, the magnetization recovery in the *z* direction following a 180° inversion pulse. Also shown is the juxtaposition of the irradiation and gradient sequence that would be used to discriminate against the tissue type with a long T_1.

5.4.3. Spin-Echo Imaging

In order to exploit the transverse relaxation time, T_2, as a discrimination parameter, one employs spin-echo (SE) sequences. The preparation part of an SE sequence must then be such as to enable the read part to generate these echoes, and possibly many of them. An MSE sequence is one with multiple spin echoes. The preparation part therefore takes the form of a 90° rotation of the net *z* magnetic moment into the *xy* plane and is usually carried out in conjunction with the slice-defining operation. The single or successive read parts take the form of 180° pulses, each of which gives rise to an echo that is encoded with information concerning the remaining two coordinates. Figure 8 illustrates the basic sequence required. If no transverse relaxation takes place, then the 180° flipping of the *xy* magnetic moment distribution should lead to perfect refocusing. However, transverse relaxation does take place; therefore refocusing is not perfect and successive echoes decrease progressively in amplitude. For a single component system, this decrease will progress with the characteristic time constant T_2. Within the heterogeneous slice of a living body, there will be many component T_2 values, and the contribution to the echoes from those regions with a short

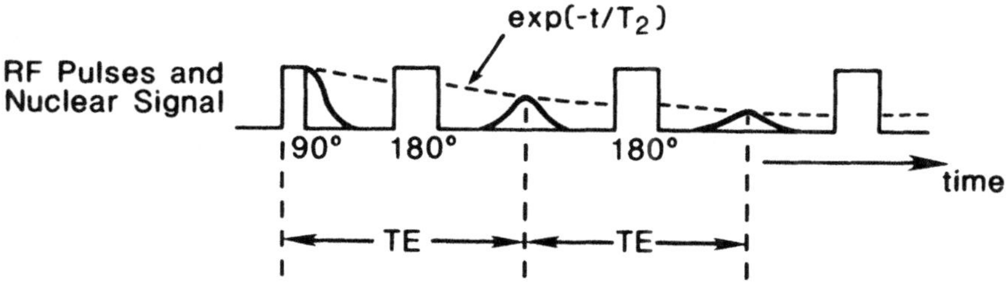

Fig. 8. A simple multiple spin-echo generating sequence in which the 90° pulse, which is usually also the slice-defining pulse, turns the net z magnetic moment into the xy plane. Following the dephasing of the subsequent signal in a field gradient, successive 180° pulses, which may or may not be slice-selective, turn over the dephasing disk of magnetic moments to ensure refocusing in the same gradient. The separation of the 180° pulses, *TE*, is twice the interval between the 90° pulse and the first 180° pulse, and the echoes occur midway between the 180° pulses. This is called a Carr-Purcell sequence.

T_2 will decrease faster than the contributions from regions with a long T_2. Thus, by generating images from distant echoes well separated from the initial 90° pulse in an MSE sequence, those contributions from short T_2 regions will be depressed, and long T_2 regions will be highlighted. The regions in which one expects T_2 to be long are those in which the water concentration is high and consequently give rise to a high molecular mobility. The spin-echo technique can therefore be expected to find utility in the demarcation of CSF and in the assessment of edema and, often, though not always, of tumors.

5.5. Paramagnetic Agents to Enhance Tissue Contrast

The presence in NMR of more than one intrinsic contrast mechanism, namely the two nuclear relaxation mechanisms in addition to the nuclear spin density, provides NMR imaging with a level of sensitivity to pathological change that is superior to that of CT. Nevertheless, several important diagnostic problems are still difficult to resolve, even with such a level of sensitivity. For example, the separation of edematous cerebral tissue from neoplastic tissue or the identification of alterations in the blood–brain barrier are either unlikely or impossible using the intrinsic relaxation mechanisms to produce image contrast. As a result, the well-trodden path of introducing exogenous contrast media to enhance tissue discrimination is also being followed by NMR imagers.

In contrast to the X-ray situation, the exogenous contrast medium is not itself detected directly in NMR imaging. Instead, it is used to modify the relaxation characteristics of the protons of water molecules, which are themselves detected during an imaging procedure and which thereby betray the presence of contrast medium in their vicinity. The type of substance that can modify the relaxational characteristics of the protons is one that is itself magnetic and that can produce time-varying magnetic fields at the proton sites. Usually, the substances used for contrast media contain paramagnetic ions. These ions can be from any of the transition-element series, but to date only the 3d iron series or the 4f rare earth series have been exploited. In particular, much effort is being devoted to the evaluation of the Gd^{3+} ion.

When employed as a contrast agent, a paramagnetic ion may be administered either in a form that dissociates in aqueous solution to give rise to free paramagnetic ions or alternatively in a chelated form that tends to isolate the paramagnetic ion somewhat from its aqueous environment. Using a chelated form affects both the toxicity and the relaxation efficiency of the paramagnetic ion. For example, free Gd^{3+} ions produce acute toxic effects because, it is thought, they compete for binding sites with Ca^{2+}. When chelated with diethylenetriaminepento-acetic acid (DTPA), a three-dimensional molecular framework encages the Gd^{3+} ion, which modifies its excretory pathway, giving rise to preferential urinary excretion, and therefore poor tissue absorption, a short tissue half-life, and much lower toxicity.

The molecular framework of the chelate, which leads to reduced toxicity, also results in modifications to the ability of the paramagnetic ions to relax the water protons. These modifications arise first from the fact that the closest distance of approach of the water molecules to the paramagnetic ion is increased by the presence of the framework, and second, from the reduced motional rate of the Gd^{3+} ion through the aqueous solvent because of the size of the molecular framework. The closest distance of approach governs the strength of the time-varying magnetic field that the water molecules experience from the paramagnetic ions. This time-varying field falls off as the inverse cube of the distance from the ion, and the relaxation efficiency varies as the square of the time-varying field strength. The motional rate, which in turn determines the time variations of the magnetic field produced by the paramagnetic ion, is a factor that can cause the relaxation efficiency to change by several orders of magnitude. When the mean motional rate matches the nuclear Larmor frequency, the relaxation efficiency is increased dramatically. Thus, when the chelating framework slows down the motion of the paramagnetic ion from the very rapid rates of motion in aqueous solution, it causes a closer matching with the proton Larmor frequency, giving rise to an increase in the relaxation efficiency.

Possibly the most common paramagnetic contrast agent being evaluated at the moment is Gd DTPA. It circulates in the vascular system, is excreted unchanged through the kidneys, and only crosses an abnormal blood–brain barrier. Early clinical applications of this contrast agent to depict the margin between brain tumor and cerebral edema have been encouragingly reported by Carr et al. (1984). Later reports (Curati et al., 1986) claim an increase in the diagnostic sensitivity of acoustic neuromas, which establishes a clear improvement over contrast-enhanced CT in demonstrating the lesions, but only a marginal improvement over air-CT. Recent studies with animal models show that the vascular permeability in a canine model of osmotic blood–brain barrier disruption could be consistently demonstrated by NMR imaging following iv administration of 0.25 mmol/kg Gd DTPA (Runge et al., 1985). However, only a marginal improvement over normal NMR imaging was observed in a feline model of acute cerebral ischemia with roughly the same dose (McNamara et al., 1986). Notwithstanding these encouraging data, the establishment of an optimal dose (probably about 0.1 mmol/kg in humans) and time of administration still have to be finalized.

6. NMR Imaging of the Central Nervous System

6.1. *Introduction and Normal Anatomy*

The fundamental objective of NMR imaging of the CNS is to provide either more or as much information as X-ray CT or myelography (with or without CT), but without the risks and discomfort of invasive procedures, radiation, and the possible exogenous contrast materials. The multiparameter nature of the NMR image appearance provides the flexibility and sensitivity to achieve this goal in many circumstances.

In the normal brain, excellent gray-white matter contrast can be obtained in several guises by means of different sequences and timings that exploit differences in the relaxation rates and in the

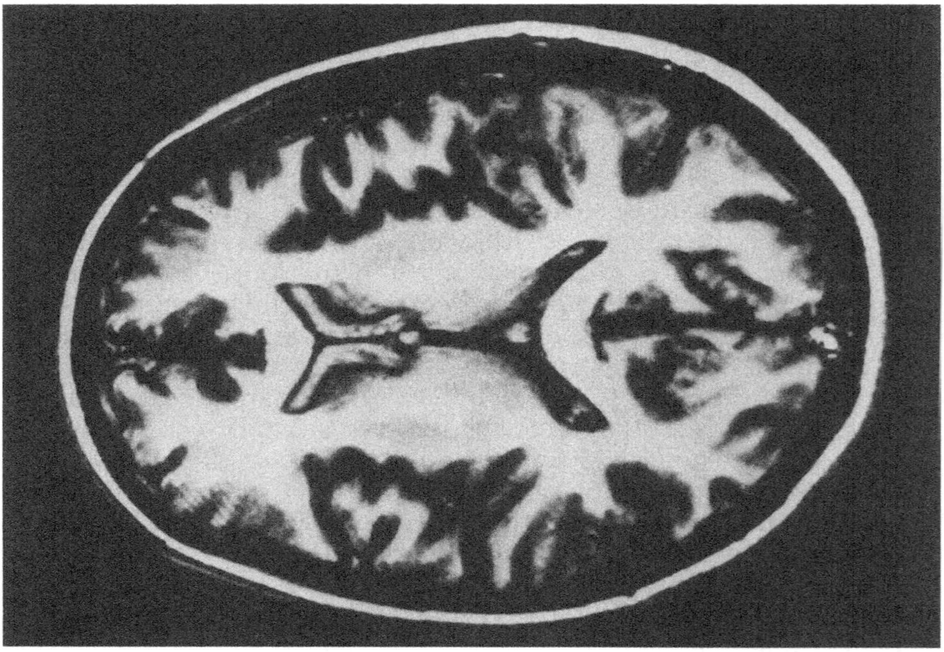

Fig. 9. A transverse section through a normal brain taken with an IR sequence using TI = 800 ms and TR = 3000 ms. The field strength was 1.5T.

mobile proton densities (*see*, for example, Figs. 9 and 10, which illustrate IR or MSE sequences). For example, gray matter is about 4 g water per g dry weight and has a long T_1 of typically ~600 ms at 0.3T. White matter, in contrast, is about 2 g water per g dry weight and has, therefore, a shorter T_1 of typically ~400 ms at 0.3T. Thus, water content and the difference in the water-to-dry weight ratio markedly affects the relaxation rates of the water fraction because the relative amount of macromolecular surface to which the water molecules can absorb is much different in the two types of brain tissue. Inversion recovery sequences, which exploit differences in T_1, are probably the best for emphasizing gray-to-white contrast, and these sequences enable both the cerebral cortex and the sub-

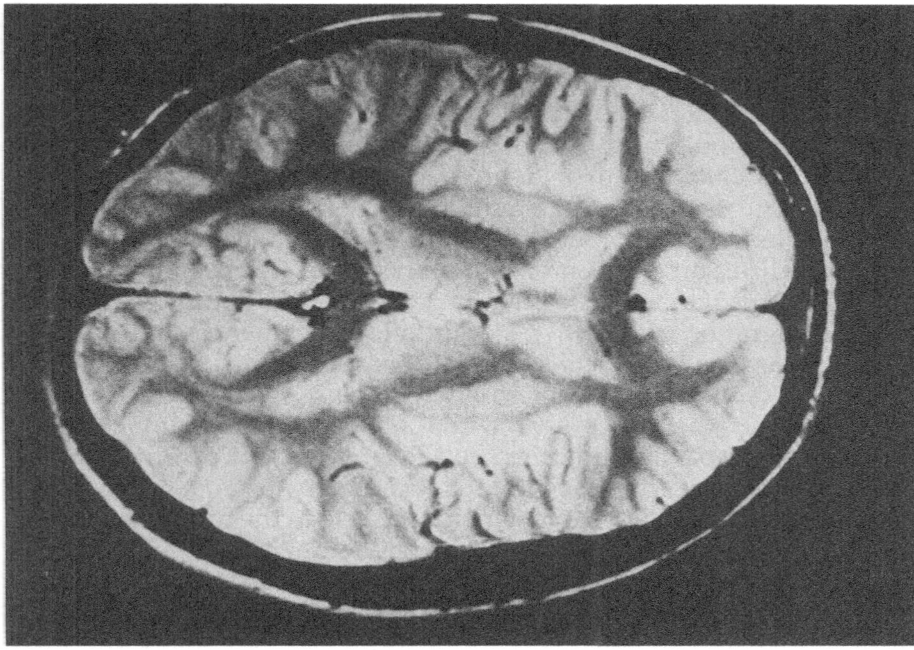

Fig. 10. A transverse section through a normal brain taken with an SE sequence using $TE = 30$ ms and $TR = 3000$ ms. The field strength was 1.5T.

cortical gray matter structures to be readily distinguished from white matter. CSF, because it has the highest water content, has by far the longest T_1 and T_2 (typically 1200 and 150 ms, respectively at 0.3T). It is therefore very easy to highlight the CSF spaces by obtaining images from the later echoes of an MSE sequence, in which the relative signal intensity from the brain tissue with shorter T_2 values will have been substantially reduced.

In contrast to the situation with X-ray CT, bone artifacts are not a problem in NMR imaging. This is particularly important when imaging the posterior fossa region, in which CT images are degraded by bone artifacts, but where brain stem structures can be clearly delineated with NMR. In fact, the whole of the CNS can be observed down to the conus medullaris. The filum terminale and the nerve fibers below the conus medullaris cannot yet be clearly identified, however. Contrast within the spinal cord can be manipulated by varying the pulse sequences in exactly the same way as for the brain, and thus the CSF signal from the central canal and the subarachnoid space can be depressed by discriminating against its long T_1 using an IR sequence or highlighted in the later echoes of an MSE sequence by exploiting its long T_2. The clinical use of IR, SR partial saturation (PS), and SE sequences has also been described didactically in a pair of papers by Bydder and Young (1985a,b).

Recognition of the effects of individual sequences on the image appearance is particularly important for the diagnosis of pathology. It is essential to be aware of which pulse sequences are likely to fail or to be most successful with lesions of a particular type and therefore of a particular anticipated T_1 and T_2 change from normal tissue. For example, edema, infarcts, and many tumors give rise to an increase in mobile proton density because of the influx of water. If this water influx is not accompanied by serum proteins, it leads to a lengthening of T_1 and T_2, which can in turn cause changes in the image that are greater than those caused by the relative change in proton density. If, however, serum proteins accompany the increase in water content, as for example, in vasogenic edema, the increase in macromolecular surface onto which the extra water molecules can absorb, can, as far as changes in relaxation rates are concerned, offset the effects of an increase in water, as explained in section 5.3. Decreases of T_1 are also observed in other pathologic processes, e.g., fat-containing lesions, in which increases in macromolecules can be anticipated.

6.2. Cerebrovascular Disease

The site of cerebral infarction, whether embolic, thrombotic, or lacunar, is usually disclosed by an area of increased T_1 and T_2 in a location of well-known vascular distributions (DeWitt et al., 1984; Bydder, 1984). Moreover, the absence of bone artifacts permits the observation of any involvement of the cerebral cortex. The increases in T_1 and T_2, respectively, manifest themselves as a reduction in signal intensity on IR scans or a relative increase in intensity in the later echoes of an MSE sequence. In some cases, the occlusion of vessels may be inferred from the lack of signal from these vessels when employing an SR procedure. A collection of such sequences provides images that enable the site, the shape, and the margins of the lesion to be determined, as well as the absence of any mass effect; all of these are important for diagnosing an infarct.

Intracranial hemorrhage in general shows three distinct regions (DeWitt et al., 1984; Bydder, 1984) that can change in intensity and distribution with the passage of time following the onset of symptoms. A region of short T_1 characteristic of a reduced rate of molecular motion is usually associated with a clot. This is often accompanied by a central region of prolonged T_1 that, if present, is likely to result from a liquefaction. The edematous rim of the hemorrhage often has an intermediate T_1 value. The time course of the changes in these regions is not yet particularly well established, and it has been suggested (Dooms et al., 1986) that NMR will be the most useful with chronic hematomas. In addition to these three regions, the presence of mass effect is also an important characteristic in diagnosing hematoma.

Clots within aneurysms are also visible in NMR images, because of their short T_1 region (Worthington et al., 1983). The wall of the aneurysm, made visible in NMR images by a thin layer of mural thrombus, has also been reported (DeWitt et al., 1984). It must be borne in mind, however, that NMR, though readily able to demonstrate giant aneurysms, does not yet provide the level of anatomical detail required prior to surgical intervention. Because it depicts the hemorrhagic and vascular foci very well, NMR imaging is also proving superior to CT in differentiating arteriovenous malformations from small neoplasms (Lemme-Plaghos et al., 1986).

6.3. Tumors of the CNS

Early claims by Damadian (1971) that the T_1 of tumor tissue was longer than that of normal tissue raised expectations that in vivo measurement of proton relaxation rates would provide a specific index for neoplastic tissue (Goldsmith et al., 1978). Although this expectation is unlikely ever to be fulfilled, the contrast enhancement afforded by the differences in the relaxation rates of water protons between tumors and their surrounding tissue usually enables one to delineate the tumor, though this is not universally true. For example, NMR imaging is highly diagnostic for brain stem tumors (Peterman et al., 1985), which are well demonstrated because of their elevated T_1 and T_2 and because of brain stem mass effects.

It must be borne in mind, however, that the proton relaxation efficiency is not always lowered (T_1 lengthened) in tumors. For example, it has been reported that lipomas and dermoid tumors (Davidson et al., 1985) display an increased longitudinal relaxation efficiency (shorter T_1), and the presence of fat could explain this. Malignant melanomas, which may contain paramagnetic free radicals, and tumors in which there has been hemorrhage are also prone to shorter proton T_1 values (MacKay et al., 1985). The relaxation times of some tumors, e.g., meningiomas, overlap with the range of relaxation times of normal tissue and are thus difficult to discriminate (MacKay et al., 1985). A discrimination problem also often exists between a tumor and the peritumor edema. Initial results with Gd DTPA (Carr et al., 1984) indicate that the use of paramagnetic contrast agents removes this difficulty and enables NMR imaging to compete favorably with contrast-enhanced X-ray CT when diagnosing meningiomas, a feat that unenhanced NMR imaging is unable to achieve (Bradley et al., 1984).

6.4. Assessment of the State of Myelination

6.4.1. Multiple Sclerosis

The detection of demyelinated plaques in multiple sclerosis (MS) is an exercise in which NMR imaging shows considerably more promise than X-ray CT (Young et al., 1981; Runge et al., 1984; Maravilla et al., 1985; Sheldon et al., 1986). MS generates an abundance of discrete lesions that can be detected by NMR and whose anatomical distribution, e.g., periventricular white matter, is very characteristic, as is the asymmetry of disease involvement and the lack of mass effect. Nevertheless, MS plaques can be difficult to distinguish from white matter infarcts. The seriousness of this problem, however, is reduced because the clinical group of patients in which a significant incidence of such infarcts occurs, namely the elderly, is somewhat different from that which exhibits MS plaques. Care must also be exercised not to confuse partial volume effects with lesions. This source of confusion is to some extent minimized by employing SE sequences that do not give rise to a high contrast between normal gray and white matter (Bydder, 1984) rather than IR sequences that, because of the high gray-white contrast that they produce, maximize partial volume effects. Although SE sequences are generally preferred for this reason, it has been reported (Runge et al., 1984) that in regions such as the brain stem, internal capsule, and corona radiatia, IR are the sequences of choice. In addition, early echo SE sequences are most useful to maintain a low relative intensity signal from CSF in order to allow distinction of periventricular lesions from CSF.

6.4.2. Development of Myelination

Studies have also been made (Johnson et al., 1983) to assess the development of myelination in infants and in children and to detect delays or deficits as compared with the normal process, although the normal variability in white matter development is still not known. Bearing in mind the much longer relaxation times in the neonatal brain and also the fact that as white matter becomes myelinated, its T_1 changes from being longer than, to being shorter than, that of gray matter, modified IR sequences are used to obtain the best gray-white matter contrast and delineation of white matter regions.

6.5. Craniovertebral Junction and the Spinal Cord

NMR imaging provides a continuous, albeit segmented, view from the brain stem and the cervicomedullary junction to the conus medullaris. Although slice thickness can be a limitation when delineating the finer structures in sagittal and coronal sections, these sections provide excellent visual images of the anatomical relationships required for surgical planning, particularly at the craniovertebral junction. For example, Chiari I and II malformations are readily observed (Spinos et al., 1985) in sagittal or coronal sections, but transverse sections are also needed (Lee et al., 1985) for the delineation of hydromyelia and syringomyelia. When imaging the spinal cord, one can also take advantage of the increased sensitivity of surface-coil radio frequency antenna. The increase in sensitivity afforded by these antenna, if they are circular, extends a distance into the tissue of about one coil radius, and the spinal cord can therefore be readily accommodated. Spinal cord anomalies and defects, such as those associated with spina bifida, are therefore prime candidates for evaluation with surface coil NMR. A discussion of these and other pathological entities in the spinal column and craniovertebral junction can be found in a review by Han et al. (1984).

6.6. Imaging with the Sodium Nucleus, ^{23}Na

The possibility of imaging the spatial distribution of ^{23}Na nuclei in the brain was realized quite early in the development of

NMR imaging (DeLayre et al., 1981; Maudsley et al., 1982). Because the vast majority of the sodium ions in normal tissue resides in the intravascular and interstitial spaces, a breakdown of the ATP-driven sodium pump (precipitated by some disease process such as a stroke) leads to an increase in the intracellular Na^+ concentration by factors of up to 2 or 3. This accumulation of ^{23}Na in the local region of pathology should, in principle, enable the location, extent, and severity of that pathology to be estimated from a spatial mapping (a term used for low-resolution imaging) of the ^{23}Na concentration.

Unfortunately the ^{23}Na nucleus suffers from two serious handicaps in so far as NMR imaging or mapping is concerned. First, it suffers from a low signal strength when compared to the proton because of its low concentration and because of its smaller magnetic moment. Second, the much shorter decay time of its transverse magnetization (T_2) means that if typical 2DFT imaging sequences are used, the majority of this already weak signal has decayed before it can be space encoded and then acquired. Techniques using volume acquisition of the FID and projection reconstruction are being developed in an attempt to alleviate these problems (Ra et al., 1986). Nevertheless, the net result is a coarser spatial resolution than one has become used to with proton imaging.

Notwithstanding the low intensity of the ^{23}Na signal, it is, in fact, of all the signals from physiologically occurring magnetic nuclei, the next most intense after that of the proton. This fact makes it a reasonable candidate with which to search for and locate the breakdown of the energy metabolism. The ^{31}P signal from ATP is divided between three spectral lines and would, even if spatial encoding ambiguities were sorted out, give rise to an even coarser spatial distribution than that for ^{23}Na. A comparison of the relative imaging performance of ^{23}Na and ^{1}H can be found in the article by Feinberg et al. (1985), which evaluates quantitatively the merits of sodium imaging. Turski et al. (1986) have used it clinically to identify acute vasogenic edema and show course sodium maps as well as high-resolution proton images of a patient with a falx meningioma.

Experiments on small animals and excised organs nevertheless continue to raise the expectations of sodium NMR. For example, by means of compartment-specific shift reagents and at magnetic field strengths six times those of current clinical magnets, Balschi et al. (1986) have succeeded in charting the time course of the Na^+ concentrations in both the intra- and extracellular compartments during the onset of ischemia induced in a rat muscle. Before anything approaching this can be achieved clinically, not only must nontoxic shift reagents be developed, but imaging strategies must be perfected that can accommodate the rapidly decaying component of the ^{23}Na signal ($0.7 \rightarrow 3$ ms), and the effective signal-to-noise ratio must be substantially improved.

7. Summary

The general aim of this chapter has been to provide an understanding of the way in which the NMR technique gives rise to images of the central nervous system that display, not only normal human morphology, but also soft tissue pathology, where and when it occurs. The NMR technique is one in which atomic nuclei are used as probes of matter. The signals that one can derive from these nuclear probes are so sensitive to the magnetic environment that the nuclear probes experience, that, by judiciously generating a controlled spatial variation in that magnetic environment, i.e., a magnetic field gradient, one can cause the nuclear signals to tell one from where in space they were emitted. This encoding of spatial information in the NMR signal is essential for imaging. It is, moreover, what makes imaging with NMR somewhat different from the NMR that has been practiced in basic science laboratories since 1946. In order to provide contrast between tissues, however, e.g., between gray and white matter, and to understand the sensitivity of NMR to soft tissue pathology, e.g., the discrimination of MS plaques from surrounding white matter, one must come to terms with the phenomenon of nuclear relaxation, a phenomenon that is well understood from the basic science NMR work of the last 40 years. That tissue contrast is not simply a function of tissue density is a fundamental difference between NMR and the x-ray methods. NMR possesses the additional parameters of longitudinal and transverse relaxation rates, with which to adjust tissue contrast to maximum advantage. Notwithstanding the sensitivity of NMR image contrast to the intrinsic proton relaxation rates in tissue, contrast between two tissue types is sometimes insufficient, e.g., between tumor tissue and peritumor edematous tissue. In such cases, the introduction of exogenous paramagnetic centers, in order to modify the relaxation rates of their neighboring water protons, can often provide the required contrast, if the delivery vehicles for those paramagnetic centers discriminate between the two tissue types.

It is with capabilities such as those outlined above, which were dealt with more fully in sections 2 to 5, that the vast majority of studies of the central nervous system are carried out. In discussing this imaging methodology, we have chosen to omit a detailed discussion of the separation of the fat proton signal from the water proton signal. The need for this separation arises because protons in the CH_2 groups of lipids resonate at a slightly higher magnetic field than the water protons and, if care is not taken, give rise to a fat image slightly displaced from the water image. This separation is much more important in parts of the body other than the CNS (Pykett and Rosen, 1983), e.g., the abdomen, and it can be overcome by a variety of methods (Dixon, 1984; Hasse and Frahm, 1985).

Although no comparison has been given of the efficacy of NMR imaging in various parts of the body, the initial promise of NMR is probably best fulfilled in the CNS. This stems from the lack of detrimental effects from bodily motion and from the predominance of the water signal, neither of which holds for the other major organs. The absence of bone artifacts, which are a substantial problem with the X-ray CT modality, enables striking soft tissue resolution to be obtained in the region of the posterior fossa and along the spinal chord, as well as permitting cortical involvement to be delineated more precisely in lesions in the cerebral hemispheres. The ease with which high resolution sagittal, coronal, and oblique sections can be obtained with NMR methods markedly improves surgical and radiation therapy planning. Moreover, beyond establishing the presence, the location, and the extent of a lesion, together with any accompanying mass effect, one still has with NMR other cards up one's sleeve. For example, one can often remove further diagnostic ambiguities by exploiting one's control over the RF sequence timings, relative to postulated relaxation times, in order to determine the fluidity of a lesion or parts of it. This is not to say, however, that NMR imaging is a perfect noninvasive imaging modality. It is not. Although the frequency of invasion with contrast agents is greatly reduced with NMR, as compared to x-ray methods, paramagnetic contrast agents that target specific tissue types can still enhance the diagnostic power of imaging by NMR.

Acknowledgment

The author is grateful to the Alberta Heritage Foundation for Medical Research for a Medical Scientist Award.

References

Abragam A. (1961) *The Principles of Nuclear Magnetism,* Oxford University Press, London, UK.

Allen P. S. (1983a) *Progress in Medical and Environmental Physics* vol. 2 (Jackson D. F., Kouris K., and Spyron, N. M., eds.) Blackie, Glasgow.

Allen P. S. (1983b) *Nuclear Magnetic Relaxation and Its Possible Exploitation in NMR Imaging,* Proceedings of the UN Conference of Appl. Phys. Med. Biol., Trieste, Italy, World Publishing, Singapore.

Balschi J. A., Bitte J. A., and Ingwall J. S. (1986) Ischemia in the intact rat by interleaved ^{31}P and ^{23}Na NMR and using a shift reagent to discriminate intra and extracellular sodium. *Proc. 5th Ann. Meet. Soc. Magn. Res. Med.* Montreal, p. 343–344.

Bradley W. G., Waluch V., Yadley R. A., and Wycoff R. R. (1984) Comparison of CT and MR in 400 patients with suspected disease of the brain and cervical spinal cord. *Radiology* **152,** 695–702.

Bydder G. M. (1984) Magnetic-resonance imaging of the brain. *Radiol. Clinics N. Am.* **22,** 779–793.

Bydder G. M. and Young I. R. (1985a) MR imaging—clinical use of the inversion recovery sequence. *J. Comp. Assist. Tomogr.* **9,** 659–675.

Bydder G. M. and Young I. R. (1985b) Clinical use of the partial saturation and saturation recovery sequences in MR imaging. *J. Comp. Assist. Tomogr.* **9,** 1020–1032.

Carr D. H., Brown J., Bydder G. M., Weinmann H. J., Speck U., Thomas T. J., and Young I. R. (1984) Intravenous chelated gadolinium as a contrast agent in NMR imaging of cerebral tumors. *Lancet* **i,** 484–486.

Curati W. L., Graif M., Kingsley D. P. E., Niendorf H. P., and Young I. R. (1986) Acoustic neuromas—Gd DTPA enhancement in MR imaging. *Radiology* **158,** 447–451.

Damadian R. (1971) Tumor detection by nuclear magnetic resonance. *Science* **171,** 1151–1152.

Davidson H. D., Ouchi T., and Steiner R. E. (1985) NMR imaging of congenital intracranial germinal layer neoplasms. *Neuroradiology* **27,** 301–303.

DeLayre J. L., Ingwall J. S., Malloy C., and Fossel E. T. (1981) Gated ^{23}Na NMR images of an isolated perfused working rat heart. *Science* **212,** 935–936.

DeWitt L. D., Buonanno F. S., Kistler J. P., Brady T. J., Pykett I. L., Goldman M. R., and Davis K. R. (1984) Nuclear magnetic imaging in evaluation of clinical stroke syndrome. *Ann. Neurol.* **16,** 535–545.

Dixon W. T. (1984) Simple proton spectroscopic imaging. *Radiology* **153,** 189–194.

Dooms G. C., Uske A., Brant-Zawadzki M., Kurcharozyk W., Lemme-Plaghos L., Newton T. H., and Norman D. (1986) Spin-echo MR imaging of intracranial hemorrhage. *Neuroradiology* **28,** 132–138.

Farrar T. C. and Becker E. D. (1971) *Pulse and Fourier Transform NMR.* Academic, New York.

Feinberg D. A., Crookes L. A., Kaufman L., Brant-Zawadski M., Posin J. P., Arakawa M., Watts J. C., and Hoenninger J. (1985) Magnetic resonance imaging performance: A comparison of sodium and hydrogen. *Radiology* **156,** 133–138.

Goldsmith M., Koutcher J. A., and Damadian R. (1978) NMR in cancer. 13. Application of NMR malignancy index to human mammary tumors. *Br. J. Cancer* **38,** 547–554.

Haase A. and Frahm J. (1985) Multiple chemical shift selective NMR imaging. *J. Mag. Reson.* **64,** 94–102.

Han J. S., Benson J. E., and Yoon Y. S. (1984) Magnetic resonance imaging in the spinal column and craniovertebral junction. *Radiol. Clinics N. Am.* **22,** 805–827.

Johnson M. A., Pennock J. M., Bydder G. M., Steiner R. E., Thomas D. J., Howard R., Bryant D. R. T., Payne J. A., Levene M. I., Whitelaw A., Dubowitz L. M. S., and Dubowitz V. (1983) Clinical NMR imaging of the brain in children—normal and neurologic disease. *Am. J. Neuroradiol.* **4,** 1013–1026.

Kumar A., Welti D., and Ernst R. R. (1975) NMR Fourier zeugmatography. *J. Mag. Res.* 18, 69.

Lee B. C. P., Zimmerman R. D., Manning J. J., and Deck M. D. F. (1985) MR imaging of syringomyelia and hydromyelia. *Am. J. Roentgen* **144,** 149–156.

Lemme-Plaghos L., Kucharczyk W., Brank-Zawadzki M., Uske A., Edwards M., Norman D., and Newton T. H. (1986) MRI of angiographically occult vascular malformations. *Am. J. Roentgen* **146,** 1223–1228.

Locher P. R. (1980) Computer simulation of selective excitation in NMR imaging. *Phil. Trans. Roy. Soc. Lond.* **B289,** 537.

MacKay I. M., Bydder G. M., and Young I. R. (1985) MR imaging of central nervous system tumors that do not display increase in T_1 or T_2. *J. Comp. Assist. Tomogr.* **9,** 1055–1061.

Mansfield P. and Morris P. G. (1982) NMR Imaging in Biomedicine, in *Advances in Magnetic Resonance* (Waugh J. S., ed.) Academic, New York.

Mansfield P., Maudsley A. A., Morris P. J., and Pykett I. L. (1979) Selective pulses in NMR imaging: A reply to criticism. *J. Mag. Res.* **33,** 261–274.

Maravilla K. R., Weinreb J. C., Suss R., and Nunnally R. L. (1985) Magnetic resonance demonstrations of multiple-sclerosis plaques in the cervical cord. *Am. J. Roentgen* **144,** 381–385.

Maudsley A. A., Hilal S. K., Simon H. E., and Perman W. H. (1982) Multinuclear NMR imaging. *Proc. 1st Ann. Meet. Soc. Mag. Res. Med.* Boston, Massachusetts, p. 102.

McNamara M. T., Brant-Zawadzki M., Berry I., Pereira B., Weinstein P., Derugin N., Moore S., Kucharczyk W., and Brasch R. C. (1986) Acute experimental cerebral ischemia: MR enhancement using Gd-DTPA. *Radiology* **158,** 701–705.

Peterman S. B., Steiner R. E., Bydder G. M., Thomas D. J., Tobias J. S., and Young I. R. (1985) Nuclear magnetic resonance imaging (NMR), (MRI), of brain-stem tumors. *Neuroradiology* **27,** 202–207.

Pykett I. L. (1982) NMR imaging in medicine. *Sci. Am.* **246,** 78.

Pykett I. L. and Rosen J. R. (1983) NMR: In-vivo proton chemical shift imaging. *Radiology* **149,** 197–201.

Ra J. B., Hilal S. K., and Cho Z. H. (1986) A method for in vivo MR imaging of the short T_2 component of ^{23}Na. *Magn. Res. Med.* **3,** 296–302.

Runge V. M., Price A. C., Kirshner H. S., Allen J. H., Partain C. L., and James A. E. (1984) Magnetic resonance imaging of multiple sclerosis—a study of pulse-technique efficacy. *Am. J. Neuroradiol.* **5,** 691–702.

Runge V. M., Price A. C., Wehr C. J., Arkinson J. S., and Tweedle M. F. (1985) Contrast enhanced MRI—evaluation of a canine model of osmotic blood-brain barrier disruption. *Invest. Radiol.* **20,** 830–844.

Sheldon J. J., Siddharthan R., Tobias J., Sheremata W. A., Siola K., and Viamonte M. (1986) MR imaging of multiple sclerosis: Comparison with clinical CT examination in 74 patients. *Am. J. Roentgen* **145,** 957–964.

Slichter C. P. (1964) *Principles of Magnetic Resonance* Harper and Row, New York.

Spinos E., Laster D. W., Moody D. M., Ball M. R., Witcofski R. L., and Kelly D. L. (1985) MR evaluation of Chiari-I malformation at 0.15T. *Am. J. Roentgen* **144,** 1143–1146.

Turski P. A., Perman W. H., Hald J. K., Houston L. W., Strother C. M., and Sackett J. F. (1986) Clinical and experimental vasogenic edema: In vivo sodium MR images. *Radiology* **160,** 821–825.

Worthington B. S., Kean D. M., and Hawkes R. C. (1983) NMR imaging in the recognition of giant intracranial aneurysms. *Am. J. Neuroradiol.* **4,** 835–836.

Young I. R., Hall A. S., Pallis C. A., Bydder G. M., Legg N. J., and Steiner R. E. (1981) Nuclear magnetic resonance imaging of the brain in multiple sclerosis. *Lancet* **ii,** 1063–1066.

Measurement of Cerebral Ions

Neil M. Branston and Robert J. Harris

1. Introduction

1.1. Historical—Why Cerebral Ions Are Measured

A knowledge of how ions are involved in the brain's normal and pathophysiological activity and of their concentrations, fluxes, and interactions with variables such as cerebral blood flow (CBF) is essential for understanding and treating cerebrovascular disease and other neurological disorders.

The facts that in the brain energy is supplied mainly by oxidative metabolism of glucose, that there is so little storage of such energy sources in the brain, and that energy failure is accompanied by major movements of cerebral ions are central features of experimental brain research and neurological practice. Several examples of the importance of cerebral ions may be mentioned. Anoxia, ischemia, or hypoglycemia lead to energy failure in the brain associated with, or preceded by, loss of dynamic electrical activity and movements in ionic species such as K^+, Ca^{2+}, and H^+ into or out of the extracellular compartment. In focal ischemia of the cerebral cortex, the infarct is typically surrounded by a region in which blood flow, although reduced to a level at which electrical activity such as the EEG is abolished and the extracellular potassium activity (K_e) is slightly elevated, is still high enough to maintain ATP concentration at close to nomral. This region, termed the ischemic penumbra (Astrup et al., 1981a), embodies a critical state of balance for metabolism and tissue perfusion, and is therefore an important target for the application of therapeutic measures to produce some degree of functional recovery. Measurement of K_e in progressive ischemia was an essential technique in forming the concept of the ischemic penumbra, and ion measurements will be important in evaluating the effects of treatment. Understanding the triggering and pathophysiology of epileptic seizures requires a knowledge of the associated ionic changes in both the extra- and intracellular compartments, and the role of glial cells in buffering K_e is one of the factors implicated in spreading depression, which may be involved in the neurological symptoms of classical migraine (Gardner-Medwin, 1981). Brain edema following subarachnoid hemorrhage and vasospasm, or following other forms of stroke, is a major contributor to neurological deficit and death (Ng and Nimmannitya, 1970; Yatsu and Coull, 1981); the onsets of ischemic and vasogenic cerebral edema are strongly associated with ion movements between the intracellular, extracellular, and intravascular compartments (Hossmann et al., 1977).

The principal ions to be considered in this chapter are K^+, Na^+, Cl^-, H^+, Ca^{2+}, Mg^+, and HCO_3^-. The first six of these can be measured in vivo using ion-sensitive microelectrodes by methods to be described in detail later. K^+, Ca^{2+}, H^+, and Mg^+ in particular have powerful effects on the excitability of neurons, transmitter release from presynaptic terminals, and tissue blood flow. It should be kept in mind that in the normal brain, functional activity involves much smaller fluctuations in ionic concentrations than under many experimental or pathophysiological conditions, and for proper functioning a stable background of extracellular ionic levels must be maintained against which these small changes can take place.

1.2. Ions in Normal Cerebral Physiology

1.2.1. Electrical Stimulation

When the normal cerebral or cerebellar cortex is activated by repetitive electrical stimulation, either directly to its surface or via an afferent pathway, K_e increases from normal stable values of about 3 m*M* to an upper limit of about 10 m*M* (Lothman et al., 1975; Heinemann et al., 1977; Branston et al., 1978; Leniger-Follert et al., 1978; Nicholson et al., 1978; Stockle and Ten Bruggencate, 1978). At the same time, the local extracellular potential moves in a negative direction and oxidation of NADH increases (Lothman et al., 1975), these correlations indicating that much of the increase in oxidative energy turnover of stimulated cortex is expended in the active transport of ions. In cortex subjected to mild hypoxia or ischemia, when the baseline value of K_e is slightly elevated, local stimulation can provoke much larger, temporary increases of K_e, up to 80 m*M* (Branston et al., 1982), that may occur even without apparent stimulation and that resemble the well-known increases seen in spreading depression (Bures et al., 1974; Nicholson and Kraig, 1981) elicited by other triggering stimuli such as anoxia, mechanical trauma, and chemical stimuli.

These increases in K_e are associated with decreases in Ca_e

(Heinemann et al., 1977; Harris et al., 1981), from resting levels of 1.2–1.5 m*M* to about 1 m*M* if K_e rises no higher than 6 m*M*, but to 0.1 m*M* or lower if K_e increases to above 30 m*M*. Associated with these ionic and oxidative metabolic changes is an increase in local CBF. The nature of the coupling between electrical activation and flow changes, and the involvement of ionic fluxes in this coupling, are controversial and are discussed briefly below.

1.2.2. Vascular Control of CBF

There is abundant evidence that the extracellular activity of H^+ ions influences cerebrovascular resistance; that is, the diameters of cerebral vessels. Acid solutions applied to the surface of the brain cause dilatation of pial arterioles, whereas alkaline solutions cause constriction (Elliott and Jasper, 1949; Wahl et al., 1970; Kuschinsky et al., 1972; Betz, 1977), and cerebral blood flow correlates with pH_e rather than $PaCO_2$ (Severinghaus et al., 1966). The well-known influence of $PaCO_2$ on CBF is caused not by the direct action of CO_2 itself, but by the ready permeability of the arteriolar vessel wall to CO_2 and the impermeability of the vessel wall to H^+ and HCO_3^- ions (Lassen, 1968). However, it is questionable whether pH_e contributes significantly by itself to the regulation of CBF. The doubt arises from experiments demonstrating that the expected association between flow and pH_e (recorded together in the same region of brain) does not always hold, either in normal brain or under pathophysiological conditions such as seizure activity or hypoxia (Astrup et al., 1976; *see* Siesjo, 1984, for further review).

Results obtained both in vivo and in vitro have shown that relatively small increases in K_e cause vasodilatation, with vasoconstriction occurring at higher levels of K_e (Kuschinsky et al., 1972; Knabe and Betz, 1972; Cameron et al., 1976). Again, it is doubtful that K^+ ions exert a general vasoactive influence in brain, but they may reinforce vascular changes that have already occurred, such as hyperemia in epileptic activity (Astrup et al., 1979) and in activation produced by local or orthodromic brain stimulation (Leniger-Follert and Lubbers, 1976; Leniger-Follert, 1984). Similar restrictions apply to the interpretation of the observed decreases in Ca_e with stimulation and seizure activity (Heinemann et al., 1977). The mechanisms of CBF control and their linkages to metabolic demands and electrical activity probably involve several factors, including the movement of several ionic species between extracellular and intracellular compartments, the involvement of each factor being variable and dependent on the nature of the disturbance or control task. As emphasized by Siesjo (1984), the key event in the contraction or relaxation of the smooth muscle of the vessel wall is the interaction of free cytosolic Ca^{2+} with the actin–myosin complex; contraction is initiated by an increase in Ca^{2+} activity, and any factors that change the fluxes of Ca^{2+} through membrane channels or its binding within the cell will, in turn, influence cerebrovascular resistance.

1.3. Ions in Cerebral Pathophysiology

1.3.1. Ischemic Lesions

A fruitful line of research involving ion measurement has been the study of ischemic lesions. The research has centered upon the measurement of extracellular ions, primarily K^+, but also including Ca^{2+}, Na^+, and H^+, with respect to time and depth of ischemia. To produce partial ischemia, the main animal models have used middle cerebral artery occlusion (baboon and cat) and bilateral carotid artery occlusion (rat), all with added hypotension.

With a reduction in CBF from normal values, the first pathophysiological change observed is a reduction in the rate of clearance of K^+ following a transient increase of K_e evoked by local electrical or orthodromic stimulation (Branston et al., 1982). This change has been associated with flows of 25–30% of normal, and becomes progressively greater as flow is reduced to lower levels. Mayevsky (1978) showed a diminution of the size of the oxidation cycle evoked by spreading depression in rats after bilateral carotid artery occlusion, which indicates that energy metabolism has been compromized, although at this stage of ischemia, spontaneous and evoked electrical activity are still essentially normal (Siesjo, 1984).

During progressive hypotension in the rat, the next change is a deepening acidosis (Harris and Symon, 1984) that proceeds rapidly with the reduction in CBF. In the primate the acidosis begins at a lower level of CBF—around 20 mL/100 g/min—but deepens more quickly (Astrup et al., 1977). As pH begins to fall, there is a loss of electrical activity, but still no effect upon resting levels of K_e, Ca_e, or Na_e.

As flow is reduced further to around 16 mL/100 g/min in the rat and cat, or to 10 mL/100 g/min in primates, a slow, steady increase in K_e commences. In the primate, this proceeds to about 13 m*M* and is followed by a rapid rise to around 50 m*M* (Astrup et al., 1977; Branston et al., 1977; Harris et al., 1981). In the rat, the slow rise in

K_e proceeds from 3 to 6 m*M* and then rises rapidly to 40 m*M* (Harris and Symon, 1984).

As K_e reaches 13 m*M*, there is a rapid fall in Ca_e to 15–20% of control levels (Harris et al., 1981; Harris and Symon, 1984), where it reaches a plateau.

Studies of ion changes after cardiac arrest, made predominantly in the rat, have shown a similar sequence (Hansen, 1977; Hansen and Zeuthen, 1981; Kraig et al., 1983; Mutch and Hansen, 1984). In addition, extracellular sodium activity (Na_e) falls markedly with Ca_e (Nicholson et al., 1977; Hansen and Zeuthen, 1981).

The changes of K_e after cardiac arrest have been the subject of pharmacological manipulation. Hansen (1977, 1978) has shown that the time from cardiac arrest to the onset of the fast K_e increase is longer in juvenile rats in comparison to adult rats and in hyperglycemic rats in comparison to normoglycemic rats. Both sets of results were attributed to the preservation of anaerobic metabolism and thus of ionic homeostasis for a longer period in the juvenile and hyperglycemic rats. With a similar interpretation, Astrup et al. (1980) showed that the time to the fast K_e increase was inversely proportional to the preischemic metabolic rate.

1.3.2. Generation of Cerebral Edema

Early work on the formation of ischemic edema utilized measurements of total brain content of sodium and potassium in addition to the presence of other markers such as albumin, pertechnate, and antipyrine. The results showed that there was a marked increase in the sodium content of gray matter detectable from 3 h after onset of ischemia, and that there was a concomitant decrease in potassium content (Watanabe et al., 1977; Pappius, 1979; Schuier and Hossmann, 1980). In long-term studies, sodium content peaked at 3–7 d and was starting to recover by 20 d postischemia (O'Brien et al., 1974).

After a 1-h ischemic insult, sodium and water content changes showed two patterns. Either they returned to normal by around 20 h (Zimmermann and Hossmann, 1975; Pappius, 1979), or when there was no recovery of electrical function there was also no recovery of sodium and potassium levels (Zimmermann and Hossmann, 1975).

In further support of the role of sodium and potassium in determining the resolution of ischemic brain edema, the measurement of activity of the Na^+, K^+-ATPase enzyme by MacMillan (1982) showed that the enzyme was not damaged by 30 min of severe ischemia in the rat, and Mrsulja and Mrsulja (1981) showed that the restoration of Na^+, K^+-ATPase activity to normal levels was coincident with the normalization of brain water content.

1.3.3. Hypoglycemia

Astrup and Norberg (1976) showed that there was a rapid, large increase in K_e at around the onset of the isoelectric EEG. This was confirmed by Harris et al. (1981), who showed also that Ca_e fell rapidly to levels similar to those seen in ischemia when K_e had risen to 13 m*M*. It was also found that the marked changes in high-energy phosphates seen at the onset of isoelectric EEG (e.g., Agardh et al., 1978; Ghajar et al., 1982) occurred after the ion changes and not before, at least as observed within the areas of ion recording. Similarly, the rapid increase in free fatty acid content, thought to result from raised intracellular calcium levels, also followed the ion changes rather than preceding them (Wieloch et al., 1984).

1.3.4. Mechanisms of Ischemic Cell Death

Of the many proposed mechanisms leading to ischemic cell death in the brain, two specifically implicate calcium and hydrogen ions (pH).

Dead cells always contain much calcium, but it is difficult to prove whether the calcium accumulation is the cause or effect of cell death. Schanne et al. (1979) have produced compelling evidence in favor of a role for calcium. They showed that rat hepatocytes incubated in low-calcium media (around 10^{-7} *M*) revealed low mortality rates in comparison with cells incubated in normal calcium media when both were challenged with membrane-active toxins. Some proposed mechanisms whereby calcium could cause cell damage involve changes in membrane function and energy metabolism. Raised intracellular free calcium (Ca_i) will increase plasma membrane permeability (Meech, 1978) and will increase the concentration of free fatty acids (e.g., Ruszczewski, 1978; Gardiner et al., 1981) through stimulation of phospholipases (Edgar et al., 1982) and other lipolytic enzymes. This condition will seriously affect the functioning of membrane-bound proteins (Sun and Sun, 1976). Increases in Ca_i will also reduce the formation of useful energy as a result of futile cycling of calcium (Peng et al., 1977; Siesjo, 1981).

A fall of pH in anoxia was one the earliest ion changes to be recorded (e.g., Holmes, 1932). However, there is still little evidence supporting a causal link between low pH and ischemic cell death. Friede and Van Houten (1961) showed in cat brain slices that continued glycolysis was associated with histopathological changes. Myers (1981) and Rehncrona et al. (1981) have shown that animals accumulating more than 16–17 μmol/g of lactic acid revealed brain injury, whereas those accumulating less recovered well. In addition, Rosner and Becker (1984) demonstrated that some protection from fluid percussion injury occurs in animals treated with an intracellular alkalinizing agent. Acidosis is proposed to act through effects upon membrane function and enzyme activities. Results from work on hypoglycemia show that cell damage, similar to that found in ischemia, can be produced by 30 min of hypoglycemia, but this is not associated with a fall in pH (Agardh et al., 1980; Pellegrino and Siesjo, 1981). Siesjo (1981) accordingly concluded that pH may exacerbate cell damage, but is not sufficient to determine its extent.

2. General Methods

2.1. Introduction

The methods for measuring cerebral ions fall into five broad groups: the use of ion-selective membranes (glass or ion-exchange resins), isotopes, spectroscopy, X-ray microanalysis, and fluorescence methods.

2.2. Ion-Selective Electrodes (ISE)

2.2.1. Glass Electrodes

Glass membranes were the first to be developed and are still the most selective membranes available for some ions. Currently, glass electrodes are used predominantly for intracellular recordings in vitro in preparations such as the snail neuron. Glass electrodes are also still used for in vivo recordings of cerebral pH_e.

Construction of glass electrodes is achieved in two main designs: the spear-type electrode (Hinke, 1961; Heuser et al., 1975a; Astrup et al., 1976, 1977; Cragg et al., 1977; Siemkowicz and Hansen, 1981; Morris et al., 1983) and the recessed-tip-type electrode (Thomas, 1970, 1974). These and other designs have been described in detail by Thomas (1978).

In constructing the spear-type electrode, a single capillary of borosilicate or aluminosilicate glass is drawn to produce a short shank (the shank is the portion of the electrode from the shoulder, where the barrel starts to get thinner, to the tip). The tip is broken back sufficiently to enable 10–100 μm of the tip of the longer and finer shank of a piece of ion-selective glass tube to pass through. When in position, the inside of the ion-selective glass tube is placed under pressure and a heating coil is advanced toward the point at which the two glasses meet. As the inner glass melts, it is blown outward and fuses with the outer insulating glass. The tip of the ion-selective glass is then sealed, the glass proximal to the seal is removed, and the electrode is boiled in distilled and deionized water to hydrate the glass. The electrode is then filled with an appropriate internal reference solution, after which it is ready for use.

To make the recessed-tip type of electrode, the same procedure as described above is followed, except that the ion-selective glass does not protrude from the insulating glass and is sealed 5–10 μm proximal to the tip.

The glass electrodes are not easy to make; however, once constructed, they are hardy, reliable, and can be used for many experiments. A major disadvantage of the spear type of electrode is that on implantation it is difficult to confirm that all of the ion-selective glass is within the brain. If any part of the ion-selective glass remains in CSF or buffered superfusion fluid, this will influence the electrode response, which is derived from the average of the ion activity surrounding the whole of the exposed surface of sensitive glass. The recessed-tip design, although excellent for recording inside cells, has a slow response time and so can reliably record only steady-state ion levels.

2.2.2. Liquid Membrane Electrodes

Construction of the liquid membrane electrode is more art than science, a fact demonstrated by the profusion of different techniques employed. Detailed descriptions of the construction of single- and double-barreled electrodes of this type have appeared in primary publications, reviews, and symposia (Durst, 1969; Moody and Thomas, 1971; Walker, 1971; Khuri et al., 1972a,b; Vyskocil and Kriz, 1972; Lux and Neher, 1973; Berman and Hebert, 1974; Meier et al., 1977; Thomas, 1977; Sykova et al., 1980; Koryta, 1980). Multiparametric electrodes of this type were first described in 1976 (Silver, 1976; Kessler et al., 1976) and were later used successfully in measurement of ion changes in hypoxia (Silver, 1977a,b; 1978) and in ischemia (Harris et al., 1981). Fujimoto and Honda (1980) have developed a triple-barreled electrode for intracellular use, and Dufau et al. (1982) have described their own version of the extracellular triple-barreled electrode.

The history of ISEs goes back to 1906 (*see* Koryta, 1975), but it was not until 1953 that their miniaturization for extracellular use was described (Sonnenschein et al., 1953), in this case for measuring pH. Much of the subsequent work done on ion changes in cerebral ischemia was performed using surface electrodes (Thorn and Heitmann, 1954; Meyer and Denny-Brown, 1957; Meyer et al., 1962). Later, intracellular glass electrodes were used (*see* Thomas, 1978, for a review); the first liquid ion exchanger was developed (Sollner and Shean, 1964), and highly selective liquid ion exchangers were subsequently introduced for K^+, Na^+, Ca^{2+}, and pH (Walker, 1971; Ammann et al., 1975, 1981; Steiner et al., 1979). The double-barreled electrode, effectively recording both ion and reference potentials at the same place in the tissue and thus eliminating artifacts caused by variable slow potentials across the cortex, was developed by Khuri et al. (1972a) and Vyskocil and Kriz (1972). Triple-barreled electrodes for recording ion potentials from two ion species simultaneously have also been successfully used (Silver, 1978; Harris et al., 1981). Such electrodes provide data that, in conjunction with local measurements of cerebral blood flow by the technique of hydrogen clearance (Pasztor et al., 1973) and selective local tissue sampling, enable the dynamic sequence of ion changes that follow the onset of ischemia or drug infusion (for instance) to be monitored and related to associated flow and metabolic changes occurring at the same place in the tissue. Surface recordings and superperfusion techniques for ion measurement, by comparison, only provide data averaged over a much wider area and longer time. Measurements of total tissue content of particular ions are valuable in delineating overall uptake or loss, but cannot distinguish extracellular from intracellular compartmental concentrations or account for the dynamic fluxes of ions.

The principles of construction of a glass electrode employing liquid ion exchange resins (IER) will be given in detail in section 3.

2.2.3. Solid Membrane Electrodes

In this type of electrode the IER is incorporated into a solid matrix, generally of polyvinyl chloride (PVC), but sometimes of silicone rubber. The membrane is normally about 1 mm in diameter and, consequently, these electrodes are typically used for measurements from the surface of the brain rather than within it.

Detailed descriptions of the production of these membranes have been published by various authors (Fiedler and Ruzicka, 1973; Ruzicka et al., 1973; Hoper et al., 1976; Bard et al., 1978), and details of how to make and use the whole electrode have been published by Crowe et al. (1977) and Strong et al. (1983). Astrup et al. (1977, 1981b) have used a commercially available electrode (Radiometer), suitably modified to incorporate a reference electrode. The commercially available ion electrodes have been reviewed by Meier et al. (1977, 1980).

To make the electrode, the membrane must be fixed to the end of a piece of tubing, normally Pyrex glass or PVC (1 mm id, 2 mm od). This is achieved in two main ways. First, a 1-mm diameter section of the membrane is prepared (cut from the master disk) and glued into its housing using epoxy cement (Araldite, Ciba Geigy) or other adhesives such as silicone rubber or PVC in tetrahydrofuran. Second, a small amount of the master disk is redissolved and either painted onto the end of the glass, or introduced into the end of the glass, where it is left to dry. This latter method results in a fine membrane. The internal filling solution is then introduced using the back-filling method (*see* section 3.1), and electrical contact is made by chlorided silver wires.

Reference electrodes have been made in two ways. Either a chlorided silver wire is wrapped around the active electrode concentrically, down to the tip, or a second glass tube is prepared. The tube is plugged with a porous ceramic cylinder and back-filled with 150 m*M* NaCl.

The complete electrode array may be applied to the measurement site in three ways; by fixing it in position with cement, using a guide cannula (Crowe et al., 1977; Mayevsky, 1978), by holding it in place with a clamp, or by suspending it from a spring system (Strong et al., 1983).

2.2.4. Verification of Electrode Function

Before using an ISE in vivo it is necessary to confirm in vitro that it has the appropriate time constant and selectivity.

The time constant of electrodes is measured using a step change in the primary ion concentration. This is achieved either by adding a large volume of concentrated solution to a small volume of low concentration, or using an overflow system in which a small chamber, fed from underneath, is flushed through from one concentration to another. Under dynamic conditions, electrodes can show a change in calibration in comparison to static conditions. The electrode time constant is of the order of 100 ms for glass and liquid membrane electrodes, except recessed tip designs, and 2–3 s for solid membrane electrodes.

The selectivity of the IER is very important in determining the accuracy of any recording. For instance, a sodium IER is available that is only suitable for intracellular use, in which calcium levels are very low, but is unsuitable extracellularly, where the calcium is much higher. The EMF output of an ion-selective membrane is generated across the IER and is proportional to the activity of the primary ion in the test solution. However, the presence of additional ions is accounted for by the Nicolsky-Eisenmann equation (Meier et al., 1977)

$$\mathrm{EMF} = E_o + s\log\left[a_i + \sum_{j \neq i} K^{\mathrm{pot}}_{ij}[a_j]^{z_i/z_j}\right] \tag{1}$$

where E_o is the intercept (see below), s is the slope that is defined according to the Nernst equation as 2.303 RT/z_iF, a_i and a_j are the activities, and z_i and z_j the charges, of primary and interfering ions, respectively, and k^{pot}_{ij} is the selectivity coefficient of the ions. The intercept E_o is given by

$$E_o = E^o_i + E_R + E_D \tag{2}$$

where E^o_i is the potential difference between the internal filling solution and the IER, E_R that between the metallic leads and the two reference solutions, and E_D the liquid junction potential between the reference electrolyte and the test solution. E^o_i and E_R are constant for any given ion-selective membrane, whereas E_D will alter with different test solutions.

In practice, once the selectivity of the IER has been proved sufficient to allow the influence of interfering ions to be ignored, the EMF may be considered to be defined by the Nernst equation

$$\mathrm{EMF} = E_o + s\log(a^o_i/a^i_i) \tag{3}$$

where a^o_i and a^i_i are the activities of the primary ion in the test and reference solutions, respectively.

There are two main methods of assessing the selectivity coefficient; the single-solution method and the fixed-interference method. Both of these have been described in detail (Meier et al., 1977; Simon et al., 1978; Ammann et al., 1979; Guilbault, 1979; Lee,

1980). Briefly, the single-solution method requires measurement of the electrode EMF in chloride solutions of each ion at graded concentrations, one ion at a time. The values are then inserted in an equation that yields the selectivity coefficient. The fixed-interference method requires the plotting of EMF against primary ion activity measured with the interfering ion at a fixed activity; the resulting activity values are then inserted in the appropriate equation to obtain the selectivity coefficient. There remains, however, some doubt about the validity of these two methods (Tsien and Rink, 1980; Harris, 1985); for instance, it is known that IERs react less to interfering ions in constant ionic strength solutions than in single-salt solutions (Kriz and Sykova, 1980). An improved method has been described by Harris (1985).

2.2.5. Comparison of Glass, Liquid and Solid Membrane ISEs

The choice of measurement technique will, of course, be determined by the type of recording desired. For instance, although glass and liquid membrane electrodes are used routinely for the measurement of cytoplasmic ion activities in, for example, invertebrate preparations (Thomas, 1970, 1974, 1976, 1977, 1978; Aickin and Thomas, 1975; Ellis and Thomas, 1976; Evans and Thomas, 1983), they are not used in vivo for intracellular recordings; these electrodes cannot be used to give values for ion activities in subcellular organelles. In terms of selectivity, pH-sensitive glass is better than the available resins, although the pH resin is excellent and more than adequate for biological use; but the other glasses are not as good as available resins. There is little to choose between solid and liquid membranes regarding selectivity, since the same resins are used.

The effective volume in which recording actually takes place varies considerably among different electrodes, ranging from a pocket of extracellular fluid of 2–30 μm diameter for liquid membrane electrodes, through one of 20–200 μm diameter for glass electrodes, to 1 mm diameter for pial surface electrodes constructed from solid membranes. To record rapidly occurring ion changes, diffusional distances between cells and electrode should be small enough to ensure that the associated diffusional delays are small compared to the lowest component time constant in the ion signal. Solid membrane electrodes effectively make a spatial average over a large area and so are more appropriate for steady-state recordings of changes occurring uniformly over homogeneously reacting areas.

The response time of liquid membrane electrodes is of the order of 0.1 s, whereas for solid membrane electrodes it is 2–3 s, and up to several minutes for recessed tip glass electrodes. In the case of recordings made at the surface of the brain with solid membrane electrodes, the pia acts as a further barrier to diffusion, slowing and averaging ion changes, as well as reducing peak values (Astrup et al., 1979).

The glass membrane electrode is the most reliable, and can be used over many months. The solid membrane electrode lasts over a period of weeks, whereas the liquid membrane electrode lasts for only a matter of days.

Both glass and liquid membrane electrodes are invasive to a degree dependent upon the tip size, whereas the solid membrane electrodes are placed on the tissue surface and are thus relatively noninvasive. Solid membrane electrodes can be sealed in the skull without risk of damage to electrode or animal and have been used in chronic experiments.

One great advantage of liquid membrane electrodes over the other two types is the ease with which multibarreled electrodes can be made. Such electrodes enable the response of at least two ionic species to be measured simultaneously in response to a stimulus or ischemic insult, the changes occurring in the same pocket of ECF. A further advantage of double- or multi-barreled glass and liquid membrane electrodes is that the active and reference electrodes lie at the same point in the tissue. Signal amplification uses a differential configuration (see below) so that any separation of active and reference electrodes would tend to produce artifactual ion changes from any locally generated dc potential shifts.

The value of K_e in normal brain is about 3.5 m*M*, ranging from 2.7 to 9.0 m*M* in the literature (Vyskocil et al., 1972; Lux and Neher, 1973; Prince et al., 1973; Heinemann and Lux, 1975; Heuser et al., 1975a; Lothman et al., 1975; Dora and Zeuthen, 1976; Mutsuga et al., 1976; Astrup et al., 1976, 1977, 1979; Branston et al., 1977; Nicholson et al., 1977, 1978; Hansen, 1981; Harris et al., 1981; Siemkowicz and Hansen, 1981; Takahashi et al., 1981; Hubschmann and Kornhauser, 1982; Urbanics et al., 1982; McCreery and Agnew, 1983; Harris and Symon, 1984).

Resting levels of Ca_e are more variable than those of K_e, averaging around 1.0 m*M*, with a published range of 0.3–2.2 m*M* (Heinemann et al., 1977; Nicholson et al., 1977, 1978; Hansen and Zeuthen, 1981; Harris et al., 1981; Siemkowicz and Hansen, 1981; McCreery and Agnew, 1983; Harris and Symon, 1984).

Extracellular pH measured using a liquid IER yields resting levels in the range of 7.2 to 7.49 units (Kraig et al., 1983; Mutch and Hansen, 1984; Harris and Symon, 1984). Results using glass membrane electrodes with varying tip diameters give an average pH_e of about 7.3 units (Heuser et al., 1975a; Astrup et al., 1976, 1977; Betz, 1977; Silver, 1977a,b; Hansen, 1981; Siemkowicz and Hansen, 1981; Urbanics et al., 1982).

Values of Mg^{2+}, measured intracellularly in invertebrate neurons by means of liquid membrane ISEs, have been reported as averaging 0.76 m*M* (Alvarez-Leefmans et al., 1983).

2.3. Spectrophotometry

Atomic flame emission spectrophotometry (e.g., *see* Williams and Wilson, 1981) has long been used to measure absolute levels of sodium and potassium in pieces of excised brain (West and Matsen, 1972; Zimmermann and Hossmann, 1975; Watanabe et al., 1977; Pappius, 1979; Schuier and Hossmann, 1980). The methods have been described in detail (Bourke et al., 1965; McBroom et al., 1971; McDonald et al., 1977) and are now standard. Essentially, the tissue must be solubilized in strong acid or alkali and diluted appropriately before analysis in standard apparatus. Tissue calcium cannot be measured accurately with emission photometry because the signal from the low levels of calcium is swamped by those from the high levels of sodium and potassium.

Atomic absorption spectrophotometry is a more sensitive technique that has been used to measure changes in brain calcium levels (Yanagihara and McCall, 1982). The samples for atomic absorbtion spectrophotometry are prepared in the same way as for flame photometry (McBroom et al., 1971).

Spectrophotometric techniques can, by definition, only be used to make one measurement in an experiment, and they average the total ion content, both free and bound, within each sample. These techniques are primarily suited to analyzing the gross movements of ions into and out of tissue.

2.4. Electron Probe X-Ray Microanalysis

X-Ray microanalysis is used to make qualitative or quantitative measurements of elemental concentrations in biological material ranging in size from blocks to subcellular organelles. To record concentrations in subcellular organelles, the method uses ultrathin sections of the material to avoid the problems associated with homogenization and centrifugation normally used to isolate these organelles. The technique is based upon the fact that when material is bombarded with a stream of high-energy electrons, an X-ray spectrum is emitted that is characteristic for the quantity of the ions present (Lechene, 1980; Gupta and Hall, 1981; Moreton, 1981; Morgan, 1984). The method measures the total of each ion present, whether in free or bound form. Correct tissue sampling and section preparation are vital. Samples must be rapidly frozen to avoid ionic redistribution across membranes, and the subsequent procedures must be carefully controlled to retard both morphologic change and ionic redistribution (Morgan et al., 1975; Lenglet et al., 1984).

Qualitative analysis is normally performed using the scanning mode on tissue blocks, whereas quantitative analysis requires ultrathin sections of the material for either scanning/transmission or transmission-alone analysis. Cryoultramicrotomy has been described (Roomans et al., 1982) with details of the use of cryoprotectants, sectioning of protected and unprotected tissue at temperatures between –40°C and –120°C to achieve sections from 16 μm to 70 nm in thickness, the freeze-drying of cryosections, the effects of freezing generally, and alternatives to cryoultramicrotomy. The calibration methods required for quantitative analysis, described by Kendall et al. (1985), are based upon the work of Hall (1979) and Hall and Gupta (1979), and use gelatin standards.

Although X-ray microanalysis has been used for many biological applications (*see* Roomans et al., 1982), direct application to brain tissue has not, so far, been extensive. Zs-Nagy et al. (1977) described a technique for quantitative scanning mode analysis of large samples taken from rat brain, and Cameron et al. (1978) began the quantitative analysis of ionic distributions between cell types and subcellular organelles.

2.5. Isotope, Fluorescence, and Other Methods for Measuring pH

In clinical applications, extracellular pH (pH_e) may be measured from samples of lumbar or cisternal CSF (Posner and Plum, 1967; Siesjo et al., 1972). This technique gives, at best, an estimate of global pH_e. The pH_e of CSF follows tissue pH changes only slowly, in lumbar samples more so than in cisternal, and the CSF values are very likely underestimates of the extent of regional cerebral tissue changes.

The many approaches to the measurement of intracellular pH have been recently reviewed (Roos and Boron, 1981). Six methods will be outlined here, three of which may be used for repeated measurements in vivo: positron emission tomography (PET), ^{31}P nuclear magnetic resonance spectroscopy (^{31}P-NMR), and umbelliferone fluorescence. The other methods are ^{14}C-labeled 5,5-dimethyl-2,4-oxazolidine dione (^{14}C-DMO) autoradiography, neutral red spectroscopy, and biochemical analysis based upon the bicarbonate/carbonic acid equilibrium. These last three methods require either sectioning or discrete sampling of the brain; umbelliferone fluorescence may also be used in this way.

All the methods mentioned above, except for biochemical analysis, measure predominantly the intracellular pH (pH_i). From a study of serial sections, the pH changes within many different brain structures may be assessed. The methods effectively produce a spatial average of pH, pH_i being calculated after allowances have been made for the influence of pH_e based upon assumptions for both the actual pH_e and the size of the extracellular space at the time of measurement. These and other assumptions used in the calculation of pH_i must be taken into account when interpreting and discussing the data obtained.

PET is a complex and sophisticated technique with which recordings may be made from any part of the brain. Its use is restricted to those groups of investigators having the required local facilities for isotope generation. PET scanning has been the subject of many reports and reviews (Jones, 1980; Phelps et al., 1982; Rhodes et al., 1983; Lammertsma, 1984; Wise et al., 1984). The technique has been used to measure pH changes in a variety of situations using $^{11}CO_2$ (Syrota et al., 1983; Buxton et al., 1984; Rottenberg et al., 1984; Brooks et al., 1984) and ^{11}C-DMO (Berridge et al., 1982). ^{31}P-NMR spectroscopy, also an expensive technique, has been used to record hemispheric changes of pH and high-energy phosphates during unilateral ischemia in the gerbil (Thulborn et al., 1982). ^{14}C-DMO has been used with quantitative autoradiography to determine tissue pH in various structures of cryostat sections of the rat brain after middle cerebral artery occlusion (Kobatake et al., 1984).

Umbelliferone fluorescence has been used for in vivo recording of brain surface pH changes (Sundt et al., 1978; Sundt and Anderson, 1980; Anderson and Sundt, 1983) and for measurement of pH within brain structures on cryostat sections (Welsh et al.,

1982; Csiba et al., 1983). Kogure et al. (1979, 1980) have used neutral red spectroscopy to measure pH in brain structures on cryostat sections.

The bicarbonate/carbonic acid equilibrium has been utilized to measure intracellular pH in samples of brain tissue under a variety of experimental conditions (Siesjo and Messeter, 1971; Siesjo et al., 1972; Pellegrino et al., 1981; Mabe et al., 1983; Siesjo et al., 1985). pH_i is calculated from a modified Henderson-Hasselbach equation, and the complete procedure, using a minimum of assumptions, has been described by Siesjo et al. (1985).

2.6. Other Methods for Measuring Ca_e

The localization of ^{45}Ca, using autoradiography, has been used to monitor routes of entry, accumulation, and retention of calcium after an ischemic insult (Dienel, 1984). Meldrum and coworkers (Meldrum, 1981; Griffiths et al., 1982, 1983, 1984; Simon et al., 1984) have used the oxalate-pyroantimonate technique to localize deposits of calcium in brain tissue. Griffiths et al. (1983) have given details of the method. Animals are perfusion-fixed with neutralized 3% glutaraldehyde/90 m*M* potassium oxalate. Brain slices (50–80 nm sections mounted in epoxy) are prepared and stained with 0.5% uranyl acetate for 8 min and 0.4% lead citrate for 10 min, the calcium then appearing as electron dense deposits under the electron microscope. That the deposits are calcium has been confirmed by depleting the calcium and thereby the deposit frequency with a chelator, and by X-ray microanalysis (Griffiths et al., 1983); but these tests do not address the problem of how much, if any, calcium is lost during the procedure (Morgan et al., 1975; Morgan, 1984).

3. Making and Calibrating a Common Type of Extracellular Ion-Selective Microelectrode

This section will deal specifically with the manufacture and calibration of liquid ion-exchanger microelectrodes, together with the associated electronics and recording techniques. Examples of recordings obtained using electrodes of this design are given in section 4.

3.1. Manufacture

A method for making triple-barreled double ISEs will be described; it can obviously be adapted to produce a double-barreled electrode, one with a smaller tip diameter, or to measure different pairs of ions. The electrodes are reliable and have been used successfully in many experiments.

The electrodes are made from borosilicate glass, 2.0 mm od and 1.16 mm id (obtained in the UK from Clark Electromedical). The glass tubes are first cleaned by drawing acetone through them for extracellular electrodes, or fuming nitric acid followed by distilled water washes for intracellular electrodes. The barrels are then stuck together using cyanoacrylate glue, and the electrode is pulled in a conventional vertical electrode puller (CF Palmer, UK) by heating it for 25–30 s, turning it through 180° (to ensure that the tips are pulled together), and pulled using the solenoid to produce shanks of length 1.0–1.5 cm. The lower half of the blank is discarded since the filling capillary cannot reach the tip because of the twist.

The upper half is kept and the tip broken back, under magnification, until its total diameter is 10–15 μm. At this stage, the electrode is ready for silanization by the saturated vapor method, using either dimethyldichlorosilane or trimethylchlorosilane. This renders the active barrels hydrophobic, while leaving the reference barrel hydrophilic (Munoz et al., 1983). Silanization ensures that the IER will remain at the tip, and reduces the electrical leakage between the test solution and the internal reference solution along the path between the IER and surrounding glass. Selectivity is also improved by good silanization. One of the short barrels is chosen as the reference barrel, marked, and plugged with a small piece of putty or wax to keep out silane. A small amount of dimethyldichlorosilane (Hopkin and Williams, BDH, in the UK) is poured into a conical flask, the top of which is covered with two layers of parafilm secured with a rubber band. Three holes are cut through the parafilm and the electrodes suspended, tip uppermost, through the holes. A stream of air is passed over the electrodes for 60–90 s while the flask is gently warmed. The shoulder of the electrode array is then strengthened, using a small amount of epoxy cement, and the electrode placed in a wooden block and heated in an oven at a temperature above 100°C for at least 20 min.

A ball of epoxy cement with a diameter of about 1 mm is then applied near the tip, 0.7–1.0 mm from the tip if the electrode is to be

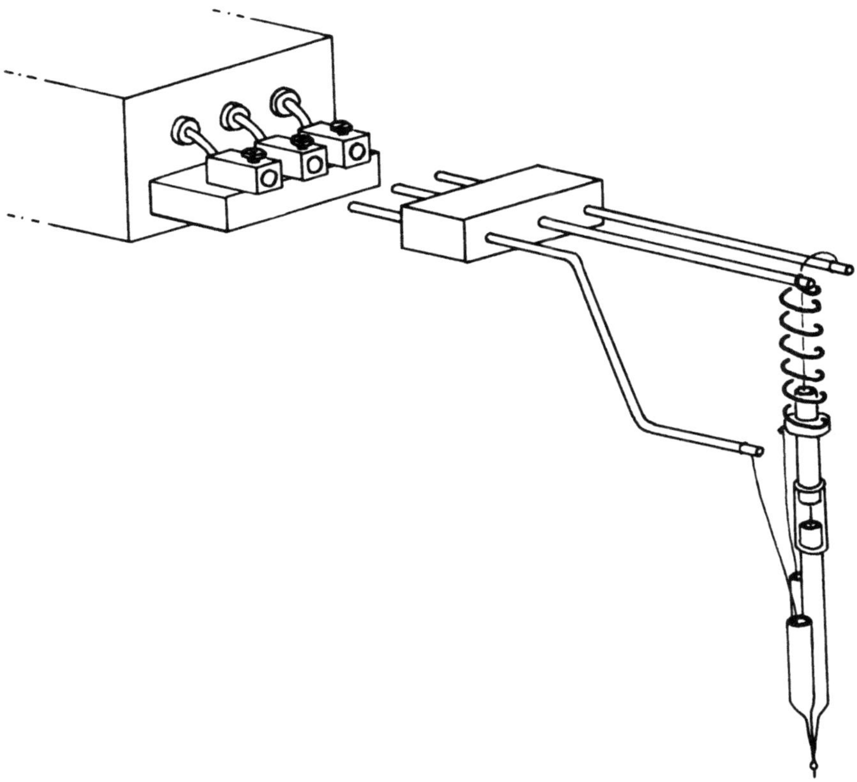

Fig. 1A. Diagram of suspension system for triple-barrelled ISEs. At upper left is the box housing the preamplifier into which the three horizontal wires shown are inserted and fixed by means of small screw terminals, as shown. These wires, spaced apart by about 8 mm, are made of stout copper and (from left to right) carry the first ISM, reference, and second ISM input signals to the preamplifier. The block on which the screw terminals are mounted and the central reinforcing block are both made of polytetrafluorethylene (PTFE, Teflon). The spring soldered to the end of the middle wire is made of 125 μm insulated copper wire; its lower end is glued to a piece of glass tube (2 mm od) that fits inside a short length of rubber tube and takes the weight of all three barrels. The spring forms the electrical connection for the reference barrel.

used for recording from primate cortex and 0.3–0.5 mm if for rat experiments. The purpose of the ball (Fig. 1) is to limit and define the depth of insertion of the electrode in brain tissue and to distribute any weight applied to the tissue by the spring suspension (see below).

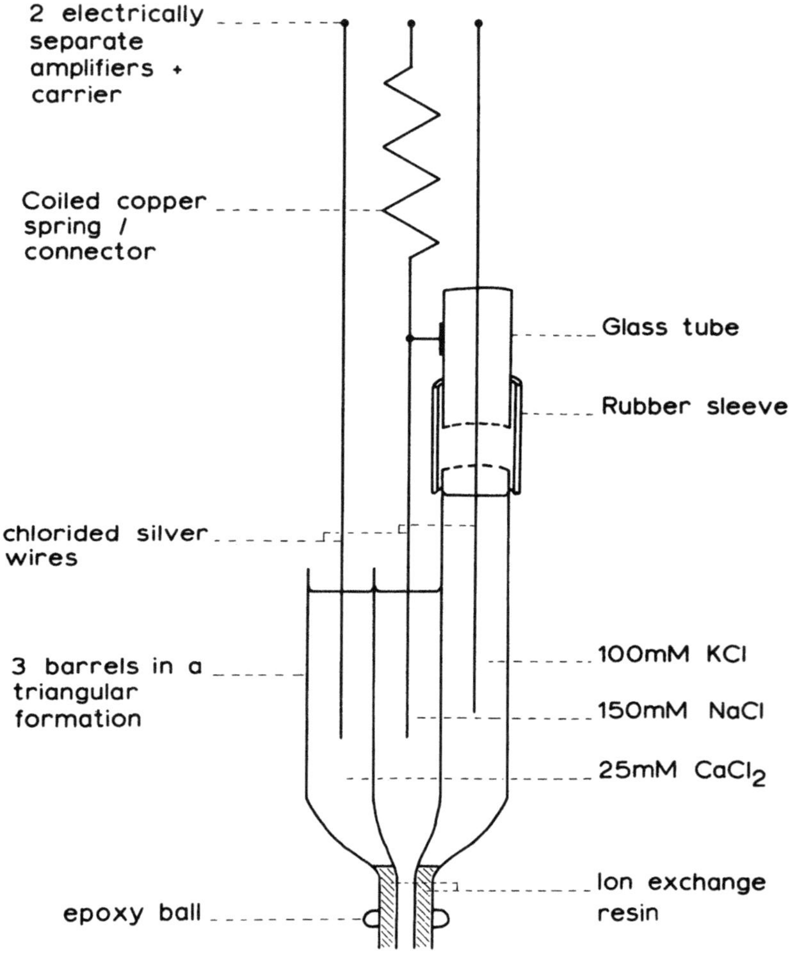

Fig. 1B. The triple-barreled ISE in close-up view, with distortion of the dimensions of the various components to clarify the details of construction.

The electrode is filled by the "back-fill" method using fine tubes drawn from the same tubing as the electrodes, but with a shank length of 4 cm. First, a small amount of the IER is introduced into the tip of one active barrel. The IER used is Corning 477317 for K^+ (Walker, 1971), ETH 1001 in nitrophenyl octyl ether for Ca^{2+} (Oehme et al., 1976), and tridodecylamine for pH (Amman et al., 1981). This is then followed by the internal reference solution, which should be 100 m*M* KCl for K^+ and 25 m*M* $CaCl_2$ for Ca^{2+}, whereas for pH a special buffer is used consisting of 10 m*M* 2-(*N*-morpholino)-ethanesulfonic acid in 150 m*M* NaCl/3 m*M* KCl, brought to a pH of 6.7 with 1.0 m*M* NaOH.

Next, the other active barrel is similarly filled, and finally the reference barrel is filled with isotonic saline. During filling, the electrode is held in a jig and the filling capillary advanced using a standard three-dimensional micromanipulator.

The filled electrode is suspended by a spring assembly, which is shown in detail in Fig. 1. The differential preamplifers (Fig. 2) are mounted directly above the electrode and their outputs are connected by cables to the remaining circuitry shown in Fig. 2.

3.2. Calibration

To calibrate the electrode, its tip is immersed in solutions close in composition and ionic strength to that of CSF, and a plot made of output voltage (relative to the reference barrel) against the logarithm of ionic concentration as indicated by the basic Nernst equation. K^+ electrodes have linear calibration curves at K^+ concentrations of greater than 6 m*M*. Below this value, there is a loss of selectivity caused by interference from Na^+ ions. Ca^{2+} electrodes have been shown to have a linear calibration over the concentration range found extracellularly (Ammann et al., 1979; Tsien and Rink, 1980), and so calibration over the range of 0.25–2.5 m*M* is sufficient with extrapolation, when necessary, of the calibration curves to values outside this range. pH electrodes have also been shown to possess a linear calibration over the ranges expected experimentally (Ammann et al., 1981), so that calibration over the range of 6.3–7.5 pH is again sufficient.

For in vivo measurements, the electrodes are lowered into brain tissue, using micromanipulators, through a small hole made in the pia. Both the head of the animal and the manipulator must be firmly held to minimize vibration-induced movement; such movement will undoubtedly increase tissue damage and may produce artifactual ion changes.

In the experiments described below, the electrodes were implanted into brain close to two or three other electrodes designed for recording cerebral blood flow using the technique of hydrogen clearance (Pasztor et al., 1973). This association brings out an important feature of these ISEs: their capacity to generate accurate recordings of the changes of up to two ionic species as a result of a specific physiological change, in this case an ischemic insult characterized by focal reductions in tissue blood flow.

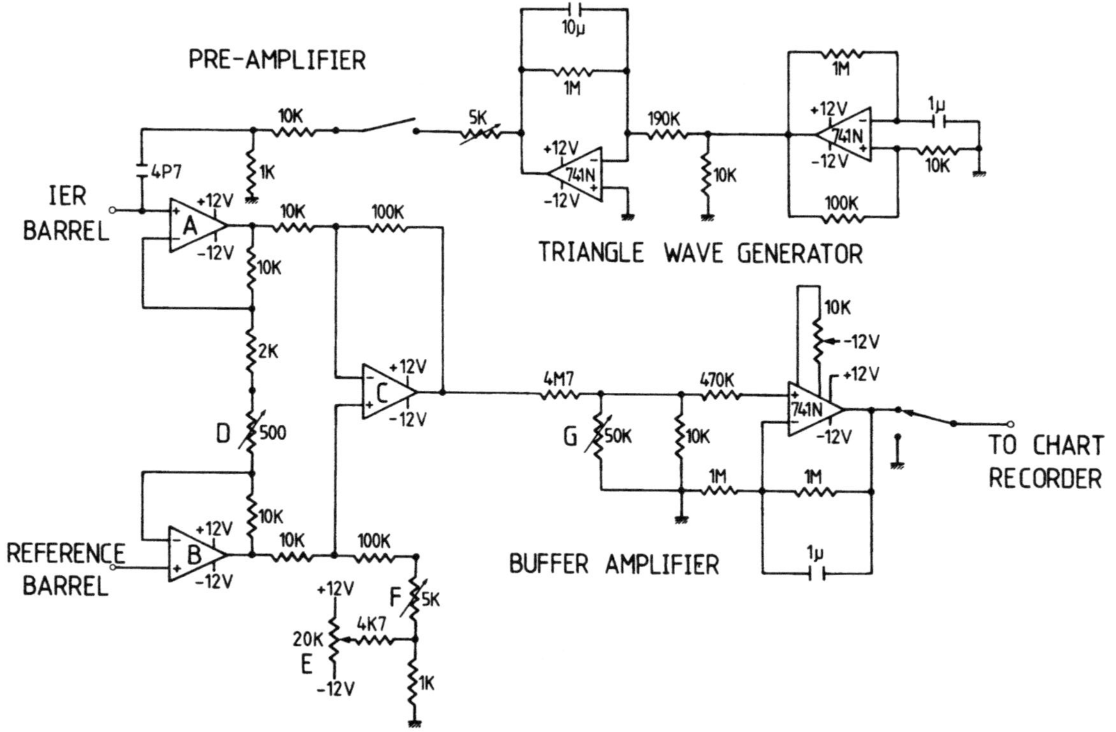

Fig. 2. Circuit diagram of an amplifier system for use with ISEs. For a triple-barreled ISE, two independent preamplifiers and buffer amplifiers would be required, with the reference barrel signal common to both. A and B are the input operational amplifiers; suitable ones are AD515L and LM308, respectively. Amplifier C can be a 741N. Variable resistors D, E, F, and G are used to adjust preamplifier gain, output offset voltage, common mode rejection, and overall system gain, respectively. Adjustment of F is most important.

3.3. Electrode Signal: Generation, Amplification, and Recording

An ion-selective microelectrode, as described above, is essentially an ion-selective barrier (ion-exchange resin) confined by glass together with an internal reference solution and an external test solution that is physically identified with the extracellular space. The signal voltage generated by the electrode system is given, theoretically and under ideal conditions, by the Nernst equation. However, the insertion of any electrode into the brain creates an artificial space in the tissue that is an extension of, but inevitably much larger than, the true extracellular space. This artificial extracellular space presents an additional diffusional path for ions moving between the real extracellular space of adjacent normal tissue and the tip of the microelectrode, and it is important to consider to what extent the dynamic changes in ionic activities are distorted before they actually reach the electrode and are transduced into signal voltages.

The artificial space surrounding the electrode is probably up to three times the electrode diameter (Herz et al., 1969; Neher and Lux, 1973; Lothman et al., 1975; Astrup et al., 1976), and the average diffusion coefficient of an ion through this space will have a value somewhere between that associated with relatively unrestricted movement through aqueous solution and that for the intact tissue. If we consider the K^+ ion alone and suppose, as the worst case, that diffusion is through cerebral cortex with a coefficient $D = 3.9 \times 10^{-6}$ cm^2/s and that the time for an ion to diffuse along a distance r meters is $t = r^2/2D$ (Lux and Neher, 1973), we obtain a value for t of 187 ms if the value of r is taken as 12 μm (a reasonable asumption for the radius of the dead space around the electrode). Similar considerations will apply to other ionic species. This time lag introduced by diffusion is several times less than the time constants associated with the fastest ionic changes observed experimentally (Fisher et al., 1976; Nicholson and Kraig, 1981; Harris et al., 1981). With careful electrode construction and insertion, then, ionic fluctuations should be essentially unaffected by the presence of the artificial extracellular space.

The initial task of the signal amplification system is to subtract the potential generated by the reference electrode from that of the associated active electrode (the ISE). Figure 2 shows the differential amplifier forming the first part of a circuit developed by us for this purpose and used routinely in this laboratory (Harris et al., 1981). The design takes into account the very high resistance of the ion electrodes, which is typically of the order of 10^{10} Ω for K^+, Ca^{2+}, and H^+ electrodes. The input resistance of the amplifier A to which the electrode is connected should be at least 100 times greater than this. The reference barrel, containing only electrolyte and no IER, has a much lower resistance and so the specifications of the second amplifier B are not so stringent.

Common mode rejection (the subtraction of unwanted dc or ac signals common to both amplifier inputs) is optimized by adjusting the gains of the two amplifiers to be equal; this is achieved by connecting both inputs to a square wave voltage source (e.g., 4 V peak, 10 Hz) and adjusting the variable resistor F until the signal observed using an oscilloscope at the output is minimum. The output offset voltage (that present at the output when the inputs are the same) may be adjusted to zero using the variable resistor, E. There is also provision for injecting a small current of square waveform into the ISE (from a triangular waveform voltage generator via a small capacitor, as shown in the top part of Fig. 2), in order to measure its resistance; the amplifier output voltage will be proportional to the voltage generated across the ISE and thus proportional to the electrode's resistance. The amplitude of this output may then be compared with that observed when the ISE is replaced with a high resistor of known value.

The whole differential amplifier shown in Fig. 2 should be mounted as close to the electrodes as possible to minimize stray capacitance and noise pickup. The overall gain of the preamplifier is adjusted to be 100, using the variable resistor, D. The output is connected to the input of the buffer/filter amplifier shown at the right of Fig. 2, the purpose of which is to filter out unwanted ac noise before the signal is displayed or otherwise recorded. The time constant associated with the filter shown is 500 ms. A simple resistive attenuator, G, is used to match the output to the chart recorder. If detailed measurements are to be made from the chart recorder traces, it is advisable to use a recorder with a pen traverse of at least 6 inch, such as the Rikadenki 6-channel (model 360B), the 10-channel flatbed (model RIX), or equivalent.

4. Use of Ion-Selective Electrodes in Studies of Focal Ischemia

4.1. Introduction

A widely used general experimental protocol involves the production of a focal ischemic lesion in the cerebral cortex, and the procedures described below illustrate the use of ISEs in such experiments. The techniques are obviously applicable to many other experimental formats.

First, it is worth comparing the different effects produced by complete and incomplete ischemia. In complete ischemia, produced for instance following cardiac arrest, tissue pO_2 rapidly falls to zero. The sequence of changes in extracellular ions, recorded with ion-selective microelectrodes during the first three minutes of ischemia in the rat, is illustrated in Fig. 3 and is a summary of typical results obtained by many workers (e.g., Heuser et al., 1975b; Hossmann et al., 1977; Nicholson et al., 1977; Hansen and Zeuthen, 1981; Urbanics et al., 1982). Immediately blood flow ceases, there is a steady and pronounced decrease in pH_e, which reaches a plateau at about 6.5, an accompanying slow increase of K_e to about 10 m*M*, and slight shifts in other ion activities and in the extracellular dc potential. When pH_e reaches its plateau, K_e rapidly rises to 60 m*M* and there are similar abrupt and major shifts in the dc potential and other ion levels. These changes are related to the rapid reduction of high-energy phosphates, starting at the onset of ischemia, which is almost complete at the time of the major ion shifts.

With incomplete ischemia, however, the residual blood flow provides a continuing, although reduced, supply of oxygen and metabolic substrates such as glucose, and allows some clearance of the products of metabolism, notably lactate, to occur from the tissue. As a result, the degree of metabolic derangement and movement of ions at a given place in the tissue will depend on the severity of ischemia occurring there. A focal ischemic lesion, such as that produced by occluding the middle cerebral artery, is characterized by a central region of dense ischemia, in which the major changes described above will have occurred, surrounded by regions of less dense ischemia—including the so-called ischemic penumbra discussed in section 1—in which changes in ion activities may be minimal.

4.2. Examples of Measurements

The first example of measurements made with ISEs in ischemia is taken from data summarized in Harris et al. (1981). In this study, K_e and Ca_e were measured in the cerebral cortex at a total of 12 sites in six baboons anesthetized with alpha-chloralose. As part of the experimental protocol, the middle cerebral artery was occluded on one side, and Fig. 4 illustrates the effects on the

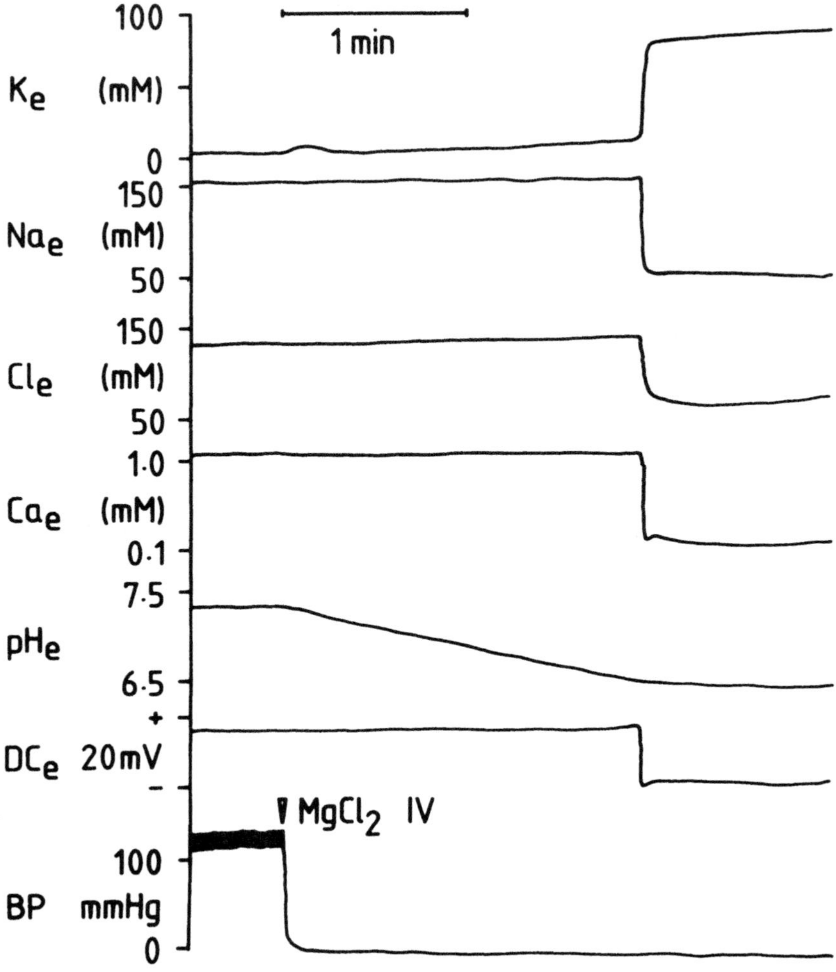

Fig. 3. Illustration of the sequence of changes in extracellular ions recorded with ISEs during the first few minutes after arrest of the cerebral circulation in the rat. DC_e is the extracellular steady potential recorded by the reference barrel relative to a remote common electrode.

extracellular ion activities. There was an initial slow increase in K_e followed by a sharp rise to a plateau of about 50 m*M*; this has also been described by other workers (Astrup et al., 1977; Branston et al., 1977) and, in this case, occurred when local blood flow fell to 9 mL/100 g/min, as measured with adjacent cortical hydrogen clear-

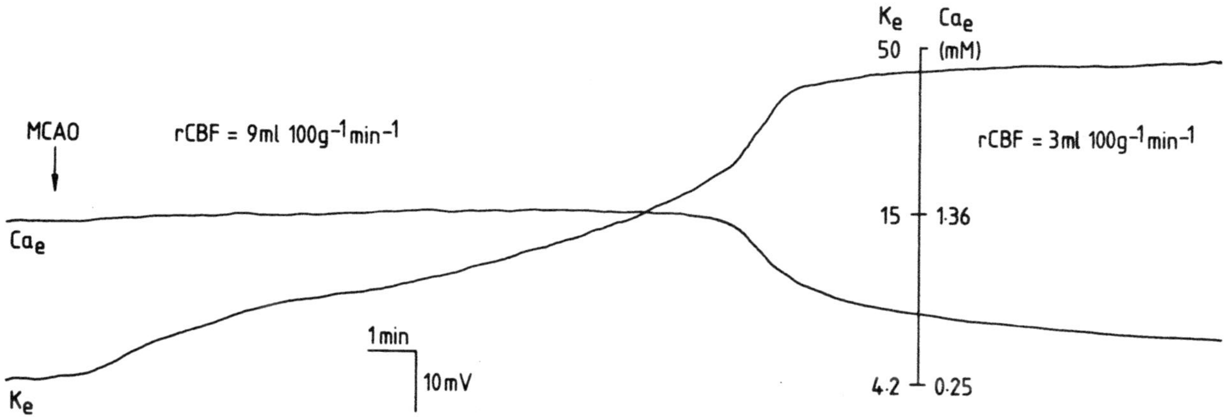

Fig. 4. The effect of middle cerebral artery occlusion upon K_e and Ca_e. When local regional cerebral blood flow (rCBF) is reduced to the critical level shown, there is a steady rise in K_e followed by a fast rise at the time when Ca_e falls. The trace shows the amplifier output voltage levels as recorded; the scale of molarities has been added to aid interpretation.

ance electrodes (Pasztor et al., 1973). Ca_e, on the other hand, was essentially unaffected during the slow rising phase of K_e, but when K_e reached a mean value of 13.4 m*M*, decreased abruptly to a mean value of 0.28 m*M*. Such results indicate that, although K^+ ions certainly move first, the major K^+ and Ca^{2+} shifts occur at the same time and at the same density of ischemia.

The second example (Figs. 5 and 6) is taken from Harris and Symon (1984), in which the relationships between changes in pH_e, K_e, Ca_e, and the local dc potential were studied in the rat cerebral cortex subjected to progressive ischemia. Triple-barreled microelectrode arrays, to measure two ions simultaneously, were used in conjunction with locally placed flow electrodes. With bilateral carotid artery occlusion followed by progressive controlled exsanguination to reduce blood flow (Fig. 5), K_e increased slowly without appreciable change in Ca_e and was critically dependent on local blood flow with a subsequent major shift in both ions when flow reached a critical threshold, as in the baboon experiments outlined above. However, the results obtained in the rat cortex differed from those in the primate in two major ways. First, the average value that K_e had to attain before the major ion shifts occurred was 13 m*M* in the primate, but only about 6 m*M* in the rat. Second, K_e in the rat began to rise rapidly before the fall in Ca_e, whereas in the primate the two events could not be distinguished in time. These interesting species differences may relate to the higher metabolic rate of the rat (Siesjo, 1978).

In six animals, K_e and pH_e were recorded together at the same site; an example of the concurrent changes in these variables, together with the dc potential, is shown in Fig. 6. Unlike Ca_e and the dc potential, but like K_e, the value of pH_e was labile during the initial phase of ischemia and varied with blood pressure. After the onset of the major shift in K_e and dc potential, there was a brief alkalotic shift (averaging 0.14 pH units), followed by an acidosis to levels of pH averaging about 6.5. The acidosis is very likely caused by the accumulation of lactic acid and other acidic products as a result of continued, although greatly reduced, substrate supply. The transient alkalotic shift has been noted by several workers (e.g., Kraig et al., 1983; Mutch and Hansen, 1984) as preceding an ischemic acidotic shift. It has not yet been satisfactorily explained; but since it is abolished by Mn^{2+}, a calcium channel blocker, it may be mediated by Ca^{2+} (Kraig et al., 1983).

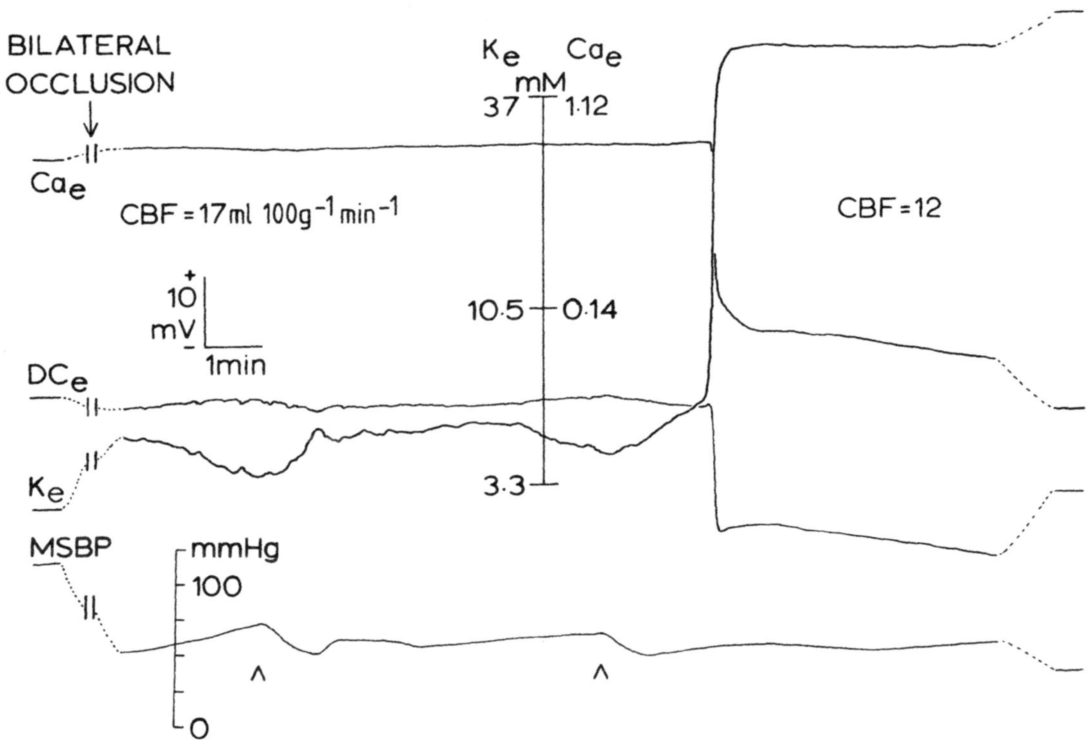

Fig. 5. The effect of bilateral carotid artery occlusion and hypotension on K_e, Ca_e, and DC_e in the rat. The bottom trace is the mean systemic blood pressure. Arrowheads indicate the start of periods of exsanguination.

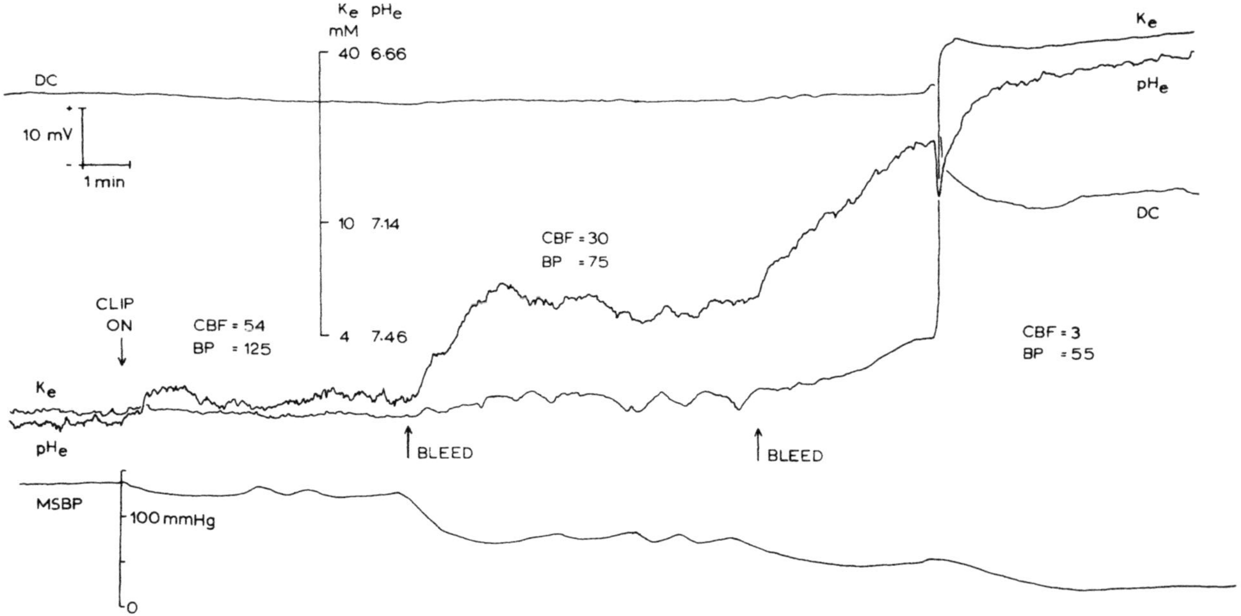

Fig. 6. Results obtained using a similar rat preparation to that used to obtain the data of Fig. 5, but with pH_e being recorded instead of Ca_e.

5. Conclusions

The techniques for the measurement of cerebral ions are both many and varied, giving the experimenter the capability of recording over the entire range of regional sizes from subcellular organelles to the whole brain. It is likely, therefore, that for any study in which the measurement of cerebral ions is indicated, a suitable method will be available. However, it will still be important to establish at the outset that the method chosen is appropriate in terms of factors such as adequate selectivity between ionic species and electrode response time (if ISEs are to be used).

In interpreting experimental and clinical results, a distinction should be drawn between the properties of intracellular and extracellular ions, since the concentrations and roles of ions within the cell are quite different from those outside. Also, whenever a redistribution of ionic species is observed to occur between tissue compartments, it is not always easy to see to what degree the ionic shifts and associated intracellular changes are causative, as distinct from effects, of the pathophysiology and sequelae of energy failure and irreversible cell damage. The references given below should be helpful in the interpretation of results, as well as providing sufficient further details to enable any of the techniques described in this chapter to be pursued.

References

Agardh C-D., Kalimo H., Olsson Y., and Siesjo B. K. (1980) Hypoglycemic brain injury. 1. Metabolic and light microscopic findings in rat cerebral cortex during profound insulin-induced hypoglycemia and in the recovery period following glucose administration. *Acta Neuropathol.* (Berl.) **50**, 31–41.

Agardh C-D., Folbergrova J., and Siesjo B. K. (1978) Cerebral metabolic changes in profound, insulin-induced hypoglycemia, and in the recovery period following glucose administration. *J. Neurochem.* **31**, 1135–1142.

Aikin C. C. and Thomas R. C. (1975) Microelectrode measurement of the internal pH of crab muscle fibres. *J. Physiol.* (Lond.) **252**, 803–815.

Alvarez-Leefmans F. J., Gamino S. M., and Rink T. J. (1983) Cytoplasmic free magnesium in neurones of *Helix aspersa* measured with ion-selective micro-electrodes. *J. Physiol.* (Lond.) **345**, 104P.

Ammann D., Guggi M., Pretsch E., and Simon W. (1975) Improved calcium ion-selective electrode based on a neutral carrier. *Anal. Lett.* **8,** 709–720.

Ammann D., Meier P. C., and Simon W. (1979) Design and Use of Ca^{2+} Selective Microelectrodes, in *Detection and Measurement of Free Ca^{2+} Ions in Cells* (Ashley C. C. and Campbell A. K., eds.), Elsevier, Amsterdam.

Ammann D., Lanter F., Steiner R. A., Schulthess P., Shijo Y., and Simon W. (1981) Neutral carrier-based hydrogen ion-selective microelectrode for extra- and intracellular studies. *Anal. Chem.* **53,** 2267–2269.

Anderson R. E. and Sundt T. M. (1983) Brain pH in focal cerebral ischemia and the protective effects of barbiturate anesthesia. *J. Cereb. Blood Flow Metab.* **3,** 493–497.

Astrup J. and Norberg K. (1976) Potassium activity in cerebral cortex in rats during progressive severe hypoglycemia. *Brain Res.* **103,** 418–423.

Astrup J., Heuser D., Lassen N. A., Nilsson B., Norberg K., and Siesjo B. K. (1976) Evidence Against H^+ and K^+ as the Main Factors in the Regulation of Cerebral Blood Flow During Epileptic Discharges, Acute Hypoxemia, Amphetamine Intoxication and Hypoglycemia. A Microelectrode Study, in *Ionic Actions on Vascular Smooth Muscle* (Betz E., ed.), Springer-Verlag, Berlin.

Astrup J., Symon L., Branston N. M., and Lassen N. A. (1977) Cortical evoked potential and extracellular K^+ and H^+ at critical levels of brain ischemia. *Stroke* **8,** 51–57.

Astrup J., Blennow G., and Nilsson B. (1979) Effects of reduced cerebral blood flow upon EEG pattern, cerebral extracellular potassium, and energy metabolism in the rat cortex during bicuculline-induced seizures. *Brain Res.* **177,** 115–126.

Astrup J., Rehncrona S., and Siesjo B. K. (1980) The increase in extracellular potassium concentration in the ischemic brain in relation to the pre-ischemic functional activity and cerebral metabolic rate. *Brain Res.* **199,** 161–174.

Astrup J., Siesjo B. K., and Symon L. (1981a) Thresholds in cerebral ischemia—the ischemic penumbra. *Stroke* **12,** 723–725.

Astrup J., Skovsted P., Gjerris G., and Sorensen H. R. (1981b) Increase in extracellular potassium in the brain during circulatory arrest: Effect of hypothermia, lidocaine and thiopental. *Anesthesiology* **55,** 256–262.

Bard D. M., Fry C. H., and Treasure T. (1978) An ion selective electrode for the determination of calcium activity. *J. Physiol.* (Lond.) **276,** 1P–2P.

Berman H. J. and Hebert N. C. (1974) *Ion-Selective Microelectrodes.* Plenum, New York.

Berridge M., Comar D., Roeda D., and Syrota A. (1982) Synthesis and *in vivo* characteristics of [2-^{11}C]5,5-dimethyl-oxazolidine-2,4-dione (DMO). *Int. J. Appl. Radiat. Isot.* **33,** 647–651.

Betz E. (1977) Ionic and metabolic control of local cerebral blood flow. *Acta. Clin. Belg.* **32,** 119–128.

Bourke R. A., Greenberg E. A., and Tower D. B. (1965) Variations of cerebral cortex fluid spaces *in vivo* as a function of species brain size. *Am. J. Physiol.* **208,** 682–692.

Branston N. M., Strong A. J., and Symon L. (1977) Extracellular potassium activity, evoked potential and tissue blood flow: Relationships during progressive ischaemia in baboon cerebral cortex. *J. Neurol. Sci.* **32,** 305–321.

Branston N. M., Symon L., Strong A. J., and Hope D. T. (1978) Measurements of regional cortical blood flow during changes in extracellular potassium activity evoked by direct cortical stimulation in the primate. *Exp. Neurol.* **59,** 243–253.

Branston N. M., Strong A. J., and Symon L. (1982) Kinetics of resolution of transient increases in extracellular potassium activity: Relationships to regional blood flow in primate cerebral cortex. *Neurol. Res.* **4,** 1–19.

Brooks D. J., Lammertsma A. A., Beaney R. P., Leenders K. L., Buckingham P. D., Marshall J., and Jones T. (1984) Measurement of regional cerebral pH in human subjects using continuous inhalation of $^{11}CO_2$ and positron emission tomography. *J. Cereb. Blood Flow Metab.* **4,** 458–465.

Bures J., Buresova O., and Krivanek J. (1974) *The Mechanism and Applications of Leao's Spreading Depression of Electroencephalographic Activity.* Academic, New York.

Buxton R. B., Wechsler L. R., Alpert N. M., Ackerman R. H., Elmaleh D. R., and Correia J. A. (1984) Measurement of brain pH using $^{11}CO_2$ and positron emission tomography. *J. Cereb. Blood Flow Metab.* **4,** 8–16.

Cameron I. L., Sheridan P. J., and Smith N. R. (1978) An X-ray microanalysis study of differences in concentration of elements in brain cells due to opiates, cell type and subcellular localisation. *J. Neurosci. Res.* **3,** 397–410.

Cameron I. R., Caronna J., Miller R., and Linton R. A. F. (1976) The Action of K^+ at the Cerebral Vessels, in *Ionic Actions on Vascular Smooth Muscle* (Betz E., ed.), Springer, Berlin.

Cragg P., Patterson L., and Purves M. J. (1977) The pH of brain extracellular fluid in the cat. *J. Physiol.* (Lond.) **272,** 137–166.

Crowe W., Mayevsky A., and Mela L. (1977) Application of a solid membrane ion-selective electrode to *in vivo* measurements. *Am. J. Physiol.* **233,** C56–C60.

Csiba L., Paschen W., and Hossmann K.-A. (1983) A topographic quantitative method for measuring brain tissue pH under physiological and pathophysiological conditions. *Brain Res.* **289,** 334–337.

Dienel G. A. (1984) Regional accumulation of calcium in postischemic rat brain. *J. Neurochem.* **43,** 913–925.

Dora E. and Zeuthen T. (1976) Brain Metabolism and Ion Movements in the Brain Cortex of the Rat During Anoxia, in *Ion and Enzyme Electrodes in Biology and Medicine* (Kessler M., Clark L. C., Lubbers D. W., Silver I. A., and Simon W., eds.), Urban and Schwarzenberg, Munich.

Dufau E., Acker H., and Sylvester D. (1982) Triple barrelled ion-sensitive microelectrode for simultaneous measurements of two extracellular ion activities. *Med. Prog. Technol.* **9,** 33–38.

Durst R. A. (1969) Ion-selective Electrodes. *Natl. Bur. Std. Spec. Publ. 314* Washington, DC.

Edgar A. D., Strosznajder J., and Horrocks L. A. (1982) Activation of ethanolamine phospholipase A2 in brain during ischemia. *J. Neurochem.* **39,** 1111–1116.

Elliott K. A. C., and Jasper H. H. (1949) Physiological salt solutions for brain surgery. Studies of local pH and pial vessel reactions to buffered and unbuffered isotonic solutions. *J. Neurosurg.* **6,** 140.

Ellis D. and Thomas R. C. (1976) Microelectrode measurement of the intracellular pH of mammalian heart cells. *Nature* (Lond.) **262,** 224–225.

Evans M. G. and Thomas R. C. (1983) The effects of acid solutions on intracellular pH and Na in snail neurones. *J. Physiol.* (Lond.) **341,** 60P.

Fiedler U. and Ruzicka J. (1973) Selectrode—The universal ion-selective electrode. VII. A valinomycin-based potassium electrode with nonporous polymer membrane and solid state inner reference system. *Anal. Chim. Acta* **67,** 179–193.

Fisher R. S., Pedley T. A., Moody W. J. Jr., and Prince D. A. (1976) The role of extracellular potassium in hippocampal epilepsy. *Arch. Neurol.* **33,** 76–83.

Friede R. L. and Van Houten W. H. (1961) Relations between post mortem alterations and glycolytic metabolism in the brain. *Exp. Neurol.* **4,** 197–204.

Fujimoto M. and Honda M. (1980) A triple-barrelled microelectrode for simultaneous measurements of intracellular Na^+ and K^+ activities and membrane potential in biological cells. *Jpn. J. Physiol.* **30,** 859–875.

Gardiner M., Nilsson B., Rehncrona S., and Siesjo B. K. (1981) Free fatty acids in the rat brain in moderate and severe hypoxia. *J. Neurochem.* **36,** 1500–1505.

Gardner-Medwin A. R. (1981) Possible roles of vertebrate neuroglia in potassium dynamics, spreading depression and migraine. *J. Exp. Biol.* **95,** 111–127.

Ghajar J. B. G., Plum F., and Duffy T. E. (1982) Cerebral oxidative metabolism and blood flow during acute hypoglycemia and recovery in unanesthetised rats. *J. Neurochem.* **38,** 397–409.

Griffiths T., Evans M. C., and Meldrum B. S. (1982) Intracellular sites of early calcium accumulation in the rat hippocampus during status epilepticus. *Neurosci. Lett.* **30,** 329–334.

Griffiths T., Evans M. C., and Meldrum B. S. (1983) Intracellular calcium accumulation in rat hippocampus during seizures induced by bicuculline or L-allylglycine. *Neuroscience* **10,** 385–395.

Griffiths T., Evans M. C., and Meldrum B. S. (1984) Status epilepticus: The reversibility of calcium loading and acute neuronal pathological changes in the rat hippocampus. *Neuroscience* **12,** 557–567.

Guilbault G. G. (1979) Recommendations for publishing manuscripts on ion-selective electrodes. *Ion-Selective Electrode Rev.* **1,** 139–143.

Gupta B. L. and Hall T. A. (1981) The X-ray microanalysis of frozen-hydrated sections in scanning electron microscopy: An evaluation. *Tissue Cell* **13,** 623–643.

Hall T. A. (1979) Problems of the Continuum-Normalisation Method for the Quantitative Analysis of Sections of Soft Tissue, in *Microbeam Analysis in Biology* (Lechene C. P. and Warner R., eds.), Academic, New York.

Hall T. A. and Gupta B. L. (1979) EDS Quantitation and Application to Biology, in *Introduction to Analytical Electron Microscopy* (Hren J. J., Goldstein J. I., and Joy D. C., eds.), Plenum, New York.

Hansen A. J. (1977) Extracellular potassium concentration in juvenile and adult rat brain cortex during anoxia. *Acta Physiol. Scand.* **99,** 412–420.

Hansen A. J. (1978) The extracellular potassium concentration in brain cortex following ischemia in hypo- and hyper-glycemic rats. *Acta Physiol. Scand.* **102,** 324–329.

Hansen A. J. (1981) Extracellular Ion Concentration in Cerebral Ischemia, in *The Application of Ion-Selective Microelectrodes* (Zeuthen T., ed.), Elsevier, Amsterdam.

Hansen A. J. and Zeuthen T. (1981) Extracellular ion concentrations during spreading depression and ischemia in the rat brain cortex. *Acta Physiol. Scand.* **113,** 437–445.

Harris R. J. (1985) Extracellular Ion Activity Changes in Cerebral Ischaemia. Doctoral thesis, University of London, London, UK.

Harris R. J. and Symon L. (1984). Extracellular pH, potassium, and calcium activities in progressive ischaemia of rat cortex. *J. Cereb. Blood Flow Metab.* **4,** 178–186.

Harris R. J., Symon L., Branston N. M., and Bayhan M. (1981) Changes in extracellular calcium activity in cerebral ischaemia. *J. Cereb. Blood Flow Metab.* **1,** 203–209.

Heinemann U. and Lux H. D. (1975) Undershoots following stimulus-induced rises of extracellular potassium concentration in cerebral cortex of cats. *Brain Res.* **93,** 63–76.

Heinemann U., Lux H. D., and Gutnick M. J. (1977) Extracellular free calcium and potassium during paroxysmal activity in the cerebral cortex of the cat. *Exp. Brain Res.* **27,** 237–243.

Herz A., Zieglgansberger W., and Farber G. (1969) Microelectrode studies concerning the spread of glutamic acid and GABA in brain tissue. *Exp. Brain Res.* **9,** 221–235.

Heuser D., Astrup J., Lassen N. A., and Betz E. (1975a) Brain carbonic acidosis after acetazolamide. *Acta Physiol. Scand.* **93,** 385–390.

Heuser D., Hossmann K. A., Schindler U., and Betz E. (1975b) Changes in extracellular ion activities and brain volume during prolonged cerebral ischemia and recovery. *Pflugers Arch.* **355** (suppl. 198), R99.

Hinke J. A. (1961) The measurement of sodium and potassium activities in the squid axon by means of cation-selective glass microelectrodes. *J. Physiol.* (Lond.) **156,** 314–335.

Holmes E. G. (1932) Observations on the variation of pH of brain tissue. *Biochem. J.* **26,** 2010–2014.

Hoper J., Kessler M., and Simon W. (1976) Measurements with Ion-Selective Surface Electrodes (pK, pNa, pCa, pH) During No Flow Anoxia, in *Ion and Enzyme Electrodes in Biology and Medicine* (Kessler M., Clark L. C., Lubbers D. W., Silver I. A., and Simon W., eds.), Urban and Schwarzenberg, Munich.

Hossmann K.-A., Sakaki S., and Zimmermann V. (1977) Cation activities in reversible ischemia of the cat brain. *Stroke* **8,** 77–81.

Hubschmann O. R. and Kornhauser D. (1982) Effect of subarachnoid hemorrhage on the extracellular microenvironment. *J. Neurosurg.* **56,** 216–221.

Jones T. (1980) Positron emission tomography and measurements of regional tissue function in man. *Br. Med. Bull.* **36,** 231–236.

Kendall M. D., Warley A., and Morris I. W. (1985) Differences in apparent elemental composition of tissues and cells using a fully quantitative X-ray microanalysis system. *J. Microscopy,* in press.

Kessler M., Hajek K., and Simon W. (1976) Four Barrelled Microelectrode for the Measurement of Potassium, Sodium and Calcium Ion Activities, in *Ion and Enzyme Electrodes in Biology and Medicine* (Kessler M., Clark L. C., Lubbers D. W., Silver I. A., and Simon W., eds.), Urban and Schwarzenberg, Munich.

Khuri R. N., Hajjar J. J., and Agulian S. K. (1972a) Measurement of intracellular potassium with liquid ion-exchange microelectrodes. *J. Appl. Physiol.* **32,** 419–422.

Khuri R. N., Agulian S. K., and Kalloghlian A. (1972b) Intracellular potassium in cells of the distal tubule. *Pflugers Arch.* **335,** 297–308.

Knabe U. and Betz E. (1972) The Effect of Varying Extracellular K^+, Mg^{2+}, and Ca^{2+} on the Diameter of Pial Arterioles, in *Ionic Action on Vascular Smooth Muscle* (Betz E., ed.) Springer, Berlin.

Kobatake K., Sako K., Izawa M., Yamamoto Y. L., and Hakim A. M. (1984) Autoradiographic determination of brain pH following middle cerebral artery occlusion in the rat. *Stroke* **15,** 540–547.

Kogure K., Busto R., Santiso M., Martinez E., Alonso O. F., Halsey J. H., and Strong E. (1979) Paradoxical pH shift in the core of developing cerebral infarction. *Acta Neurol. Scand.* (suppl. 72), 52–53.

Kogure K., Alonso O. F., and Martinez E. (1980) A topographic measurement of brain pH. *Brain Res.* **195,** 95–109.

Koryta J. (1975) *Ion-Selective Electrodes.* Cambridge University Press, Cambridge, New York.

Koryta J. (1980) *Medical and Biological Applications of Electrochemical Devices.* J. Wiley, Chichester.

Kraig R. P., Ferreira-Filho C. R., and Nicholson C. (1983) Alkaline and acid transients in cerebellar microenvironment. *J. Neurophysiol.* **49,** 831–850.

Kriz N. and Sykova E. (1980) Sensitivity of K^+-Selective Microelectrodes to pH and Some Biologically Active Substances, in *Ion-Selective Electrodes and Their Uses in Excitable Tissues* (Sykova E., Hnik P., and Vyklicky L., eds.), Plenum, New York.

Kuschinsky W., Wahl M., Bosse O., and Thurau K. (1972) Perivascular potassium and pH as determinants of local pial arterial diameter in cats: A microapplication study. *Circ. Res.* **31,** 240–247.

Lammertsma A. A. (1984) Positron emission tomography of the brain: Measurement of regional cerebral function in man. *Clin. Neurol. Neurosurg.* **86,** 1–11.

Lassen N. A. (1968) Brain extracellular pH: The main factor controlling cerebral blood flow. *Scand. J. Clin. Lab. Invest.* **22,** 247–251.

Lechene C. (1980) Electron probe microanalysis of biological soft tissues: Principle and technique. *Fed. Proc.* **39,** 2871–2880.

Lee C. O. (1980) Determination of Selectivity Coefficients of Ion-Selective Microelectrodes, in *Ion-Selective Electrodes and Their Uses in Excitable Tissues* (Sykova E., Hnik P., and Vyklicky L., eds.), Plenum, New York.

Lenglet W. J., Bos A. J., v.d. Stap C. C., Vis R. D., Delhez H., and v.d. Hamer C. J. (1984) Discrepancies between histological and physical methods for trace element mapping in the rat brain. *Histochemistry* **81,** 305–309.

Leniger-Follert E. (1984) Mechanisms of regulation of cerebral microflow during bicuculline-induced seizures in anaesthetised cats. *J. Cereb. Blood Flow Metab.* **4,** 150–165.

Leniger-Follert E. and Lubbers D. W. (1976) Behavior of microflow and local pO_2 of the brain cortex during and after direct electrical stimulation. A contribution to the problem of metabolic regulation of microcirculation in the brain. *Pflugers Arch.* **366,** 39–44.

Leniger-Follert E., Urbanics R., and Lubbers D. W. (1978) Behaviour of Extracellular H^+ and K^+ Activities During Functional Hyperemia of Microcirculation in the Brain Cortex, in *Advances in Neurology,* vol. 20, *Cerebrospinal Microcirculation* (Cervos-Navarro J. and Betz E., eds.), Raven, New York.

Lothman E. W., LaManna J., Cordingley G., Rosenthal M., and Somjen G. (1975) Responses of electrical potential, potassium levels and oxidative metabolic activity of cerebral neocortex of cats. *Brain Res.* **88,** 15–36.

Lux H. D. and Neher E. (1973) The equilibration time course of $[K^+]_o$ in cortex. *Exp. Brain Res.* **17,** 190–205.

Mabe H., Blomqvist P., and Siesjo B. K. (1983) Intracellular pH in the brain following transient ischemia. *J. Cereb. Blood Flow Metab.* **3,** 109–114.

MacMillan V. (1982) Cerebral Na^+, K^+-ATPase activity during exposure to and recovery from acute ischemia. *J. Cereb. Blood Flow Metab.* **2,** 457–465.

Mayevsky A. (1978) Ischemia in the brain: The effects of carotid artery ligation and decapitation on the energy state of the awake and anesthetised rat. *Brain Res.* **140,** 217–230.

McBroom M. J., Lancaster Y. J., and Weiss A. K. (1971) Measurement of tissue calcium, magnesium, sodium and potassium by flame spectrophotometry. Aspects of methodology and instrumentation. *Anal. Biochem.* **42,** 178–190.

McCreery D. B. and Agnew W. F. (1983) Changes in extracellular potassium and calcium concentration and neural activity during prolonged electrical stimulation of cat cerebral cortex at defined charge densities. *Exp. Neurol.* **79,** 371–396.

McDonald J. M., Burns D. E., Jarett L., and Davies J. E. (1977) A rapid microtechnique for the preparation of biological material for the simultaneous analysis of calcium, magnesium and protein. *Anal. Biochem.* **82,** 485–492.

Meech R. W. (1978) Calcium-dependent potassium activation in nervous tissue. *Ann. Rev. Biophys. Bioeng.* **7,** 1–18.

Meier P. C., Ammann D., Osswald H. G., and Simon W. (1977) Ion-selective electrodes in clinical chemistry (review). *Med. Prog. Technol.* **5,** 1–12.

Meier P. C., Ammann D., Morf W. E., and Simon W. (1980) Liquid Membrane Ion-Selective Electrodes and Their Biomedical Applications, in *Medical and Biological Applications of Electrochemical Devices* (Koryta J., ed.), John Wiley, Chichester.

Meldrum B. S. (1981) Metabolic Effects of Prolonged Epileptic Seizures and the Causation of Epileptic Brain Damage, in *Metabolic Disorders of the Nervous System* (Rose F. C., ed.), Pitman Medical, London.

Meyer J. S. and Denny-Brown D. (1957) The cerebral collateral circulation. 1. Factors influencing collateral blood flow. *Neurology* **7,** 447–458.

Meyer J. S., Gotoh F., Tazaki Y., Hamaguchi K., Ishikawa S., Novailhat F., and Symon L. (1962) Regional cerebral blood flow and metabolism in vivo. *Arch. Neurol.* **7,** 560–581.

Moody G. J. and Thomas J. D. R. (1971) Selective Ion Sensitive Electrodes. Merrow, Watford.

Moreton R. B. (1981) Electron-probe X-ray microanalysis: Techniques and recent applications in biology. *Biol. Rev.* **56,** 409–461.

Morgan A. J. (1984) X-Ray Microanalysis in Electron Microscopy for Biologists. Oxford University Press, Oxford, UK.

Morgan A. J., Davies T. W., and Erasmus D. A. (1975) Changes in the concentration and distribution of elements during electron microscope preparative procedures. *Micron* **6,** 11–23.

Morris P. J., Heuser D., McDowall D. G., Hashiba M., and Myers D. (1983) Cerebral cortical extra-cellular fluid H^+ and K^+ activities during hypotension in cats. *Anesthesiology* **59,** 10–18.

Munoz J. L., Deyhimi F., and Coles J. A. (1983) Silanization of glass in the making of ion-selective microelectrodes. *J. Neurosci. Meth.* **8,** 231–247.

Mutch W. A. C. and Hansen A. J. (1984) Extracellular pH changes during spreading depression and cerebral ischemia: Mechanisms of brain pH regulation. *J. Cereb. Blood Flow Metabol.* **4,** 17–27.

Mutsuga N., Schuette W. H., and Lewis D. V. (1976) The contribution of local blood flow to the rapid clearance of potassium from the cortical extracellular space. *Brain Res.* **116,** 431–436.

Myers R. E. (1981) High Lactic Acid Not Reduced ATP: Cause of Brain Injury From Oxygen Deprivation, in *Cerebral Vascular Disease,* vol. 3 (Meyer J. S., Lechner H., Reivich M., Ott E. O., and Aranibar A., eds.), Excerpta Medica, Amsterdam.

Neher E. and Lux H. D. (1973) Rapid changes of potassium concentration at the outer surface of exposed single neurons during membrane current flow. *J. Gen. Physiol.* **61,** 385–399.

Ng L. K. Y. and Nimmannitya J. (1970). Massive cerebral infarction with severe brain swelling: A clinicopathological study. *Stroke,* **1,** 158–163.

Nicholson C. and Kraig R. P. (1981) The Behavior of Extracellular Ions During Spreading Depression, in *The Application of Ion-Selective Microelectrodes* (Zeuthen T., ed.), Elsevier, Amsterdam.

Nicholson C., Ten Bruggencate G., Steinberg R., and Stockle H. (1977) Calcium modulation in brain extracellular microenvironment demonstrated with ion-selective micropipette. *Proc. Natl. Acad. Sci. USA* **74,** 1287–1290.

Nicholson C., Ten Bruggencate G., Stockle H., and Steinberg R. (1978) Calcium and potassium changes in extracellular microenvironment of cat cerebellar cortex. *J. Neurophysiol.* **41,** 1026–1039.

O'Brien M. D., Waltz A. G., and Jordan M. M. (1974) Ischemic cerebral edema. Distribution of water in brains of cats after occlusion of the middle cerebral artery. *Arch. Neurol.* **30,** 456–460.

Oehme M., Kessler M., and Simon W. (1976) Neutral carrier Ca^{2+} microelectrode. *Chimia* **30,** 204–206.

Pappius H. M. (1979) Evolution of Edema in Experimental Cerebral Infarction, in *Cerebrovascular Diseases* (Price T. R. and Nelson E., eds.), Raven, New York.

Pasztor E., Symon L., Dorsch N. W. C., and Branston N. M. (1973) The hydrogen clearance method in assessment of blood flow in cortex, white matter and deep nuclei of baboons. *Stroke* **4,** 556–569.

Pellegrino D. and Siesjo B. K. (1981) Regulation of extra- and intracellular pH in the brain in severe hypoglycemia. *J. Cereb. Blood Flow Metab.* **1,** 85–96.

Pellegrino D., Almqvist L. O., and Siesjo B. K. (1981) Effects of insulin-induced hypoglycemia on intracellular pH and impedance in the cerebral cortex of the rat. *Brain Res.* **221,** 129–147.

Peng C. F., Kane J. J., Murphy M. L., and Straub K. D. (1977) Abnormal mitochondrial oxidative phosphorylation of ischemic myocardium reversed by Ca^{2+}-chelating agents. *J. Mol. Cell. Cardiol.* **9,** 897–908.

Phelps M. E., Mazziotta J. C., and Huang S. C. (1982) Study of cerebral function with positron computed tomography. *J. Cereb. Blood Flow Metab.* **2,** 113–162.

Posner J. B. and Plum F. (1967) Spinal-fluid pH and neurological symptoms in systemic acidosis. *N. Engl. J. Med.* **277,** 605–613.

Prince D. A., Lux H. D., and Neher E. (1973) Measurement of extracellular potassium activity in cat cortex. *Brain Res.* **50,** 489–495.

Rehncrona S., Rosen I., and Siesjo B. K. (1981) Brain lactic acidosis and ischemic cell damage. 1. Biochemistry and neurophysiology. *J. Cereb. Blood Flow Metab.* **1,** 297–311.

Rhodes C. G., Wise R. J. S., Gibbs J. M., Frackowiak R. S. J., Hatazawa J., Palmer A. J., Thomas D. G. T., and Jones T. (1983) *In vivo* disturbance of the oxidative metabolism of glucose in human cerebral gliomas. *Ann. Neurol.* **14,** 614–626.

Roomans G. M., Wei X., and Seveus L. (1982) Cryoultramicrotomy as a preparative method for X-ray microanalysis in pathology. *Ultrastruct. Pathol.* **3,** 65–84.

Roos A. and Boron W. F. (1981) Intracellular pH. *Physiol. Rev.* **61,** 296–434.

Rosner M. J. and Becker D. P. (1984) Experimental brain injury: Successful therapy with the weak base, tromethamine. With an overview of CNS acidosis. *J. Neurosurg.* **60,** 961–971.

Rottenberg D. A., Ginos J. Z., Kearfott K. J., Junck L., and Bigner D. D. (1984) *In vivo* measurement of regional brain tissue pH using positron emission tomography. *Ann. Neurol.* **15** (suppl.), S98–S102.

Ruszczewski P. (1978) Release of prostaglandin-like substances into cerebral venous blood in conditions injurious to brain in the dog. *Acta Physiol. Pol.* **29,** 489–499.

Ruzicka J., Hansen E. H., and Tjell J. (1973) Selectrode—the universal ion-selective electrode. VI. The Ca(II) selectrodes employing a new ion exchanger in a nonporous membrane and a solid-state reference system. *Anal. Chim Acta* **67,** 155–178.

Schanne F. A. X., Kane A. B., Young E. E., and Farber J. L. (1979) Calcium dependence of toxic cell death: A final common pathway. *Science* **206,** 700–702.

Schuier F. J. and Hossmann K.-A. (1980) Experimental brain infarcts in cats. *Stroke* **11,** 593–601.

Severinghaus J. W., Chiodi H., Eger E. I. II., Brandstater B., and Hornbein T. F. (1966) Cerebral blood flow in man at high altitude. Role of cerebrospinal fluid pH in normalization of flow in chronic hypocapnia. *Circ. Res.* **19,** 274.

Siemkowicz E. and Hansen A. J. (1981) Brain extracellular ion composition and EEG activity following 10 minutes ischemia in normo- and hyperglycemic rats. *Stroke* **12,** 236–240.

Siesjo B. K. (1978) *Brain Energy Metabolism.* John Wiley, Chichester.

Siesjo B. K. (1981) Cell damage in the brain: A speculative synthesis. *J. Cereb. Blood Flow Metab.* **1,** 155–185.

Siesjo B. K. (1984) Cerebral circulation and metabolism. *J. Neurosurg.* **60,** 883–908.

Siesjo B. K. and Messeter K. (1971) Factors Determining Intracellular pH, in *Ion Homeostasis of the Brain* (Siesjo B. K. and Sorensen S. C., eds.), Munksgaard, Copenhagen.

Siesjo B. K., Folbergrova J., and MacMillan V. (1972) The effect of hypercapnia upon intracellular pH in the brain, evaluated by the bicarbonate-carbonic acid method and from the creatine phosphokinase equilibrium. *J. Neurochem.* **19,** 2483–2495.

Siesjo B. K., Von Hanwehr R., Nergelius G., Nevander G., and Ingvar M. (1985) Extra- and intracellular pH in the brain during seizures and in the recovery period following the arrest of seizure activity. *J. Cereb. Blood Flow Metab.* **5,** 47–57.

Silver I. A. (1976) Multiparameter Microelectrodes, in *Ion and Enzyme Electrodes in Biology and Medicine* (Kessler M., Clark L. C., Lubbers D. W., Silver I. A., and Simon W., eds.), Urban and Schwarzenberg, Munich.

Silver I. A. (1977a) Changes in pO_2 and Ion Fluxes in Cerebral Hypoxia-Ischemia, in *Tissue Hypoxia and Ischemia* (Reivich M., Coburn R., Lahiri S., and Chance B., eds.), Plenum Press, New York.

Silver I. A. (1977b) Ion fluxes in hypoxic tissues. *Microvasc. Res.* **13,** 409–420.

Silver I. A. (1978) Cellular microenvironment in relation to local blood flow. *CIBA Foundation Symposium* **56,** 49–67.

Simon R. P., Griffiths T., Evans M. C., Swan J. H., and Meldrum B. S. (1984) Calcium overload in selectively vulnerable neurons of the hippocampus during and after ischaemia: An electron microscopy study in the rat. *J. Cereb. Blood Flow Metab.* **4,** 350–361.

Simon W. Ammann D., Oehme M., and Morf W. E. (1978) Calcium selective electrodes. *Ann. NY Acad. Sci.* **307,** 52–70.

Sollner K. and Shean G. M. (1964) Liquid ion-exchange membranes of extreme selectivity and high permeability for anions. *J. Am. Chem. Soc.* **86,** 1901–1902.

Sonnenschein R. R., Walker R. M., and Stein S. N. (1953) A microglass electrode for continuous recording of brain pH in situ. *Rev. Sci. Instrum.* **24,** 702–704.

Steiner R. A., Oehme M., Amman D., and Simon W. (1979) Neutral carrier sodium ion-selective microelectrode for intracellular studies. *Anal. Chem.* **51,** 351–353.

Stockle H. and Ten Bruggencate G. (1978) Climbing fiber-mediated rhythmic modulation of potassium and calcium in cat cerebellar cortex. *Exp. Neurol.* **61,** 226–230.

Strong A. J., Venables G. S., and Gibson G. (1983) The cortical ischaemic penumbra associated with occlusion of the middle cerebral artery in the cat. 1. Topography of changes in blood flow, potassium ion activity, and EEG. *J. Cereb. Blood Flow Metab.* **3,** 86–96.

Sun A. Y. and Sun G. Y. (1976) Functional roles of phospholipids of synaptosomal membrane. *Adv. Exp. Med. Biol.* **72,** 169–197.

Sundt T. M. Jr. and Anderson R. E. (1980) Umbelliferone as an intracellular pH-sensitive fluorescent indicator and blood–brain barrier probe: Instrumentation, calibration and analysis. *J. Neurophysiol.* **44,** 60–75.

Sundt T. M. Jr., Anderson R. E., and Van Dyke R. A. (1978) Brain pH measurement using a diffusible lipid soluble pH sensitive fluorescent indicator. *J. Neurochem.* **31,** 627–635.

Sykova E., Hnik P., and Vyklicky L. (1980) *Ion-Sensitive Micro-electrodes and Their Use in Excitable Tissues.* Plenum, New York.

Syrota A., Castaing M., Rougemont D., Berridge M., Baron J. C., Bousser M. G., and Pocidalo J. J. (1983) Tissue acid–base balance and oxygen metabolism in human cerebral infarction studied with positron emission tomography. *Ann. Neurol.* **14,** 419–428.

Takahashi H., Manaka S., and Sono K. (1981) Changes in extracellular potassium concentration in cortex and brainstem during the acute phase of experimental closed head injury. *J. Neurosurg.* **55,** 708–717.

Thomas R. C. (1970) New design for sodium-sensitive glass microelectrode. *J. Physiol.* (Lond.) **210,** 82P–83P.

Thomas R. C. (1974) Intracellular pH of snail neurones measured with a new pH-sensitive glass microelectrode. *J. Physiol.* (Lond.) **238,** 159–180.

Thomas R. C. (1976) Ionic mechanism of the H^+ pump in a snail neuron. *Nature* (Lond.) **262,** 54–55.

Thomas R. C. (1977) The role of bicarbonate, chloride and sodium ions in the regulation of intracellular pH in snail neurons. *J. Physiol.* (Lond.) **273,** 317–338.

Thomas R. C. (1978) *Ion-Sensitive Intracellular Microelectrodes.* Academic, London.

Thorn W. and Heitmann R. (1954) pH der Gehirnrinde vom Kaninchen in situ wahrend perakuter, totaler Ischamie reiner anoxie und in der Erkohlung. *Pflugers Arch.* **258,** 501–510.

Thulborn K. R., du Boulay G. H., Duchen L. W., and Radda G. (1982). A ^{31}P nuclear magnetic resonance in vivo study of cerebral ischaemia in the gerbil. *J. Cereb. Blood Flow Metab.* **2,** 299–306.

Tsien R. Y. and Rink T. J. (1980) Neutral carrier ion-selective microelectrodes for measurement of intracellular free calcium. *Biochim. Biophys. Acta* **599,** 623–638.

Urbanics R., Leniger-Follert E., and Lubbers D. W. (1982) Measurements with ion-selective electrodes in the brain cortex during a short period of ischemia and arterial hypoxia. *Z. med. Labor-Diagn.* **23,** 92–95.

Vyskocil F. and Kriz N. (1972) Modifications of single and double-barrel potassium specific microelectrodes for physiological experiments. *Pflugers Arch* **337,** 265–276.

Vyskocil F., Kriz N., and Bures J. (1972) Potassium-selective microelectrodes used for measuring the extracellular brain potassium during spreading depression and anoxic depolarisation in rats. *Brain Res.* **39,** 255–259.

Wahl M., Deetjen P., Thurau K., Ingvar D. H., and Lassen N. A. (1970) Micropuncture evaluation of the importance of perivascular pH for the arteriolar diameter on the brain surface. *Pflugers Arch.* **316,** 152–163.

Walker J. L. (1971) Ion specific liquid ion exchanger microelectrodes. *Anal. Chem.* **43,** 89A–93A.

Watanabe O., West C. R., and Bremer A. (1977) Experimental regional cerebral ischemia in the middle cerebral artery territory in primates. 2. Effects on brain water and electrolytes in the early phase of MCA stroke. *Stroke* **8,** 71–76.

Welsh F. A., O'Connor M. J., Marcy V. R., Spatacco A. J., and John R. L. (1982) Factors limiting regeneration of ATP following temporary ischemia in cat brain. *Stroke* **13,** 234–242.

West C. R. and Matsen F. A. (1972) Effects of experimental ischemia on electrolytes of cortical cerebrospinal fluid and on brain water. *J. Neurosurg.* **36,** 687–699.

Wieloch T., Harris R. J., Symon L., and Siesjo B. K. (1984). Influence of severe hypoglycemia on brain extracellular calcium and potassium activities, energy and phospholipid metabolism. *J. Neurochem.* **43,** 160–168.

Williams B. L. and Wilson K. (1981) *A Biologist's Guide to Principles and Techniques of Practical Biochemistry,* 2nd ed., E. Arnold, London.

Wise R. J., Thomas D. G., Lammertsma A. A., and Rhodes C. G. (1984) PET scanning of human brain tumors. *Prog. Exp. Tumor Res.* **27,** 154–169.

Yanagihara T. and McCall J. T. (1982) Ionic shift in cerebral ischaemia. *Life Sci.* **30,** 1921–1925.

Yatsu F. M. and Coull B. M. (1981) Stroke, in *Current Neurology,* vol. 3 (Appel S. H., ed.), Wiley, New York.

Zimmermann V. and Hossmann K.-A. (1975) Resuscitation of the monkey brain after one hour's complete ischemia. II. Brain water and electrolytes. *Brain Res.* **85,** 1–11.

Zs-Nagy I., Pieri C., Giuli C., Bertoni-Freddari C., and Zs-Nagy V. (1977) Energy dispersive X-ray microanalysis of the electrolyte in biological bulk specimens. *J. Ultrastruct. Res.* **58,** 22–33.

Immunocytochemical Procedures for the Demonstration of Putative Neurotransmitters in Cerebral Vessels, Cerebrum, and Retina

Anitha Bruun and Lars Edvinsson

1. Introduction

Immunocytochemistry as a technique has had a tremendous impact on the development of neurobiology in the last decade. It is not a new technique. The principle that antibodies can be used as histochemical reagents was recognized early by Marrack (1934), showing its applicability if suitable markers were linked to antibody molecules. Of the various markers that have subsequently been examined, only the fluorescent dyes have provided the sensitivity necessary. The first example was fluorescein isothiocyanate-labeled antibodies that were employed to localize pneumococcal antigens in infected tissues (Coons et al., 1942). Another example was the localization of adrenocorticotrophic hormone using fluorescein-labeled antibodies raised against a partially purified antigen (Marshall, 1951). Localization of endogenous antigens in tissues was thus accomplished by the "indirect" immunocytochemical technique (Coons et al., 1955; Mellors et al., 1955). Another procedure involved the use of horseradish peroxidase, which resulted in the elegant peroxidase–antiperoxidase (PAP) technique (Sternberger et al., 1970).

By 1970 all methodology necessary for the application of immunocytochemistry to more detailed studies was available. The first neural antigen to be localized by immunocytochemistry was dopamine beta-hydroxylase, which was found to be present throughout the peripheral sympathetic neuronal system (Geffen et al., 1969). Within a few years immunocytochemistry became a major tool for mapping of neurotransmitter pathways in the central and peripheral nervous systems.

In this chapter we will describe some useful aspects of immunocytochemical methodology, mainly for the inexperienced reader, to provide an insight into the type of data that can be obtained, and illustrate this with examples from cerebral blood vessels, cerebrum, and retina.

2. Immunological Basis

Many macromolecules that occur naturally in neural tissues can be extracted and used as antigens or substances that, upon introduction into a suitable host, will give rise to the formation of antibodies. Small polypeptides or other agents isolated from neural tissues or synthesized in the laboratory may be made antigenic by coupling them to a carrier compound, which might be a protein or a polysaccharide (Weir, 1973). These antigenic agents are characterized by their capacity to induce a host response and by their specificity to combine with the antibody formed. Usually, the antibody has one or more combining sites for the specific antigen and this is an immunoglobulin (IgG) composed of polypeptide chains (Weir, 1973; Sternberger, 1974). In tissue sections, antibodies bound to antigens can be visualized by labeling the antibody with a marker compound (Weir, 1973). The in vitro immunological reaction (i.e., on the section) using marker compounds constitutes the basis for the localization of antigens in, for example, neural pathways, by immunocytochemistry. The value of the technique is directly related to the specificity of the antibody for the antigenic substance. Therefore, the specificity of the antiserum used in the immunocytochemical procedure must be known in detail (Joh et al., 1963).

For small peptides, the specificity tests of the antibodies are largely based on cross-reactivity between the formed antibody and its parent peptide (Pickel et al., 1979). The choice of immunocytochemical technique for demonstrating, for example, neuronal pathways, is dependent on the availability of a purified antigen and its specific antibody, as well as on the necessity for showing the chemical nature of the pathway as opposed to the anatomical localization. The purification procedures for most neuron-specific antigens are difficult and laborious, and they are therefore available only in limited amounts. Furthermore, the antibody and antiserum used in any study should be tested for specificity, and this can be done in several ways; (1) precipitation reactions in gels, (2) inactivation, (3) immunoprecipitation and radioimmunoassay, (4) solid-phase binding, and/or (5) immunoblots. Requirements such as these are essential for the interpretation of the immunocytochemistry data, and when not available, the results are open to question (Landis, 1985).

3. Types of Immunocytochemistry Methods

Once the antibody has attached to the desired site, it is identified by one of four widely used marking systems.

3.1. Indirect Fluorescence

As can be seen in Fig. 1A, a second antibody is used that selectively binds to the primary antibody. This second antibody is linked to a fluorophore in order to recognize the binding site by fluorescence microscopy. Two antigenic sites can be simultaneously labeled in one single preparation by either of two ways:

(1) By removing the initial antibody with acid potassium permanganate (exposure around 30 s) and then by restaining with another antibody, according to Tramu et al. (1978). (2) By using a monoclonal antibody coupled to fluorescein and a different polyclonal antibody complexed with rhodamine. With appropriate filter settings on the fluorescence microscope, the distribution of both types of antigens can then be analyzed in the same part of the tissue section (Furness et al., 1984). This is of particular value when a single type of nerve cell body contains different neurotransmitter candidates.

3.2. Peroxidase–Antiperoxidase (PAP) Complex

For use of PAP complex, a secondary antibody is chosen that specifically binds to a relatively invariant region of the primary antibody. This secondary antibody then serves as a bridge between the primary antibody and the antibody portion of the PAP complex (for details, *see* Fig. 1B). Subsequently, a small amount of hydrogen peroxidase and a substrate such as diaminobenzidine are added. The peroxidase catalyzes the formation of an insoluable reaction product that has the feature of being visible both upon light and electron microscopy and, in addition, requires less of the primary antibody.

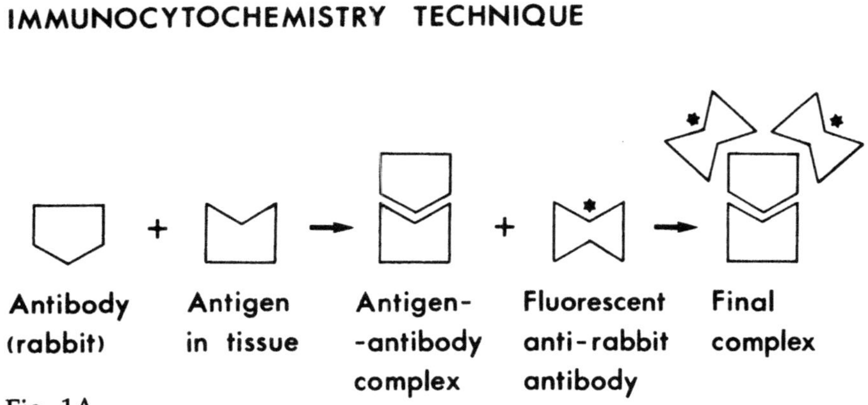

Fig. 1A

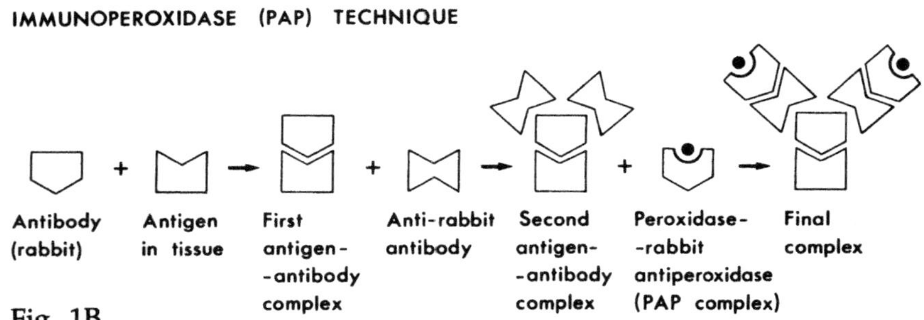

Fig. 1B

Fig. 1. Schematic drawing illustrating the principle of (A) the indirect immunofluorescence method and (B) the PAP technique. In this example, the primary antibody was raised in rabbit and the secondary fluorescent antibody was raised against rabbit using goat as the host animal.

3.3. Avidin-Biotin Procedure

This method relies on a specific and high-affinity interaction of biotin and the protein avidin. A primary antibody is supplied, and a secondary antibody is coated with biotin and allowed to bind to the primary. The site of the second antibody is identified with a complex of avidin linked to biotinylated peroxidase. Avidin thus forms a bridge between the biotinylated secondary antibody and the biotinylated peroxidase. As above the peroxidase generates an easily identifiable reaction product.

3.4. Colloidal Gold

This approach has mainly been used as a marker for electron microscopy. The colloidal gold particles are too small to be of value for light microscopic analysis. The secondary antibody is absorbed to colloidal gold, which marks the site of the primary antibody.

4. Indirect Immunofluorescence Method

In the following we will concentrate our discussion on the indirect immunocytochemical technique, since this has generally been the most widely used (Cuello, 1983; De Mey, 1983). We have for this review performed some variations of our standard procedures for immunocytochemistry in order to shed some light on technical problems that may confront the inexperienced immunocytochemistry reader.

4.1. Fixation

The three types of specimens examined were cerebral vessels, brain tissue, and retina. The tissue material was fixed either with various concentrations of formaldehyde in phosphate buffer (0.1*M*, pH 7.3) or with a buffered picric acid-formaldehyde fixative with some modification of the initial suggestions (for details, *see* Appendix I) (Stefanini et al., 1967; Zamboni et al., 1967). The specimens were obtained either following perfusion fixation for 10 min with ice-cold solutions or following immersion fixation of specimens rapidly dissected out (<5 min).

The influence of varying the mode of fixation and the composition of the fixatives is illustrated in Fig. 2. Four different antibodies were used (Table 1). The strong immunofluorescence of neuropeptide Y (NPY) in cerebral vessels was not noticeably affected by the different procedures that were used. The NPY immunoreaction was best for cerebral cortex and retina when fixation was performed with 4% formaldehyde or following perfusion fixation in the picric acid-formaldehyde fixative (Fig. 2). A good immunoreaction of the 5-hydroxytryptamine (5-HT) antibodies was seen only following fixation of the tissue in 4% formaldehyde (Fig. 2), which is in concert with information available in the literature (Steinbusch, 1981). As can be seen, only occasional fibers and cells appear in retina if minor modifications of this fixative are in-

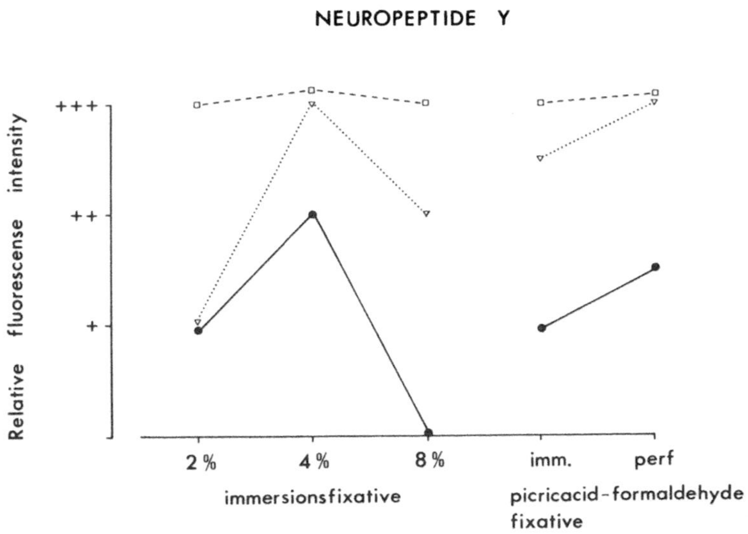

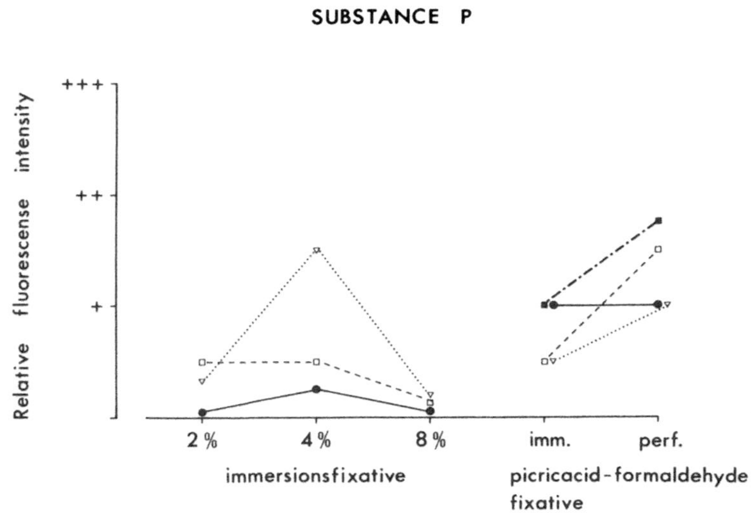

Fig. 2. Comparison of tests showing optimum fixation procedures for NPY, 5-HT, SP, and VIP in specimens from guinea pig. ■ — · — · — ■ = retina, □ – – – – □ = cerebral vessels, ▽ · · · · · ▽ = cerebrum (continued on next page).

5-HYDROXYTRYPTAMINE

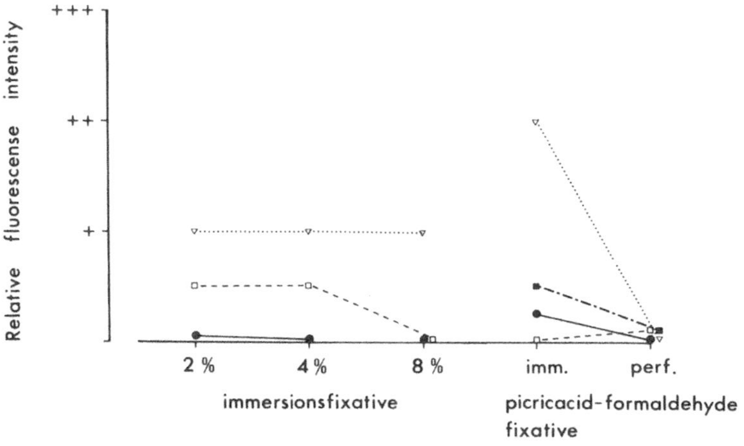

VASOACTIVE INTESTINAL PEPTIDE

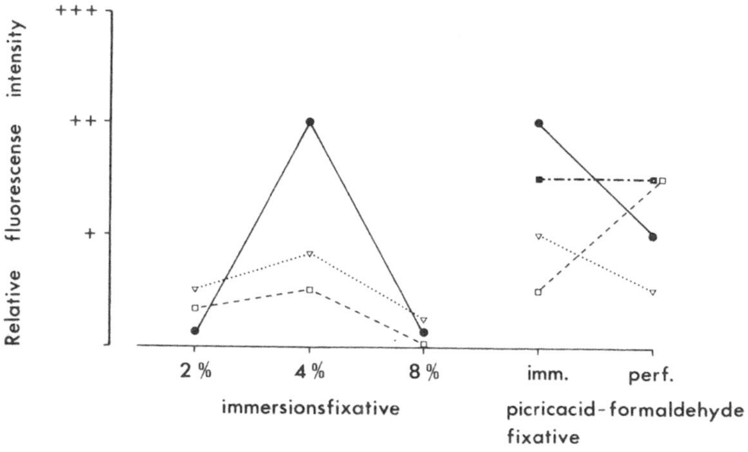

Table 1
Details on the Antibodies Used in This Study

Antibody code	Antigen	Working dilution	Produced in	Source	Reference
NPYY	NPY	1:500	Rabbit	P. C. Emson, MRC Cambridge, UK	Sundler et al., 1983
5-HT	5-HT	1:200	Rabbit	P. C. Emson, MRC Cambridge, UK	Steinbusch et al., 1978
14411	SP	1:400	Rabbit	Immuno Nuclear Co., Stillwater, Minnesota, USA	Stenfors Tornqvist, 1983
7852	VIP	1:400	Rabbit	Milab, Malmo, Sweden	Alm et al., 1980

troduced. Substance P (SP) and vasoactive intestinal peptide (VIP) fibers were most reliably demonstrated in cerebral vessels fixed by vascular perfusion with the picric acid-formaldehyde mixture (Fig. 2). The standard 4% formaldehyde fixative solution resulted in demonstration of fewer SP and VIP immunoreactive fibers.

For cortex there was no difference between immersion fixation in 4% formaldehyde or the picric acid-formaldehyde solution. The same was seen for VIP immunoreactivity in the retina. However, SP fibers were recognized with good precision only by fixatives containing picric acid. A remarkable finding was that perfusion fixation resulted in only a modestly better histochemical outcome as compared to immersion fixation.

4.2. Sections and Whole Mounts

Following fixation, the specimens were rinsed with a solution of 20% sucrose in a phosphate buffer at 4°C (for details, *see* Appendix I). Preliminary tests with 10 and 30% sucrose did not result in a better immunocytochemical outcome compared to 20% sucrose. The material was sectioned at 15 μM in a cryostat at –20°C.

In addition, whole mounts of cerebral vessels were fixed as above, although following the rinsing in the sucrose solution, the tissue was mounted on chromalum-subbed slides. The advantage with this method is that the entire nerve plexus can be visualized. However, only certain tissues are thin enough for this approach. We have used this procedure for studies on the distribution of various perivascular fibers (Edvinsson et al., 1981), whereas others have successfully applied it to various gastrointestinal tissues in which laminae of material can be dissected free in thin sheaths suitable for examination as whole mounts (Costa et al., 1980).

The immunocytochemical outcome of cryostat sections and whole mounts was compared only in cerebral vessels. Overall we noted a better immunocytochemical reaction with the whole mount procedure than with cryostat sections for 5-HT, SP, and VIP (not checked for NPY).

4.3. Immunolabeling

The antibodies used have been characterized in detail by the authors listed in Table 1, and their specificity in the concentrations used is reasonable. The site of the antigen-antibody reaction was revealed by fluorescein isothiocynate-labeled goat anti-rabbit IgG in a dilution of 1:80. Control sections were routinely incubated with antiserum inactivated by the previous addition of excess antigen (10–100 μg/ml). The absolute identity of the immunoreactive material is not certain in that cross-reactivity with possible unknown peptides containing the same immunoreactive site cannot be excluded. It is, therefore, appropriate to refer to the immunoreactive material as NPY-like, 5-HT-like, SP-like, or VIP-like. However, for brevity, the immunoreactive material is usually referred to as NPY, 5-HT, SP, or VIP in various published reports. A range of peptide antibody concentrations is compared in order to evaluate the best immunocytochemical outcome. It may vary depending on antibody, species, and tissue. It is a well-known phenomenon and should be examined in detail for each new modification of the immunocytochemical procedure.

4.4. Microscopy

The sections were examined in a fluorescence microscope (Leitz Dialux 20) with epi-illumination special filters selected to give peak excitation at 450–490 nm and barrier filter with cut-off edge at 515 nm (Leitz I 2/3 filter system).

5. Immunocytochemistry of Selected Tissues

5.1. Neuropeptide Y

Plexuses of nerve fibers showing NPY immunoreactivity were seen in the adventitia and at the adventitia-media border of cerebral blood vessels from guinea pig, frog, and chicken (Fig. 3). Such fibers have previously been observed in the cerebrovascular bed of rat, cat, and humans. The general distribution of the NPY fibers was similar to that of the noradrenaline-containing fibers. The fibers were particularly abundant around major cerebral arteries that belong to the circle of Willis (Fig. 4); they were also seen in pial arterioles on the convexity of the cerebral cortex and occasionally around penetrating arterioles (Fig. 3) (Edvinsson et al., 1983a, 1984a). NPY fibers were, in addition, associated with pial veins, but the nerve plexus was less dense in comparison with that of the arteries. Removal of the superior cervical ganglion resulted in the demonstration of only occasional NPY immunoreactive fibers. Staining of sections or whole mounts sequentially for NPY or dopamine-beta-hydroxylase revealed that NPY and noradrenaline

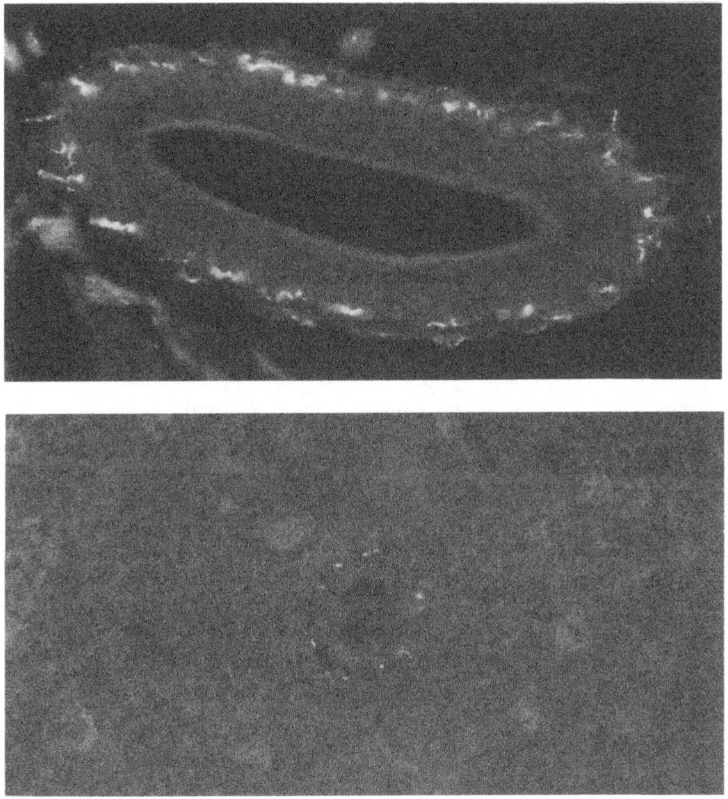

Fig. 3. Immunocytochemical demonstration of NPY-containing fibers in a pial artery (top), × 280, and in an intracerebral arteriole (bottom), × 310, from chick.

CEREBRAL ARTERIES OF THE GUINEA-PIG

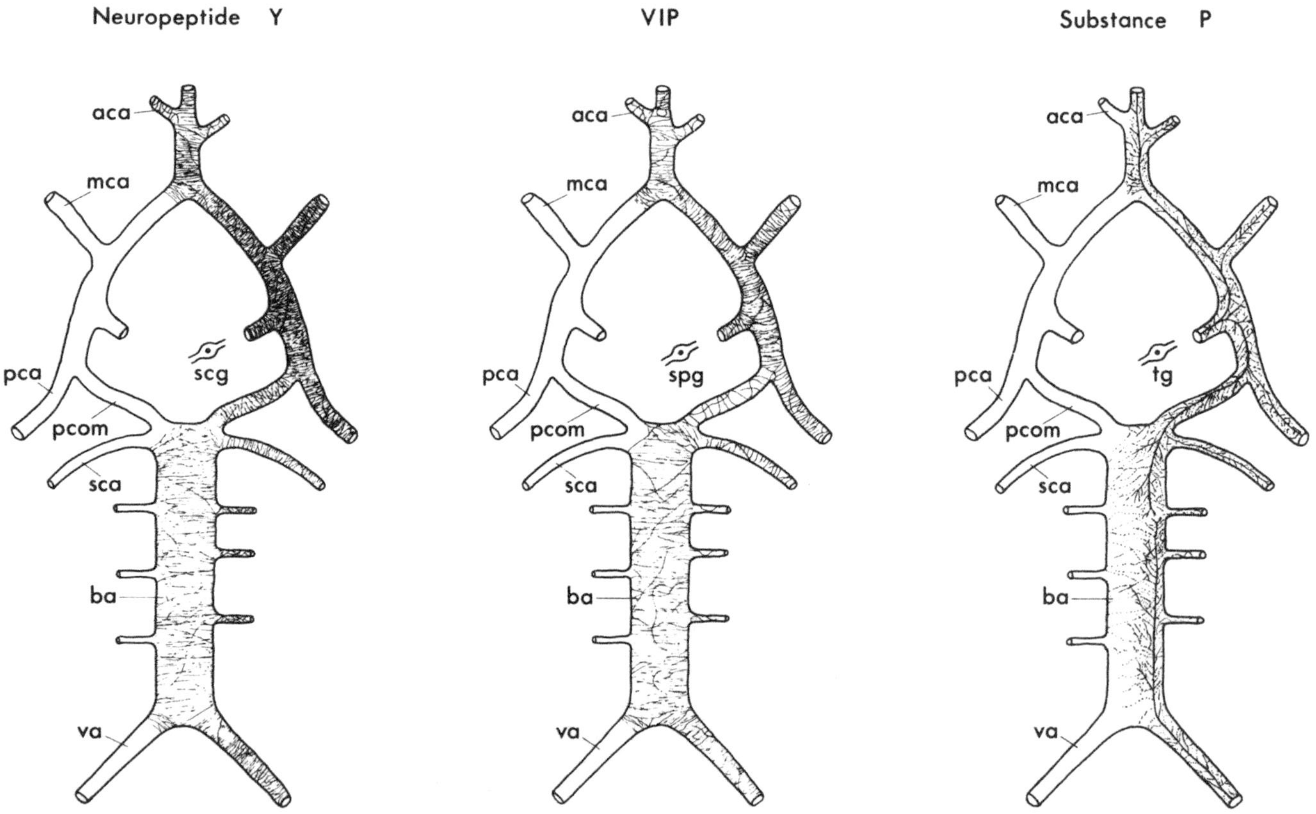

Fig. 4. Overview of peptidergic innervation of major cerebral arteries of the guinea pig. The majority of the NPY fibers originate in the superior cervical ganglion (SCG) (Edvinsson et al., 1984a). The VIP fibers have a slightly less dense distribution. The origin is not clear, but the sphenopalatine ganglion (SPG) has been suggested as a tentative localization of the nerve cell bodies (Hara et al., 1985). The arteries belonging to the circle of Willis and the cranial part of the basilar artery are all supplied with SP-containing fibers that originate in perikarya in the trigeminal ganglion (TG) (Liu-Chen et al., 1983). The caudal part of the basilar artery and the vertebral artery are thought to be supplied with SP fibers from dorsal root ganglia (Yamamoto et al., 1983). A similar distribution and origin of the CGRP fibers has been shown (Uddman et al., 1985).

coexist (Ekblad et al., 1984). NPY has been found to be a potent cerebrovascular constrictor in vitro and *in situ* (Edvinsson et al., 1984a) and to reduce cerebral blood flow without changing cerebral metabolism following intracerebral administration (Tuor et al., 1986).

NPY-immunoreactive cell bodies were regularly seen in cerebral cortex (Allen et al., 1983), particularly in the deeper layers (Fig. 5.), and to have long dendrites that sometimes traversed the corpus callosum to make direct contact with cell bodies in the caudate nucleus (Tuor et al., 1986). NPY-immunoreactive fibers and terminals were widely distributed in the brain with particularly dense arrangements in the central amygdaloid nucleus, nucleus accumbens, medial forbrain bundle, stria terminals, several preoptic and hypothalamic nuclei, and various brainstem regions (Everitt et al., 1984). Studies in various mammals have revealed that NPY is the peptide that, of those studied thus far, is present in highest concentrations in the brain (Allen et al., 1983). Its functional role is not known, but an involvement in feeding behavior and mood regulation has been advocated.

Neurons displaying NPY immunoreactivity were found among amacrine cells in retina from humans, baboon, pig, cat, guinea pig, pigeon, chicken, frog, carp, goldfish, skate, and mudpuppy (Fig. 6). The immunoreactive cell bodies were located in the inner-most cell row of the inner nuclear layer with processes forming one, two, or three more or less well-defined sublayers in the inner plexiform layer (Fig. 7) (Brecha, 1983; Bruun et al., 1984; Bruun, Ehinger, and Tornquist, unpublished results).

The function of NPY in retina is not yet clear; it may be a putative transmitter or a neuromodulator for known neurotransmitters such as glycine, GABA, dopamine, and acetylcholine. It seems that NPY is present in most adrenergic neurons. The density of the NPY immunoreactive fibers in the uvea is high and directly comparable to the density of the noradrenergic fibers. Since the sympathetic adrenergic nervous system participates in the regulation of intraocular pressure, it is possible that NPY is involved in this process.

5.2. 5-Hydroxytryptamine

5-Hydroxytryptamine-positive nerve fibers were observed around guinea pig cerebral vessels (Fig. 8a). The fibers formed plexuses of delicate nerves running in all directions. Small pial

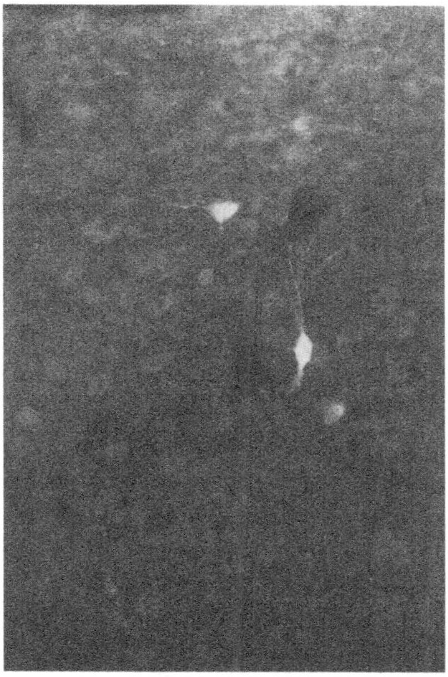

Fig. 5. Immunocytochemical demonstration of NPY-containing cell bodies and nerve fibers in the chick cerebral cortex (× 205).

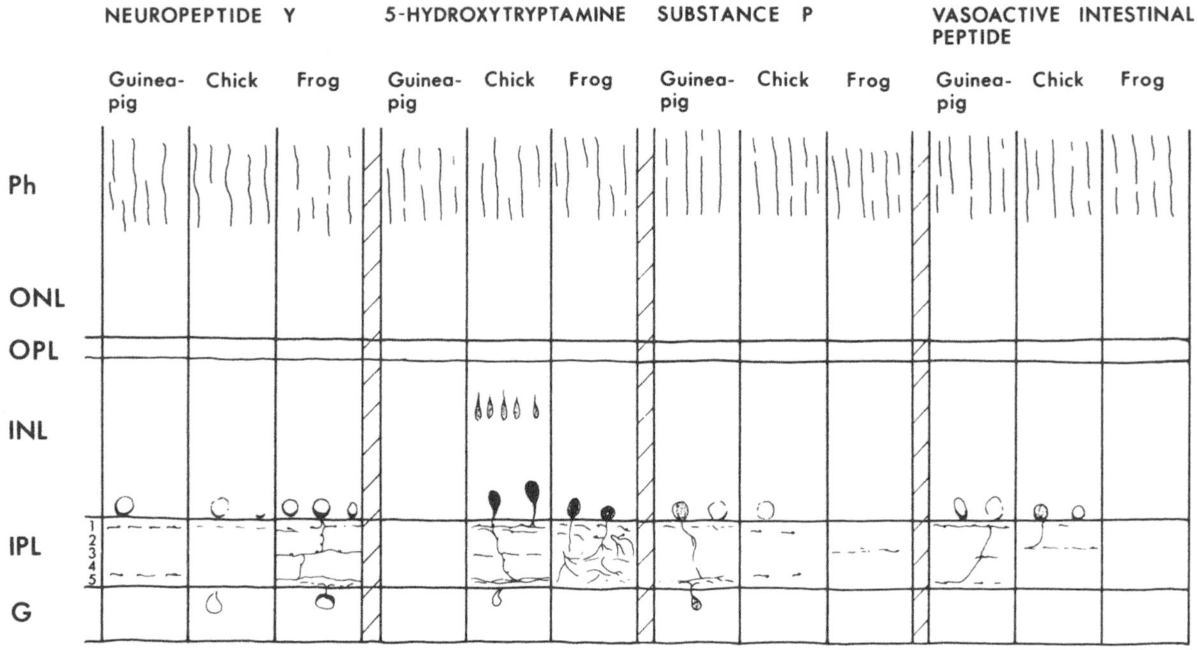

Fig. 6. Schematic drawing of retina demonstrating the morphology of NPY-, 5-HT-, SP-, and VIP-immunoreactive neurons in the various species investigated. The NPY immunoreactivity was found in amacrine cell bodies in INL and in displaced amacrine cell bodies in ganglion cell layer. The processes ramified in sublayers 1, 3, and 5 in IPL depending on the animal variations. The 5-HT immunoreactivity was found in bipolar- and amacrine cell bodies in INL and displaced amacrine cell bodies in ganglion cell layer in chick. The processes were ramifying in sublayer 1, 3, and 5 in chick and as a dense network in frog in IPL. No 5-HT immunoreactivity was found in guinea pig. The SP and VIP immunoreactivities were weaker compared to the two previous antibodies. They were, however, found in the same place in retina. In frog retina there was no or scarce immunoreactivity. Abbreviations: Ph, photoreceptors; ONL, outer nuclear layer; OPL, outer plexiform layer; INL, inner nuclear layer; IPL, inner plexiform layer; G, ganglion cell layer.

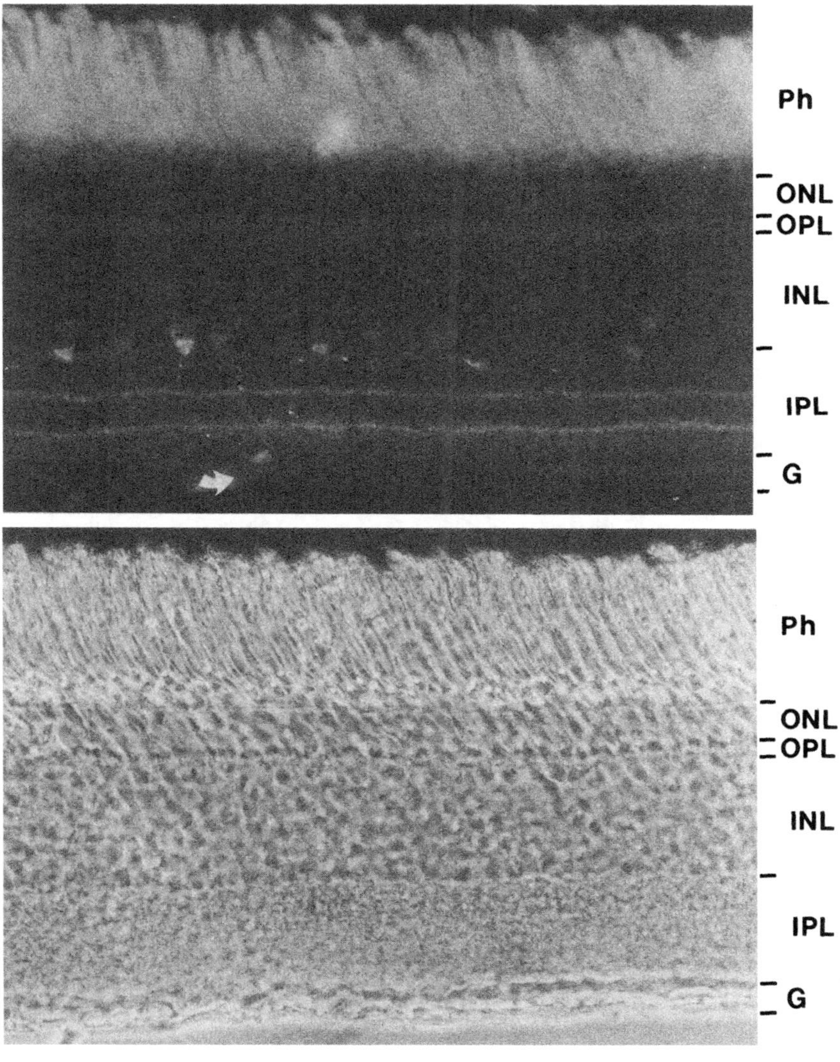

Fig. 7. Immunocytochemical demonstration of NPY immunoreactive cell bodies and fibers in retina from frog. Arrow indicates a displaced amacrine cell body. Designation of the layers as in Fig. 6 (× 260).

arterioles had only few or single 5-HT-positive nerve fibers (Edvinsson et al., 1984b). 5-Hydroxytryptamine fibers have, in addition, been observed in brain vessels from humans, cat, rabbit, rat, and mouse. The vasomotor effects of 5-HT are complex; cere-

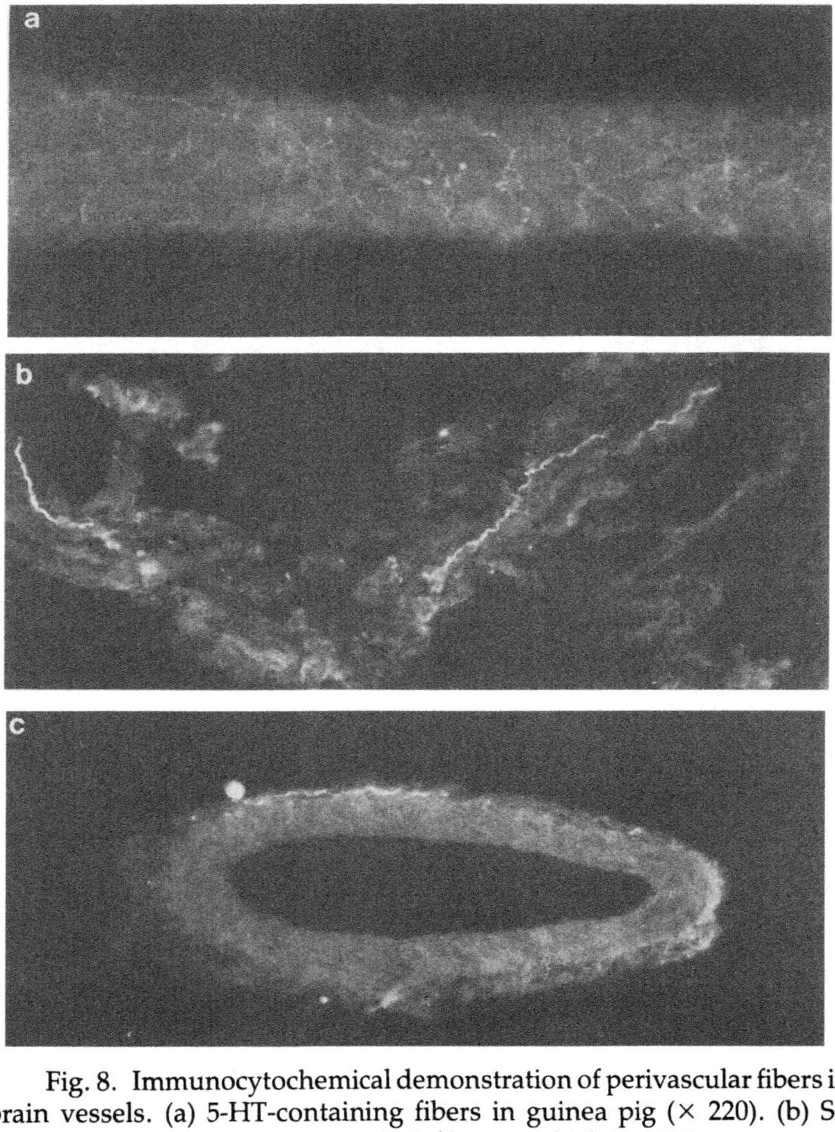

Fig. 8. Immunocytochemical demonstration of perivascular fibers in brain vessels. (a) 5-HT-containing fibers in guinea pig (× 220). (b) SP fibers in guinea pig (× 260). (c) VIP fibers in chick (× 173).

bral arterioles with diameters of less than 100 μm relax in response to 5-HT, whereas larger vessels show a constriction (MacKenzie and Edvinsson, 1980). The constriction is mediated via an unusual type of 5-HT receptor, whereas the relaxation probably occurs via a beta-adrenoceptor (Edvinsson et al., 1978). The localization of immunoreactive 5-HT in rat brain has been examined in detail by Steinbusch (1981), who showed that most parts of the central nervous system are supplied with 5-HT-positive fibers and neurons. In the guinea pig, we observed delicate 5-HT-positive fibers in the cerebral cortex and caudate nucleus, with a distribution resembling that noted previously for the rat.

5-Hydroxytryptamine-immunoreactive neurons were found in retina from chicken, pigeon, frog, and goldfish (Fig. 9a). These neurons were localized among the amacrine cell bodies, and the nerve fibers ramified in the inner plexiform layer. However, considerable interspecies variation exists. In bird retina, bipolar cell bodies have been found in the middle of the inner nuclear layer, sending processes to the inner plexiform layer and to the horizontal cells (Ehinger and Floren, 1976; Stenfors Tornqvist, 1983). In retina from guinea pig, no 5-HT-immunoreactive neurons have been observed. There are indoleamine-accumulating amacrine cells in chicken and frog retina, which thus contain endogenous 5-HT and, in addition, have an uptake mechanism for 5-HT. This supports the idea that 5-HT is a neurotransmitter in indoleamine-accumulating neurons of cold-blooded animals, and perhaps also in birds (Stenfors Tornqvist, 1983).

5.3. Substance P

A delicate network of substance P-containing nerve fibers was seen around brain arteries and veins from various mammals (Fig. 8b), including humans (Chan-Palay, 1977; Edvinsson et al., 1981; Edvinsson et al., 1983b). The major cerebral vessels in guinea pig (Fig. 4) and cat were surrounded by numerous SP fibers, whereas a moderate supply was seen in other species. As for all other perivascular fibers, the SP fibers were located in the adventitia or at the adventitia-media border. In some species the rostrally located arteries of the circle of Willis contained more SP fibers than vessels with a more caudal location. Some SP fibers were observed in cerebral vessels from frog and chicken. Capsaicin treatment has been noted to result in a substantial reduction of perivascular SP fibers (Duckles and Buck, 1982; Furness et al., 1982), indicating a sensory role of these fibers. Moreover, SP-containing cell bodies are numerous in sensory ganglia, and selective lesioning of the trigeminal ganglion causes the disappearance of SP within the eye, facial skin, and pial vessels (Unger et al., 1981; Uddman et al.,

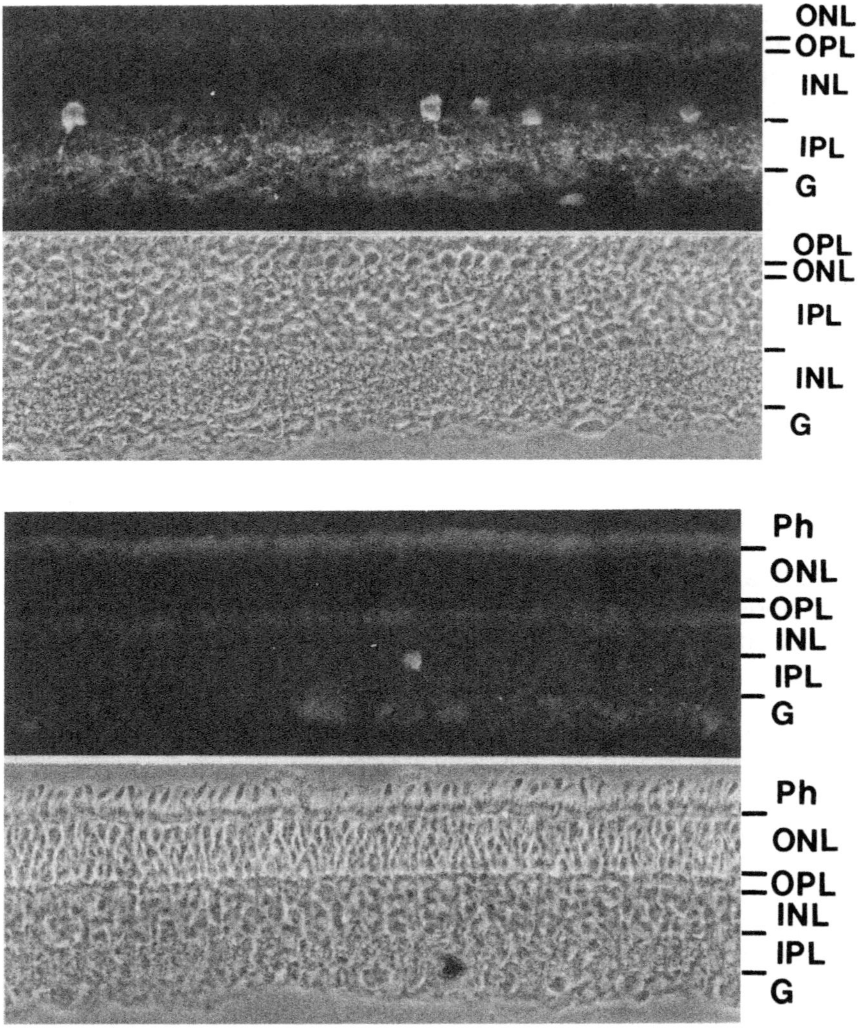

Fig. 9. Immunocytochemistry of retina. (Top) 5-HT-immunoreactive neurons in frog retina. Above, fluorescence micrograph; below, phase contrast micrograph (× 257). (Bottom) SP-immunoreactive cell body in guinea pig retina, ramifying in outer part of the inner plexiform layer. Above, fluorescence micrograph; below, phase contrast micrograph. Designation of layers as in Fig. 6 (× 280).

1985). Tracer studies have shown that SP fibers in cerebral blood vessels originate in the trigeminal ganglion (Mayberg et al., 1981; Liu-Chen et al., 1983; Yamamoto et al., 1983).

Intracortical SP-containing fibers were extremely few; a finding that is in accord with studies on the detailed distribution of SP (Hökfelt et al., 1977) and with quantitative analysis of the regional levels of SP in rat brain (Brownstein et al., 1976). However, particularly high concentrations of SP occur in the reticular part of the substantia nigra and in the interpeduncular nucleus. As one example of the functional role of SP, the nigral SP fibers have been found to exert a tonic excitatory influence on efferent GABAergic projections from pars reticulata of the substantia nigra (Melis and Gale, 1984).

SP-immunoreactivity was found in guinea pig and chicken retina, where it was observed in amacrine cell bodies and in processes that ramified in two or three layers in the inner plexiform layer (Fig. 9b) (Karten and Brecha, 1980). SP has a complicated function in retina, which has been only partly clarified (Glickman and Adolph, 1982).

5.4. *Vasoactive Intestinal Peptide*

Nerve fibers displaying VIP immunoreactivity were found in the walls of cerebral arteries of most laboratory animals and humans (Larsson et al., 1976; Edvinsson et al., 1980; Kobayashi et al., 1983a; Matsuyama et al., 1983; Edvinsson and Ekman, 1984) and were presently documented in guinea pig, frog, and chicken brain vessels (Fig. 8c). Blood vessels from guinea pig (Fig. 4), cat, and pig contained dense plexuses of VIP fibers, whereas a more sparse supply was seen in mouse, rat, humans, frog, and chicken. VIP fibers were occasionally seen to follow vessels that penetrated into the cerebral parenchyma. The origin of the perivascular VIP fibers is still unclear. It has been suggested that at least a portion of the VIP fibers to pial vessels originate in local perivascular ganglia (Gibbins et al., 1983), and others in the sphenopalatine ganglion (Hara et al., 1985). VIP is a potent dilator of cerebral arteries and veins, and this response occurs in parallel with activation of adenylate cyclase. Intraarterial infusion or local intracerebral administration of VIP causes an increase in cerebral blood flow, cerebral oxygen consumption, and glucose use, as well as signs of activation in the EEG (McCulloch and Edvinsson, 1980).

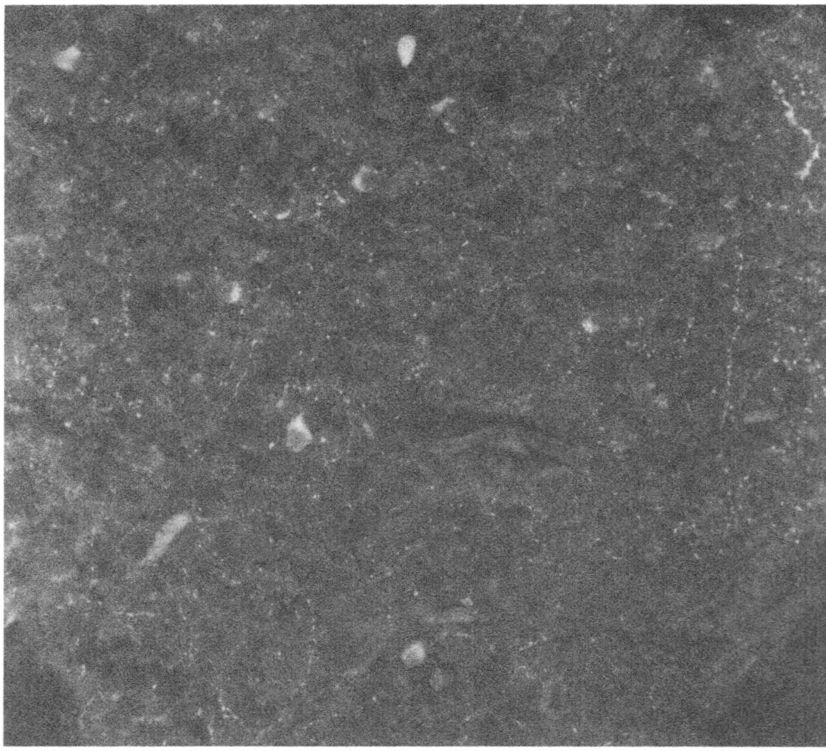

Fig. 10. Examples of VIP immunoreactive fibers in cerebral cortex of chick (× 270).

Intracortical VIP fibers were present in all cortical layers (Fig. 10). The VIP neurons were usually bipolar, with long, radially directed processes with limited branching (Loren et al., 1979). The density of the axonal varicosities was highest in cortical layers II–IV. Detailed immunocytochemical analysis has revealed that each VIP-containing neuron is confined within a radial volume of approximately 15–60 μm in diameter, with some overlapping of neighboring domains (Morrison et al., 1984). These data suggest that the VIP-neurons may play an important role in the radially oriented columns of cerebral cortex. Only scattered VIP-positive fibers were seen in the caudate nucleus.

Neurons displaying VIP immunoreactivity were detected among the amacrine cells in the retina of baboon, cynomolgous monkey, squirrel monkey, cow, pig, cat, rabbit, guinea pig, rat,

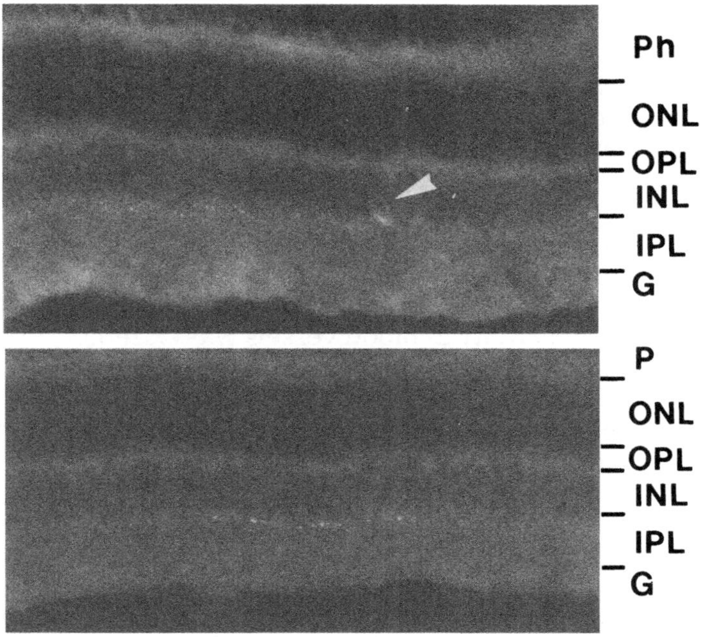

Fig. 11. VIP-immunoreactive neurons in the guinea pig retina. Above, arrow indicates a weakly immunoreactive amacrine cell body situated close to the inner plexiform layer. Below, immunoreactive nerve fibers in the inner-most part of the inner plexiform layer. Designation of layers as in Fig. 6 (× 310).

mouse, frog, and goldfish (Fig. 6). Generally, the immunoreactive cell bodies were located in the inner nuclear layer, with processes ramifying in three more or less well-defined sublayers in the inner plexiform layer (Fig. 11) (Tornqvist et al., 1982).

Little is known concerning the function of VIP in retina. VIP has been shown to stimulate retinal adenylate cyclase. The morphological localization of the retinal VIP neurons suggests that they may have a modulatory effect on signals between the retinal cells and fibers (Tornqvist et al., 1982).

6. Conclusion

Immunocytochemical techniques for the localization of specific antigens have provided the methodological basis for the microscopic identification of various agents in cell bodies and nerve fibers. The described protocol (in Appendix I) may serve as a useful starting point for researchers intending to examine the localization of, e.g., peptides. With proper attention to fixation and controls, new basic information can be obtained with high precision.

We observed that the concentration of formaldehyde in the fixative has considerable significance for the immunocomplex binding and that the presence of picric acid may change the binding capability for several antibodies toward neurotransmitters. The NPY immunoreactivity in blood vessels was extremely good for all fixatives tested, whereas in retina and cerebrum the 4% formaldehyde was preferable. For the 5-HT antibody, the best result in blood vessels and cerebrum was seen after immersion fixation with picric acid-formaldehyde. There was no observable 5-HT immunoreactivity in the retina of guinea pig. The optimum fixation procedure for SP and VIP was 4% formaldehyde immersion fixation and picric acid-formaldehyde perfusion fixation.

However, the localization of a particular antigen is entirely dependent on the specificity of the antibody; thus, it is of utmost importance that the antibodies are well characterized (Landis, 1985). A negative immunocytochemical finding does not necessarily indicate that the antigen is absent in the tissue. The antigen may be present, but because of technical limitations or low levels in the specimens, it remains undetected. Examples exist in which peptide levels are raised by ligation of a nerve or upon treatment with colchicine, enhancing the tissue levels to beyond the detection limit. It has also been demonstrated that peptide processing may vary in different regions of the organism. In addition, considerable interspecies differences exist, a fact that must be taken into consideration.

Appendix I

Immunocytochemistry Procedures

Protocol for Immunocytochemistry in Tissue Sections

1. Tissue must be obtained with a minimum period of anoxia (within minutes), to prevent degradation and diffusion of the peptide material.

2. Fix in (5 mm in thinnest dimension) 4% paraformaldehyde in Sorensen buffer, pH 7.4, or in picric acid-formaldehyde fixative, pH 7.4, at 4°C for 24 h.
3. Wash in Sorensen buffer, pH 7.4, with 20% sucrose at 4°C for 6 h × 4.
4. If necessary store in Sorensen buffer, pH 7.4, containing 20% sucrose and 0.1% sodium azide.
5. Cut in cryostat, at –20°C, 15 μm sections and mount the sections on chromalum-subbed slides. Store in freezer (–18°C).
6. Dilute primary antibody with antibody diluent to the optimum working dilution (to reduce the possibility of nonspecific and background staining). Apply about 50 μl to a slide.
7. Place in a humid environment (e.g., a plastic chamber) at room temperature for about 5 h and then at 4°C for about 20 h (usually overnight).
8. Put the humid chamber at room temperature for 1 h.
9. Rinse the slides with PBS-buffer with 0.25% Triton X, 15 min × 2.
10. Dilute the secondary antibody, 1:80, with antibody diluent and apply as in step 6.
11. Rinse as in step 9.
12. Mount in PBS:glycerol (1:1) and inspect in a fluorescence microscope with appropriate filter settings.

Protocol for Immunocytochemistry of Whole Mounts

1. Tissue must be obtained within a minimum period of anoxia to prevent degradation of the peptides and drying of the tissue.
2. Fix in 4% paraformaldehyde in Sorensen buffer, pH 7.4, or in picric acid-formaldehyde fixative, pH 7.4, for 24 h. Either mount the tissue directly on chromalum-subbed slides and fix the tissue in a jar with fixative, or immerse the specimen into the fixative solution for later mounting.
3. Wash in Sorensen buffer, pH 7.4, with 20% sucrose at 4°C for 6 h × 4 while still in a jar or still in solution.
4. When the rinsing procedure has been completed, store the slides in a freezer at –18°C, or mount the specimens on chromalum-subbed slides and store in a freezer (–18°C).

5. Before the staining procedure, immerse the slides in PBS-buffer with 0.25% Triton X for 15 min.
6. Continue the staining procedure starting with step 6 in the protocol for immunocytochemistry in tissue sections.

Appendix II

Buffers, Fixatives, and Other Solutions

Picric Acid-Formaldehyde Fixative

Dissolve 1.2 g picric acid in 300 ml 0.2*M* Na acetate with 3% sucrose, then add 300 ml 8% paraformaldehyde, prepared in a boiling waterbath until complete solution occurs.

Final concentrations:
4% Formaldehyde
0.1*M* Na acetate
0.2% Picric acid
1.5% Sucrose
pH 7.4

0.1M Sorensen Buffer, pH 7.4

Stock solution A: Dissolve 27.60 g $NaH_2PO_4 \cdot H_2O$ in 1000 ml distilled water and store in refrigerator.
Stock solution B: Dissolve 35.60 g $Na_2HPO_4 \cdot 2H_2O$ in 1000 ml distilled water and store in room temperature to avoid precipitation.
Mix: 19 ml from stock A + 81 ml from stock B + 100 ml distilled water, adjust pH to 7.4 with 1*N* HCl or 1*N* NaOH.

4% Formaldehyde in Sorensen Buffer pH 7.4

Dissolve 8 g paraformaldehyde in 100 ml 0.2*M* Sorensen buffer, pH 7.4, in a boiling waterbath until it is in complete solution, add 100 ml distilled water, let it cool to room temperature, and adjust the pH to 7.4 with 1*N* HCl or 1*N* NaOH.

Final concentrations:
4% formaldehyde
0.1*M* Na-phosphate
pH 7.4

Phosphate-Buffered Saline (PBS), pH 7.4

Stock solution A: Dissolve 1.4 g Na_2HPO_4 in 100 ml distilled water. Stock solution B: Dissolve 1.6 g NaH_2PO_4 in 100 ml distilled water. Mix 84.1 ml from A + 15.9 ml from B + 8.5 g NaCl + 900 ml distilled water, adjust pH to 7.4 with 2*N* HCl.

Antisera Diluent

PBS-buffer, pH 7.4, containing 0.25% bovine serum albumin and 0.25% Triton X.

Rinse Buffer

PBS-buffer, pH 7.4, + 0.25% Triton X.

Chromalum Gelatin Solution

1 g Gelatin
0.1 g Chromium potassium sulfate × 12 H_2O
200 ml distilled water
Dissolve in a waterbath at 60°C, and let mature overnight.
Dip the slides in the solution and allow them to dry overnight in a vertical position.

Acknowledgments

Supported by a grant from the Swedish Medical Research (5958,2321) and H. and L. Nilssons Foundation.

References

Allen Y. S., Adrian T. E., Allen J. M., Tatemoto K., Crow J., Bloom S. R., and Polak J. M. (1983) Neuropeptide Y distribution in the rat brain. Science **221,** 877–879.

Alm P., Alumets J., Håkanson R., Owman C., Sjöberg N. O., and Walles B. (1980) Origin and distribution of VIP (vasoactive intestinal polypeptide) -nerves in the genitourinary tract. *Cell Tissue Res.* **205,** 337–347.

Brecha N. (1983) Retinal Neurotransmitters: Histochemical and Biomedical Studies, in *Chemical Neuroanatomy* (Emson P. C., ed.) Raven, New York.

Brownstein M. J., Mroz E. A., Kizer J. S., Palkovits M., and Leeman S. E. (1976) Regional distribution of substance P in the brain of the rat. *Brain Res.* **116,** 299–305.

Bruun A., Ehinger B., Sundler F., Tornqvist K., and Uddman R (1984) Neuropeptide Y immunoreactive neurons in the guinea-pig uvea and retina. *Invest. Ophthalmol. Vis. Sci.* **25,** 1113–1123.

Chan-Palay V. (1977) Innervation of Cerebral Blood Vessels by Norepinephrine, Indoleamine, Substance P and Neurotensin Fibres and the Leptomeningeal Indoleamine Axons: Their Roles in Vasomotor Activity and Local Alterations of Brain blood Composition, in *Neurogenic Control of Brain Circulation* (Owman C. and Edvinsson L., ed.). Pergamon, Oxford.

Coons A. H., Creech H. J., Jones R. N., and Berliner E. (1942) The demonstration of pneumococcal antigen in tissues by the use of fluorescent antibody. *J. Immun.* **45,** 159–170.

Coons A. H., Leduc E. H., and Connolly J. M. (1955) Studies on antibody production. I. A method for the histochemical demonstration of specific antibody and its application to a study of the hyperimmune rabbit. *J. Exp. Med.*, 49–60.

Costa M., Buffa R., Furness J. B., and Solcia E. (1980) Immunohistochemical localization of polypeptides in peripheral autonomic nerves using whole mount preparations. *Histochemistry* **65,** 157–165.

Cuello A. C., ed. (1983) *Immunohistochemistry* John Wiley, Chichester.

De Mey J. (1983) A critical review of light and electron microscopic immunocytochemical techniques used in neurobiology. *J. Neurosci. Meth.* **7,** 1–18.

Duckles S. P., and Buck S. H. (1982) Substance P in the cerebral vasculature: Depletion by capsaicin suggests a sensory role. *Brain Res.* **245,** 171–174.

Edvinsson L., and Ekman R. (1984) Distribution and dilatory effect of vasoactive intestinal polypeptide (VIP) in human cerebral arteries. *Peptides* **5,** 329–332.

Edvinsson L., Hardebo J. E., and Owman C. (1978) Pharmacological analysis of 5-hydroxytryptamine receptors in isolated intracranial and extracranial vessels of cat and man. *Circ. Res.* **42,** 143–151.

Edvinsson L., Fahrenkrug J., Hanko J., Owman C., Sundler F., and Uddman R. (1980) VIP (vasoactive intestinal polypeptide)-containing nerves of intracranial arteries in mammals. *Cell Tissue Res.* **208,** 135–142.

Edvinsson L., McCulloch J., and Uddman R. (1981): Substance P: Immunohistochemical localization and effect upon cat pial arteries in vitro and in situ. *J. Physiol.* **318,** 251–258.

Edvinsson L., Rosendahl-Helgesen S., and Uddman R. (1983a) Substance P-localization, concentration and release in cerebral arteries, choroid plexus and dura mater. *Cell Tissue Res.* **234,** 1–7.

Edvinsson L., Emson P., McCulloch J., Tatemoto K., and Uddman R. (1983b) Neuropeptide Y: Cerebrovascular innervation and vasomotor effects in the cat. *Neurosci. Lett.* **43,** 79–84.

Edvinsson L., Emson P., McCulloch J., Tatemoto K., and Uddman R. (1984a) Neuropeptide Y: Immunocytochemical localization to and effect upon feline pial arteries and veins in vitro and in situ. *Acta Physiol. Scand.* **122,** 155–163.

Edvinsson L., Birath E., Uddman R., Lee T.J.-F., Duverger D., MacKenzie E. T., and Scatton B. (1984b) Indolaminergic mechanisms in brain vessels: Localization, concentration, uptake and in vitro responses of 5-hydroxytryptamine. *Acta Physiol. Scand.* **121,** 291–299.

Ehinger B. and Floren I. (1976) Indol accumulating neuron in the retina of rabbit, cat and goldfish. *Cell Tissue Res.* **175,** 37–48.

Ekblad E., Edvinsson L., Wahlestedt C., Uddman R., Håkanson R., and Sundler F. (1984) Neuropeptide Y co-exists and co-operates with noradrenaline in perivascular nerve fibres. *Reg. Pept.* **8,** 225–235.

Everitt B. J., Hökfelt T., Terenius L., Tatemoto K., Mutt V., and Goldstein M. (1984) Differential co-existence of neuropeptide Y (NPY)-like immunoreactivity with catecholamines in the central nervous system of the rat. *Neuroscience* **11,** 443–462.

Furness J. B., Papka R. E., Della N. G., Costa M., and Eskay R. L. (1982) Substance P-like immunoreactivity in nerves associated with the vascular system of guinea-pigs. *Neuroscience* **7,** 447–459.

Furness J. B., Costa M., and Keast J. R. (1984) Choline acetyltransferase and peptide immunoreactivity of submucous neurons in the small intestine of the guinea-pig. *Cell. Tiss. Res.* **237,** 328–336.

Geffen L. B., Livett B. G., and Rush R. A. (1969): Immunohistochemical localization of protein components of catecholamine storage vesicles. *J. Physiol.* **204,** 593–605.

Gibbins I. L., Brayden J. E., and Bevan J. A. (1983) Distribution and origins of VIP-immunoreactive (VIP-IR) nerves in the cranial circulation of the cat. *Reg. Pept.* **6,** 304.

Glickman R. D. and Adolph A. R. (1982) Acetylcholine and substance P: Action via distinct receptors on carp retinal ganglion cells. *Invest. Ophthalmol. Vis. Sci.* **22,** 804–808.

Hara H., Hamill G. S., and Jacobovitz D. M. (1985) Origin of cholinergic nerves to the rat major cerebral arteries: Coexistence with vasoactive intestinal polypeptide. *Brain Res. Bull.* **14,** 179–188.

Hökfelt T., Johansson O., Kellerth J.-O., Ljungdahl A., Nilsson G., Nygårds A., and Pernow B. (1977) Immunohistochemical Distribution of Substance P, in *Substance P* (von Euler U.S. and Pernow B., eds.) Raven, New York.

Joh T. H., Gegham C., and Reis D. J. (1963) Immunochemical demonstration of increased tyrosine hydroxylase protein in sympathetic ganglia and adrenal medulla elicited by reserpine. *Proc. Natl. Acad. Sci. USA* **70,** 2767–2771.

Karten H. J. and Brecha N. (1980) Localization of substance P immunoreactivity in amacrine cells of the retina. *Nature* **283,** 87–88.

Kobayashi S., Kyoshima K., Olschowka J. A., and Jacobowitz D. M. (1983) Vasoactive intestinal polypeptide immunoreactive and cholinergic nerves in the whole mount preparation of the major cerebral arteries of the rat. *Histochemistry* **79,** 377–381.

Landis D. M. D. (1985): Promise and pitfalls in immunocytochemistry. *Trends Neurosci.* **8,** 312–317.

Larsson L-I., Edvinsson L., Fahrenkrug J., Håkanson R., Owman C., Schaffalitzky de Muckadell O. B., and Sundler F. (1976) Immunohistochemical localization of a vasodilatory polypeptide (VIP) in cerebrovascular nerves. *Brain Res.* **113,** 400–404.

Liu-Chen L-Y., Han D. H., and Moskowitz M. A. (1983) Pia arachnoid contains substance P originating from trigeminal neurons. *Neuroscience* **9,** 803–808.

Lorén I., Emson P. C., Fahrenkrug J., Björklund A., Alumets J., Håkanson R., and Sundler F. (1979) Distribution of vasoactive intestinal polypeptide in the rat and mouse brain. *Neuroscience* **4,** 1953–1976.

MacKenzie E. T. and Edvinsson L. (1980) Effects of Serotonin on Cerebral Circulation and Metabolism, as Related to Cerebrovascular Disease, in *Cerebral Circulation and Neurotransmitters* (Bes A. and Geraud G., eds.) Excerpta Medica, Amsterdam.

Marrack J. (1934) Nature of antibodies. *Nature* **133,** 292–293.

Marshall J. M. (1951) Localization of adrenocorticotrophic hormone by histochemical and immunohistochemical methods. *J. Exp. Med.* **94,** 21–30.

Matsuyama T., Shiosaka S., Matsumoto M., Yoneda S., Kimura K., Abe H., Hayakawa T., Inoue H., and Tohyama M. (1983) Overall distribution of vasoactive intestinal polypeptide-containing nerves in the wall of the cerebral arteries: An immunohistochemical study using whole mounts. *Neuroscience* **10,** 89–96.

Mayberg M., Langer R. S., Zervas N. T., and Moskowitz M. A. (1981) Perivascular meningeal projections from cat trigeminal ganglia: Possible pathway for vascular headaches in man. *Science* **213,** 228–230.

McCulloch J. and Edvinsson L. (1980) Cerebral circulatory and metabolic effects of vasoactive intestinal polypeptide. *Am. J. Physiol.* **238,** H449–H456.

Melis M. R. and Gale K. (1984) Evidence that nigral substance P controls the activity of the nigrotectal GABAergic pathway. *Brain Res.* **295,** 389–393.

Mellors R. C., Siegel M., and Pressman D. (1955) Analytical pathology. I. Histochemical demonstration of antibody localization in tissue, with special reference to the antigen components of kidney and lung. *Lab. Invest.* **4,** 69–89.

Morrison J. H., Magistretti P. J., Benoit R., and Bloom F. E. (1984) The distribution and morphological characteristics of the intracortical VIP-positive cell: An immunohistochemical analysis. *Brain Res.* **292,** 269–282.

Pickel V. M., Joh T. H., Reis D. J., Leeman S. E., and Miller R. J. (1979) Electron microscopic localization of substance P and enkephalin in axon terminals related to dendrites of catecholaminergic neurons. *Brain Res.* **160,** 387–400.

Stefanini M., Demartino C., and Zamboni C. (1967) Fixation of ejaculated spermatozoa for electron microscope. *Nature* **216,** 173–174.

Steinbusch H. W. M. (1981) Distribution of serotonin-immunoreactivity in the central nervous system of the rat—cell bodies and terminals. *Neuroscience* **6,** 557–618.

Steinbusch H. W. M., Verhofstad A. A. J., and Joosten H. W. J. (1978) Localization of serotonin in the central nervous system by immunohistochemistry: Description of a specific and sensitive technique and some applications. *Neuroscience* **3,** 811–819.

Sternberger L. A. (1974) *Immunocytochemistry* Prentice-Hall, Englewood Cliffs, New Jersey.

Sternberger L. A., Hardy P. H., Cuculis J. J., and Meyer H. G. (1970) The unlabeled antibody-enzyme method of immunohistochemistry. Preparation and properties of soluble antigen-antibody complex (horseradish peroxidase-anti horseradish peroxidase) and its use in identification of spirochetes. *J. Histochem. Cytochem.* **18,** 315–333.

Stenfors Tornqvist K. (1983) 5-Hydroxytryptamine and neuropeptides in the retina. PhD thesis, Lund University, Lund, Sweden.

Sundler F., Moghimzadeh E., Håkanson R., Ekelund M., and Emson P. (1983) Nerve fibres in the gut and pancreas of the rat displaying neuropeptide-Y immunoreactivity. Intrinsic and extrinsic origin. *Cell Tissue Res.* **230,** 487–493.

Tornqvist K., Uddman R., Sundler F., and Ehinger B. (1982) Somatostatin and VIP neurons in the retina of different species. *Histochemistry* **76,** 137–152.

Tramu G., Pillez A., and Leonardelli, J. (1978) An efficient method of antibody elution for the successive or simultaneous location of two antigens by immunocytochemistry. *J. Histochem. Cytochem.* **26,** 322–324.

Tuor U., Kelly, Tatemoto K., Edvinsson L., and McCulloch J. (1986) Neuropeptide Y and the Cerebral Circulation, in *Neural Regulation of Brain Circulation* (Owman C. and Hardebo J. E., eds.) Elsevier, Amsterdam pp. 333–354.

Uddman R., Edvinsson L., Ekman R., Kingman T., and McCulloch J. (1985) Innervation of the feline cerebral vasculature by nerve fibers containing calcitonin gene-related peptide: Trigeminal origin and co-existence with substance P. *Neurosci. Lett.* **62,** 131–136.

Unger W. G., Butler J. M., Cole D. F., Bloom S. R., and McGregor G. P. (1981) Substance P, vasoactive intestinal polypeptide (VIP) and somatostatin levels in ocular tissue of normal and sensorily denervated rabbit eyes. *Exp. Eye Res.* **32,** 797–801.

Weir D. M., ed. (1973) *Handbook of Experimental Immunology* Blackwell Scientific, Oxford.

Yamamoto K., Matsuyama T., Shiosaka S., Inagaki S., Senba E., Shimizu Y., Ishimoto I., Hayakawa T., Matsumoto M., and Tohyama M. (1983) Overall distribution of substance P-containing nerves in the wall of the cerebral arteries of the guinea-pig and its origin. *J. Comp. Neurol.* **215,** 421–426.

Zamboni L. and De Martino C. (1967) Buffered picric-acid formaldehyde: A rapid fixative for electron-microscopy. *J. Cell. Biol.* **35,** 148A.

Sensory-Evoked Potentials

Betty L. Grundy

1. Introduction

Sensory-evoked potentials (EPs) are the electrophysiologic responses of the nervous system to sensory stimulation (Chiappa and Ropper, 1982a,b; Greenberg and Ducker, 1982). These responses, which reflect the condition of sensory pathways, can be used to identify abnormal or nonfunctional sensory pathways throughout the nervous system. EPs have been used for mapping the various structures that serve sensory functions of the nervous system in animals and humans (Kelly et al., 1965; Stohr and Goldring, 1969; Woolsey et al., 1979; Allison et al., 1986); for monitoring the functional integrity of specific sensory pathways and, to some extent, of adjacent pathways as well (Grundy, 1982, 1983); and for measuring the neurologic responses to experimental and therapeutic interventions (Grundy et al., 1981a,b,c; Hacke, 1985; Mackey-Hargadine and Hall, 1985). This chapter will present a classification of EPs according to types of stimulation, modes of recording, methods of signal processing, types of poststimulus latencies, and purported neural generators. Current techniques will be described and experimental and clinical applications will be summarized.

2. Classification of Evoked Potentials

2.1. Types of Stimulation

2.1.1. Somatosensory

The somatosensory system can be stimulated by electrical (Dawson, 1947; Jones, 1977), mechanical (Pratt and Starr, 1981), or thermal (Carmon et al., 1976) techniques. Electrical stimuli are delivered to cutaneous receptors through surface electrodes to test sensory function in specific dermatomes, most often to localize radiculopathies diagnostically or during surgical procedures. In clinical practice, sensory or mixed nerves are stimulated electrically by using surface or subdermal electrodes. Somatosensory-evoked potentials (SEPs) elicited by stimulating the median nerve at the wrist, the posterior tibial nerve at the ankle, or the common peroneal nerve at the knee are often used for diagnostic and monitoring applications; the sural nerve at the ankle and the ulnar nerve at the wrist also provide information. Ring electrodes are sometimes used for pure sensory stimulation of the fingers. Magnetic transients can be used to stimulate peripheral nerves by inducing an electrical current transcutaneously with no electrode contact (Barker et al., 1985).

When mixed nerves are stimulated, the most common clinical practice, the stimulation is complex and includes orthodromic sensory stimuli, antidromic motor stimuli, and a later component of sensory stimulation produced by the muscle contractions caused by orthodromic motor impulses. The information reaching the spinal cord and cortex is therefore complex and dispersed over time (Burke et al., 1980).

Electrical stimulation of mixed nerves sufficient to produce a motor twitch in muscles supplied by the nerve is called "supramaximal." Although it is true that stimulus intensities above this level have relatively little effect on the evoked potential, some increase in amplitude definitely occurs, and the stimulus level itself should be documented and kept constant during recording. An electrical current representing the sum of the motor threshold plus the sensory threshold has been suggested (Lesser et al., 1979). Some workers use the lowest current that produces a visible muscle twitch; others use the highest level comfortably tolerated by the subject.

2.1.2. Trigeminal

This nerve may be stimulated by techniques similar to those used to stimulate somatosensory nerves. Surface electrodes are placed on the face or on the gingivae (Bennett and Jannetta, 1980). Electrical stimulation of the tooth pulp is thought to represent a pure pain stimulus. Tooth pulp EPs are valuable in experimental studies of analgesic agents (Derendorf et al., 1982).

2.1.3. Auditory

The auditory system is most often stimulated by broad-band filtered clicks or pure tones. Shielded headphones are used in the diagnostic laboratory; a variety of ear pieces has been developed for intraoperative recording, when headphones would interfere with access to the surgical field. Effective stimulation depends on unimpaired air conduction of sound to the tympanic membrane. The external auditory canal must not be obstructed by cerumen, a foreign body, or a pathologic condition. Pressure from the headphones may cause the external auditory canal to collapse, particularly in patients at the extremes of age. Otoscopic examination should be performed before recording auditory-evoked potentials, and headphones must be placed carefully on infants and elderly patients (American Electroencephalographic Society, 1984).

Either rarefaction or condensation clicks may be used. Rarefaction clicks are most effective as stimulators of the cochlea (Stockard et al., 1979). Clicks of alternating polarity are sometimes used to minimize stimulus artifact when a recording electrode is near the electromagnetic transducer (Grundy et al., 1982).

The intensity of auditory stimulation is most accurately characterized as the sound pressure level (SPL), measured in decibels (dBs). Clinically, however, intensity is more often described as decibels above the mean hearing level of the normal population (dBHLs) or decibels above the sensation level in a particular subject (dBSLs). Audiologists more often use dBHLs, particularly when auditory EPs are used to measure hearing. Neurologists frequently use dBSLs to better standardize peripheral sensory input in groups of patients with some partial hearing loss.

With pure tone stimulation, particular portions of the cochlea can be tested. As the displacement wave moves along the basilar membrane of the cochlea, the point of maximal displacement depends on the frequency of the stimulus. Low-frequency signals stimulate maximally at the apex, and high-frequency signals, at the base. High-frequency fibers can also respond to low-frequency stimulation; this confuses interpretation when there is a pathologic condition. Tone bursts are symmetrical around 0 voltage and include both positive and negative components. The important characteristic clinically is the initial phase or polarity of the tone, which can be adjusted to improve recordings when stimulus artifact is excessive. With alternating polarity of pure tone stimuli, both the stimulus artifact and the electrical response of the cochlea, the cochlear microphonic, are obliterated while the later neural potentials persist. Several techniques are available for calibrating the polarity, temporal characteristics, and amplitude spectra of auditory stimuli (Gorga et al., 1985).

2.1.4. Visual

The visual system may be stimulated with flashes or with reversing patterns. Flash stimulation is most often used for electroretinography (ERG) (Armington, 1986). Full-field stimulation is used so that all parts of the retina are stimulated directly and stray light has minimal effect. The subject sits in a dark room and looks into one side of a sphere, the entire sphere being illuminated by the flashes. With adaptation to different wavelengths of light, some distinction can be made between the photopic system, which entails the response of cones to high levels of illumination, and the scotopic system, which depends on stimulation of the retinal rods by low levels of illumination. Test stimuli can be superimposed on color adaptation fields. During adaptation to long-wavelength light (red), photopic sensitivity is reduced, and the ERG reflects predominately scotopic activity. In contrast, with adaptation to short-wavelength light (blue), scotopic sensitivity is reduced, and the ERG reflects primarily photopic activity. The ERG produced by a bright flash to an eye totally adapted to the dark reflects both scotopic and photopic activity. The retina can also be stimulated regionally. Stimuli directed to the fovea have the greatest photopic effect, whereas the greatest scotopic effect is elicited when the stimulus forms a ring around the central region and the image falls on the peripheral parts of the retina, where the rods are concentrated.

Both flash and pattern-reversal stimulation are used to elicit visual EPs (Sokol, 1986). In clinical diagnostic applications, pattern reversal is used more often than flash because the former yields more reproducible EPs within and among subjects. Edge receptors in the retina are stimulated by the reversal of a checkerboard pattern on a television screen. Various check sizes may be used and redescribed as the number of degrees of visual arc, a function of the check size on a television screen used for stimulation, and the distance between the screen and the subject's eye. Corrective lenses must be worn during this test, and the subject must cooperate by focusing on a spot in the center of the television screen. Because of the need for cooperation from the subject, pattern reversal stimulation cannot be used in anesthetized patients.

Flash stimulation is provided by a bright flash or by an array of light-emitting diodes (LEDs) mounted in opaque goggles. With goggles over closed eyes, the luminance of the stimulus cannot be measured or reproduced completely because of variability in the characteristics of eyelids. Light-emitting diodes mounted in scleral contact lenses can be used for flash stimulation during anesthesia and operation; there is less variability than when the LEDs are mounted in goggles, but contact lenses introduce the risk of corneal abrasion, particularly when the eye is near the surgical field and cannot be observed directly. On the other hand, care must be taken with goggles to ensure that the goggle is not displaced onto the eye intraoperatively. Pressure on the eye may cause thrombosis of the central retinal artery, with permanent loss of vision in the affected eye. Problems in achieving adequate visual stimulation and the exquisite sensitivity of visually evoked potentials (VEPs) to anesthetic agents have limited the applications of VEPs in surgical and comatose patients.

2.1.5. Motor

Because SEPs are transmitted primarily through the dorsal columns in the spinal cord and do not always reflect the function of the anterior cord, other methods have been developed to test motor function in both animals and humans. Either the motor cortex or the spinal cord can be stimulated by percutaneous electrical discharges (Merton et al., 1982; Levy et al., 1984); magnetic transients also have been used (Barker et al., 1985). Electrical current can be delivered to the motor cortex by using bipolar scalp electrodes or by placing one electrode on the scalp over the motor cortex and another electrode on the hard palate. Magnetic stimulation is achieved by holding a magnet over the motor cortex, spinal cord, or peripheral nerve. Electrical stimulation of the motor cortex can be uncomfortable, and experience with this technique is still limited.

Theoretical concerns about scalp-evoked motor potentials include the possibility of kindling with the production of a seizure focus after prolonged repetitive stimulation, but total current delivery is less than that classically used for inducing seizures in experimental animals. Transcortical stimulation of the motor cortex can produce remarkable hypertension in surgical patients. Magnetic stimulation seems inherently safer, but this technique too may have some risk. Movement of ferromagnetic objects is a problem.

Motor nerve conduction studies are commonly used both diagnostically and intraoperatively to identify and test either cranial or peripheral nerves. In these cases, the stimulating anode is usually hand-held, and the cathode is placed nearby on the surface of the skin or in subcutaneous tissue. Electrodes placed in the epidural space either blindly or under direct vision, as well as electrodes placed directly on the spinal cord, can be used to elicit motor activity. Electrodes can also be placed invasively under fluoroscopic guidance or with computerized tomographic (CT) control.

2.1.6. *Noninvasive*

Noninvasive techniques to stimulate sensory systems are simpler, safer, and more comfortable for the patient than invasive techniques and therefore are preferable when the same information can be obtained. Little additional risk would seem to be added by placing electrodes within a surgical field intraoperatively, however.

2.1.7. *Invasive*

Techniques for invasive stimulation are discussed below, along with the description of techniques for invasive recording.

2.1.8. *Electrical*

With noninvasive electrical stimulation, electrode impedance is reduced by light abrasion of the skin and use of a conductive gel. Lower currents are needed when neural structures are stimulated directly. The negative pole of an electrode pair, or cathode, stimulates neurons by depolarization. When both electrodes are placed along the pathway of a cranial or peripheral nerve, the cathode is placed proximally for sensory stimulation and distally for motor stimulation to avoid blocking axons secondary to the hyperpolarization produced by the anode. Polarity is less important at higher stimulus intensities.

Because it is electrical current that actually depolarizes neurons, constant current stimulators are preferable to those with constant voltage. With constant current stimulators, small changes in electrode impedance are automatically compensated for by increased voltage; similar impedance changes alter the intensity of the stimulus provided by a constant voltage stimulator. Square-wave pulses of 50–300 μs may be used; impulses of approximately 100 μs are probably best. The level of current for each patient is set by testing for sensory and motor thresholds. For most patients, a current equal to the sum of motor threshold plus sensory threshold is reasonably comfortable (Lesser et al., 1979); higher levels, commonly 20 or even 30 mA, may be employed during anesthesia.

2.1.9. Mechanical

Somatosensory EPs may be elicited by repetitive, brisk tapping of a fingernail, by stretch of tendons, or by puffs of air to the skin (Pratt and Starr, 1981). These methods are not commonly used clinically.

2.1.10. Thermal

Single filtered responses to a short burst of laser radiation sufficiently intense to produce a second-degree burn may represent an electrophysiologic correlate of pain (Carmon et al., 1976). This is of particular interest because of the difficulties in finding an objective measurement of pain.

2.1.11. Olfactory

Olfactory stimulation has been used experimentally, but has no widespread clinical applications at this time.

2.1.12. Magnetic

Peripheral nerves, spinal cord, or cortex may be stimulated by magnetic transients delivered from a hand-held magnet placed over the appropriate neural structures (Barker et al., 1985).

2.1.13. Cognitive

A late positive potential with a latency varying from 250 to 400 ms or longer can be recorded over central parietal areas when cognitive tasks are performed in response to sensory stimulation (Goodin et al., 1978). This P300 may be elicited by having a subject attend to or count infrequent stimuli within a series of frequent, but related, stimuli. For example, tones of higher frequency may be randomly interspersed in a train of low-frequency tones. Single alerting stimuli, such as repetition at random intervals of the subject's name or other meaningful stimuli, can also elicit the P300.

2.2. Modes of Recording

2.2.1. Electrical

The overwhelming preponderance of data comes from electrical recording. Once electrical impulses are generated by the nervous system, they are transmitted virtually instantaneously through the body by volume conduction. Signal strength decreases as the distance between the neural generator of the potential and the recording electrode increases (Gevins, 1986).

2.2.1.1. Near-Field Potentials. In near-field recording, the recorded signal predominately reflects activity within 2–3 cm of the electrode. Amplitudes are in the range of 1–100 mV, and both configurations and amplitudes of waveforms are sensitive to slight changes in electrode location, on the order of 1–2 cm. In scalp recordings, near-field potentials reflect activity in the cerebral cortex. Signals recorded from electrodes placed over superficial nerves or plexuses, such as the electrode at Erb's point for recording activity in the brachial plexus, are also near-field potentials. Recordings from deep structures such as the spinal cord may be near-field potentials when invasive electrodes are used.

2.2.1.2. Far-Field Potentials. Far-field potentials are transmitted by volume conduction to electrodes placed inches or feet away from the neural generators. For example, activity arising in subcortical structures of the brain, spinal cord, and even peripheral nerve can be recorded from electrodes on the scalp. Poststimulus latencies of far-field potentials are longer than those of near-field potentials in scalp recordings, because time is required for neural transmission of impulses. Voltages of far-field potentials range up to a few hundred nanovolts.

2.2.1.3. Noninvasive Recording. This method is attractive because it is less traumatic than invasive recording and incurs virtually no risk of infection. Resolution of signals from deep structures is more difficult with noninvasive than with invasive recordings, however.

2.2.1.4. Invasive Recording. When they are feasible, invasive stimulation and recording techniques have specific advantages. Very specific areas of the nervous system, such as particular components of peripheral nerves (Kaplan et al., 1984) or the brachial plexus (Landi et al., 1980), or a particular cranial nerve (Møller and Jannetta, 1983) can be stimulated. Recordings are enhanced, because signal strength is greater and the signal-to-noise ratio is vastly improved. With appropriate techniques, accurately localized activity can be recorded from specific structures. Not all invasive recordings permit such specific localization, however.

Numerous invasive techniques have been described for recording somatosensory and brainstem auditory EPs (BAEPs). Most of these are for intraoperative recordings in patients or experimental recordings in animals. On occasion, implantation of deep brain electrodes for chronic stimulation has provided the opportunity for clinical observations to help clarify the origins of EP waveforms recorded from the scalp. Invasive recording of VEPs has been done by placing an electrode on the optic nerve to record its action potential during operations of the anterior cranial fossa.

Historically, several orthopedic surgeons used invasive electrodes to monitor spinal cord function intraoperatively (Brown and Nash, 1979). Japanese investigators inserted epidural or subarachnoid electrodes through Touhy needles of the kind normally used to insert epidural catheters for regional anesthesia (Tamaki, 1977). Sites of stimulation and recording could be localized radiologically with regard to vertebral level, but not with regard to specific tracts of the spinal cord. Similar localization of the recording site was possible by simply drilling an orthopedic threaded wire into the spinous process at the most rostral aspect of the wound during operations on the spine (Nordwall et al., 1979). Subsequently, other workers removed a small segment of the ligaments between vertebral arches and placed electrodes for intraoperative recording directly on the dura. During operations on the spinal cord, neurosurgeons placed electrodes in the subarachnoid space above and below the operative site to measure spinal cord conduction velocity (Macon et al., 1980). Others placed electrodes on specific tracts of the spinal cord, either under direct vision during operations on the cord or percutaneously, under fluoroscopic control, to permit recordings directly from motor pathways (Levy and York, 1983; Levy, 1983).

Numerous epilepsy centers now surgically implant epidural electrode arrays on the cerebral cortex for intensive electrophysiologic monitoring and video recording before operative intervention in patients with intractable epilepsy. The electroencephalogram (EEG) is recorded to document seizure activity; cortical stimulation is performed to localize functional areas of the cortex; and SEPs are recorded to localize the Rolandic fissure. Opportunities for clinical investigations while these electrodes are implanted have not yet been fully exploited.

Invasive stimulation of the facial nerve during operations on the parotid gland, the ear, and the structures in the posterior cranial fossa has been commonplace for decades. Traditionally, EPs have not been recorded, but the patient has simply been observed for motor responses to stimulation. This technique depends on normal neuromuscular transmission, however, thus the muscle relaxants often given during surgical procedures cannot be used. Motor EPs can be recorded from nerve trunks within or outside the surgical field, and audible monitoring of the electromyogram recorded from facial muscles may give immediate warning if the facial nerve is inadvertently stimulated during surgical manipulations (Harner et al., 1986).

Action potentials can be recorded directly from the eighth nerve or from the brainstem during operations in the posterior cranial fossa. Multimodality recordings from invasive and noninvasive electrodes during stimulation of nerves within the surgical field permit extensive clinical monitoring during operations on the skull base (Sehkar and Moller, 1986).

Despite their impressive advantages, invasive EP techniques have some drawbacks. Some risk of trauma or infection may be introduced when these techniques are employed. Decisions about the relative risks and benefits can only be made for individual situations, and these decisions grow easier as experience is gained with the various techniques.

2.2.2. Magnetic Sensory-Evoked Potentials

Magnetic sensory-evoked potentials are associated with magnetic transients that can be recorded by magnetic sensors positioned over the scalp. Recording is technically difficult, because the magnetic fields generated by other electrical devices interfere with signal acquisition. Recordings of magnetic fields elicited by sensory stimulation might supplement what is learned from electrophysiologic recordings (Reite et al., 1982).

2.3. Methods of Signal Processing

Regardless of the method of signal processing, electrophysiologic signals must be conditioned and amplified upon acquisition (Cooper et al., 1980; Gevins, 1986). Initial filtering is necesary even when digital filters are to be applied subsequently. Noise above the band width of interest must be eliminated before analog-to-digital conversion so that harmonics of the higher frequencies do not, by aliasing, contaminate the signal. Analog filters produce some phase shift, and the frequency response of the pen on a strip chart recorder limits the frequencies visible on a paper record. Note that the standardization mark on a clinical electrocardiogram is 1 mV, whereas that on an EP record ranges from a few hundred nanovolts to several microvolts. Evoked sensory activity can sometimes be seen in the spontaneous EEG record; to extract the SEP reliably from the noise of the ongoing background EEG, however, special methods must be used.

2.3.1. *Superimposition*

The first sensory EPs were recorded by superimposing multiple segments of EEG activity that was time-locked to the sensory stimulus (Dawson, 1947). As more repetitions were added, the pattern of the EEG waveform became more apparent. This primitive method enabled the initial demonstration of EPs but is no longer used.

2.3.2. *Averaging*

Background EEG activity is to some extent random, whereas EPs are associated with specific stimuli. Today's clinically available EP recording systems use computers to sum or average multiple segments of filtered and amplified EEG signals. Extracting a reproducible signal from random noise, averaging increases the signal-to-noise ratio according to the square root of the number of repetitions. By averaging 500 to several thousand epochs of electrical activity immediately following sensory stimulation, subcortical or spinal far-field potentials only a few hundred nanovolts in amplitude can be extracted from the spontaneous EEG, which is usually 5–100 μV in amplitude.

A simple microprocessor can time the delivery and duration of the sensory stimulus, the duration of the segments of electrical activity recorded (sweep time or epoch length), and the number of repetitions included in an average. In most commercially available systems, stimulus and recording parameters, settings for filters and amplifiers, and devices for measurement, display, and storage of the averaged EP also are controlled by computer. Some systems also have menus of electrode montages and other user-friendly semiautomatic features.

2.3.3. *Other Methods*

Many methods of signal analysis have been applied to electrophysiologic activity evoked by sensory stimulation. The user of advanced techniques for signal enhancement should be cautious in applying them when the signal-to-noise ratio is less than optimal. If the EP cannot be seen in the usual average, then sophisticated signal processing may well produce spurious results. Unfortunately, it is just when noise prevails that the user is most desperate for unambiguous data. Familiarity with the signal permits the development of techniques that improve on averaging, but comparing a newly processed signal with the customary average is essential to avoid erroneous interpretations (Nagelkerke et al., 1983).

Special filters have been used to optimize the signal-to-noise ratio. Under certain conditions, the Wiener theory of optimal filtering can be applied to ensure minimal mean square estimation error for the estimation of signals masked in noise (Wiener, 1949). Applied to EPs in 1969 (Walter, 1969), this data-adaptive method of filtering alternately adds and subtracts epochs of EEGs that are time-locked to the sensory stimulus so that the EP signal is eliminated and an estimate of the background noise (everything else) is obtained. Mathematical procedures are then used to develop a filter that would optimize the extraction of the particular EP signal (as seen in a normal average) from the noise in that particular situation. Problems have arisen, however. Wiener filtering is based on several assumptions that do not pertain to the recording of averaged EPs. First is the assumption that the signal and the noise are stationary. Second is the assumption that the signal and noise are additive and do not correlate. Finally, the theory as described by Wiener is based on knowledge rather than estimates of the spectra of signal and noise (deWeerd and Martens, 1978). The time-invariant Wiener filter technique has now been modified to provide a time-varying filter that applies differing filters to sequential components of the averaged EP (deWeerd 1981; deWeerd and Kap, 1981). This can be used to improve the averaged waveform or to decrease the number of repetitions in an average, but only modest reductions in the number of stimuli are recommended. The time-varying filter can introduce changes in amplitudes of EP, which makes these less reliable. The reliability of latency measurements is not affected.

A two-dimensional filter has been developed to reconstruct individual EPs or small subaverages after raw data have been recorded digitally (Sgro et al., 1985). Data are filtered in the frequency domain along the data sequence axis for each trial, then along the cross-trial sequence axis for comparable frequency coefficients in sequential trials. Successive responses are then stacked one above the other so that rapid changes, such as those during the operation can be readily detected.

Statistical techniques based on analysis of variance can be used to estimate the optimal number of trials in recording averaged EPs, determining the ratio of EP amplitude to estimated averaged background noise (Don et al., 1984). A quantitative threshold for EP detection can then be determined and automatically applied; this reduces the variability of test interpretation and maximizes efficiency by avoiding averages with either too few or too many repetitions.

Averaging itself introduces some error into the estimation of true EP signals, since the true signal varies to some extent from one repetition to the next, and, to that extent, averaging attenuates the true signal. Deconvolution procedures can be used to determine the exact minima and maxima in EP signals, assuming that the observed average waveform is a convolution of an original response function and a folding function (Sjontoft, 1980). Latency variability can be diminished by using the averaged EP as a template, then cross-correlating and shifting individual sweeps to enhance minima and maxima in the waveform (Woody, 1967; Barajas, 1985). By these methods, sweeps that are at the extremes of variability may be discarded. The greater the inherent variability of the physiologic signal from sweep to sweep within an average, the greater the amplitude enhancement produced by the signal processing. For example, BAEP amplitudes recorded from patients with Down's syndrome are usually smaller than those recorded from normal individuals. With latency compression analysis, the amplitudes of Down's syndrome BAEPs are enhanced more than the amplitudes of normal BAEPs (Galbraith, 1984). This may imply that the neural pathways are less stable in patients with Down's syndrome, but unless the latency-compensated BAEPs are compared with BAEPs recorded from the same patients by means of customary time-locked averaging, the abnormal amplitudes in patients with Down's syndrome would not be appreciated. To discard sweeps with poorly correlating waveforms, selective averaging can be done by cross-correlating the noise recorded 1 s before the stimulus with the template of the time-locked average EP. Latency variability is less and signal-to-noise ratios are higher with selective averaging and latency-corrected averaging, combined, than with either method alone (Peregrin and Valach, 1981). Weighted averaging can be done by assigning high weights to "good" sweeps and low weights to "bad" sweeps, either by using statistical assumptions about the characteristics of noise after a prewhitening filter is applied to make the noise approximately white (Gasser et al., 1983) or by assuming that the noise consists of a stationary and a nonstationary component, the nonstationary component consisting of multiples of the stationary noise source with the multiplying factor varying slowly with time (Hoke et al., 1984).

Peaks can be identified in single responses by sophisticated techniques for signal enhancement. In one approach, peaks are identified by comparison with a template, after preprocessing with a minimal mean square error digital filter. A latency histogram of the peaks is constructed, and clusters within the histogram are identified. All peaks within each cluster are then converted to the mean latency for that cluster, which produces a latency-corrected average version of the EP (Aunon, 1978). Single EPs can also be enhanced automatically without a template or any human intervention by modeling a prestimulus epoch of EEG by a canonical variate-analysis system-identification technique to produce a filter, which is then applied to the poststimulus epoch of EEG to enhance the signal-to-noise ratio (Krieger and Larimore, 1986).

Classification of single-trial late event-related potentials (ERPs) in multichannel recordings shows that single-channel recordings may be expected to give as accurate results as multichannel recordings for the decision-making paradigm used by Childers and his associates (1986). In this situation, channel averaging increases the probability of error in ERP classification, possibly because noise may be coherent across channels or because latencies of the ERP components vary slightly from channel to channel. Best results are obtained by channel polling or using the subject's single best channel.

The accuracy of single-response EP detection and classification, however, can be improved by using signal processing and statistical pattern recognition techniques. Moser and Aunon (1986) extract features from the time-frequency plane to capitalize on the nonstationary property of EP signals by using the geometric relationship between time and frequency. A powerful filter for suppressing EEG noise in single-response evoked potentials can be derived by using cross-correlation matrices, estimated from multielectrode recordings of the prestimulus EEG, to derive a multielectrode model of the EP (Westerkamp and Aunon, 1987).

Gasser et al. (1986) proposed a statistical method for comparing various estimators of evoked potential signals using real data rather than simulated signals, since simulated data rely on artificial assumptions and may not be representative of the actual signal and noise encountered in real data. Integrated mean square error (IMSE), a global criterion for curve estimation obtained by summing over latencies, was determined for five estimators of the EP waveform: averaging of brain electrophysiologic activity time-locked to repetitive stimulation (AVE); a posteriori Wiener filtering (APWF); time-varying filtering (TVF); attenuating the signal (ATTE), sometimes followed by low-pass filtering (ATLO); and smoothing by kernal estimation. These workers demonstrated that their statistical method for comparing estimators is mathematically sound when signals are homogeneous across repeated responses to individual stimuli and approximately correct in the face of moderate heterogeneity. For flash-evoked potentials in normal children, little improvement was gained over time-locked averaging with any of the other methods. TFV was better than APWF, but was more demanding of methodological insight and computer time than smoothing, which gave about the same improvements.

These investigators noted several limitations associated with TVF or APWF. Both procedures rescale the signal, depending on the particular signal-to-noise ratio, and this attenuation is greater for small numbers of sweeps, just when improvements over averaging are most important. The attenuation according to the individual signal-to-noise ratio hampers comparisons among groups or individuals, and confuses interpretation of topographic maps.

Finally, Gasser and his coworkers suggest that mean squared error, despite its almost universal acceptance, does not seem appropriate for evaluating estimators that allow attenuation of the signal. This may explain some of the controversy surrounding the merits of such estimators as APWF.

2.3.4. Steady-State Evoked Potentials

When sensory stimuli are presented rapidly and continuously with a constant interstimulus interval, the brain is assumed to approach a relatively stable state of responsiveness. A train of overlapping responses with minimally varying amplitude and constant frequency results. The sinusoidal waveform produced by summing the responses to 100–300 stimuli can be quantitatively analyzed by measuring the amplitude and the phase angle relative to the train of stimuli. Alternatively, Fourier analysis of the train of responses can be performed by computing the power spectra for the train of response waves evoked by each of several specific frequencies of stimulation. Ratios between the power spectra recorded over symmetrical areas of the right and left cortical hemispheres should be less than 2.0 in normal subjects (Celesia, 1982).

2.4. Poststimulus Latencies

The poststimulus latency of a peak or a complex in the EP waveform is described as the time in milliseconds from the onset of a sensory stimulus to the point on the waveform that indicates a characteristic response. Nominal designations of polarity and latency may be given to short-, intermediate-, and long-latency potentials of various sensory responses. Because time is required for neural transmission, shorter latencies are seen in EPs recorded closer to the point of stimulation. Poststimulus latencies of SEPs vary according to the subject's height, and those of BAEPs, according to gender.

In scalp recordings the short-latency subcortical potentials are far-field recordings and therefore of lower amplitude than the near-field cortical potentials. Also, auditory stimuli produce short-latency potentials within the first 10–15 ms, whereas so-called short-latency SEPs normally extend to about 20 ms after stimulation of the median nerve at the wrist and to about 40 ms after stimulation of the posterior tibial nerve at the ankle. The ERG is a short-latency potential (Armington, 1986); subcortical, short-latency VEPs have been described, but have not yet been made clinically useful (Harding and Rubenstein, 1980).

2.4.1. Intermediate

Intermediate-latency EPs recorded from the scalp are cortical in origin. They arise from primary sensory areas, and for the SEP, the intermediate-latency activity has been called the primary specific complex. It begins with a cortical positivity between 20 and 25 ms after stimulation of the medial nerve at the wrist and about 40 ms after stimulation of the posterior tibial nerve at the ankle. The intermediate-latency somatosensory activity is normally confined to the first 100 ms or so after stimulation. Intermediate-latency visual EPs arise from the occipital cortex and are nominally seen within a time window extending from about 70 to about 200 ms after stimulation.

2.4.2. Long

Long-latency potentials are thought to arise in association areas of the cerebral cortex and include particularly the cognitive potentials such as the P300, which is seen when a rare stimulus is selected from a series of common stimuli or as a response to a painful somatosensory stimulus. These potentials have been used in studies of psychiatric patients, drug effects, and cognitive function (Goodin et al., 1978).

2.5. Purported Neural Generators

Information gained from clinical pathologic correlations in patients and from experimental studies of animals has allowed some postulation of the neural structures associated with production of certain EPs. The purported generators are not absolutely proven, but the designations are clinically useful.

2.5.1. Potentials Generated by Sensory Receptors

2.5.1.1. AUDITORY. The electrocochleogram can be recorded from a needle or ball electrode in the external ear canal or from a transtympanic electrode. Transtympanic electrodes are inserted through the tympanic membrane to rest on the tympanic promontory of the middle ear and must be placed by a skilled otologic specialist. This electrode is not recommended for neurologic studies. The cochlear microphonic and summating potentials are of greater interest to otologists than to neurologists. The most prominent component of the so-called electrocochleogram actually arises from the primary fibers of the auditory nerve, as does wave I of the BAEP. Detection of the potential from the auditory nerve is especially important when BAEPs are recorded to test brainstem function in comatose patients. If eighth nerve function cannot be documented, absence of the BAEP may be caused by otologic dysfunction, and brainstem pathways may be normal despite an abnormal or absent BAEP (Stockard et al., 1986).

2.5.1.2. VISUAL. The ERG is often recorded as an important quality-control measure in recording VEPs (Armington, 1986). It is intrinsically useful for diagnostic studies of the retina. Optimal recordings are obtained from corneal electrodes placed with topical anesthesia. The a-wave, negative at the cornea, is followed by a larger cornea-positive b-wave. In appropriate circumstances, high-frequency oscillatory potentials can be seen superimposed on the leading edge of the b-wave. With corneal electrodes, the ERG is usually well seen in single sweeps. Averaging is needed only when potentials are of unusually low amplitude or absent. Dim, blue flashes to dark-adapted eyes elicit activity from the rod system without detectable input from cones. Strong, white flashes in the presence of background light selectively activate the cone system. Both rods and cones contribute to the

ERG evoked by strong white flashes to dark-adapted eyes. Averaged ERGs of lesser quality can be recorded from infraorbital disk electrodes.

2.5.2. Nerve-Action Potentials

Nerve-action potentials can be recorded from surface or invasive electrodes placed near nerve trunks through which evoked signals are traveling either orthodromically or antidromically. For example, when the median nerve is stimulated at the wrist, the resulting sensory nerve action potential can be recorded from an electrode placed over a median nerve in the forearm or at the antecubital fossa. Nerve-action potentials are important in studies of peripheral neuropathy and in documenting responses to a sensory stimulus when EPs arising from structures higher in the neuraxis are being recorded (Daube, 1986).

2.5.3. Plexus Potentials

Activity arising from the brachial plexus is commonly recorded from an electrode placed at Erb's point, and signals may also be recorded from the lumbar plexus. Direct recordings from components of the brachial plexus during surgical exploration help guide decision-making when questions arise about neurolysis versus resection of injured or pathologic components of the plexus (Landi et al., 1980).

2.5.4. Potentials Generated by Nerve Roots

These somatosensory potentials may be recorded directly from the spinal canal by using invasive techniques in animals or in patients during operation. Far-field recordings from surface electrodes can also demonstrate these potentials. They are small, however, and many repetitions must be averaged.

2.5.5. Spinal Cord Potentials

Potentials arising from the spinal cord may be recorded either invasively or noninvasively. In diagnostic recordings, electrodes are placed at intervals along the spinal column, and a traveling wave can be seen with increasing latencies at higher levels (Cracco et al., 1979). Electrical activity arising in the spinal cord is often monitored during operations on the spine or on the contents of the spinal canal. As described above, electrodes may be inserted into the interspinous ligament, the epidural space, or the subarachnoid space or may be placed directly on the spinal cord.

2.5.6. Brain Potentials

The origin of various potentials that arise in the cranial vault is controversial, but there is sufficient evidence to make informed, clinically useful assumptions about neurophysiologic origins of these potentials. An electrode placed on or near a peripheral nerve detects a triphasic potential when the nerve is depolarized. An initial positive deflection is recorded as the potential approaches the electrode. With the net flow of positive ions into the nerve, depolarization and a dominant negative potential occur, followed by a small positive deflection as the area of depolarization moves away from the electrode. An electrode placed on the proximal cut end of the peripheral nerve detects only a large positivity. It follows that in the somatosensory system, afferent volleys in fiber tracts such as the thalamocortical radiation, median lemniscus, and dorsal spinal column produce positive potentials at the scalp (Allison et al., 1986).

2.5.6.1. MEDULLA. Both somatosensory and auditory EPs reflect activity arising in the medulla. Medullary components of the SEP include the N12b, thought to arise in the ascending dorsal column, and the N13b, generated by the cuneate nucleus (Eisen and Aminoff, 1986). The potentials in the ascending dorsal column are predominately from second-order neurons, since most somatosensory afferents synapse in the dorsal horn of the spinal cord within a few segments of entry. A few first-order neurons are involved. The large primary afferent fibers carrying information from cutaneous receptors terminate in the cuneate nucleus. Fibers from this nucleus cross as internal arcuate fibers in the decussation of the lemniscus and continue as the medial lemniscus (Table 1).

Purported generators of the BAEP are shown in Table 2. Wave 2 of the BAEP probably includes potentials arising from the cochlear nucleus, although the primary source of this peak is probably the intracranial portion of the auditory nerve.

2.5.6.2. PONS. Pontine activity is reflected in both the BAEP and the short-latency SEP. The axons of the acoustic nerve, which arise from cell bodies in the spinal ganglion of the cochlea, enter the dorsolateral area of the medulla at the pontomedullary junction. Fibers of the second-order neurons then ascend both crossed and uncrossed to the medial geniculate body of the thalamus. Intervening synapses may occur in the reticular formation, the trapezoid body, the superior olivary nuclei, or the inferior colliculus. Wave 3 of the BAEP is pontine in origin and probably arises from the

Table 1
Purported Generators of Short-Latency Somatosensory-Evoked Potentials[a]

Peak	Generator
N9	Brachial plexus
N11	Spinal roots or dorsal columns
N13, 14	Spinal cord gray matter or dorsal columns
N14, 15	Brainstem, thalamus, or both
N20	Primary somatosensory cortex

[a]Stimulation: median nerve; recording: clavicle, mastoid process, 2nd cervical vertebra, primary somatosensory cortical area; reference electrode: FZ or noncephalic (early positive waves are obscured in frontally referenced recordings) (reproduced with permission from Grundy, 1982).

superior olivary nucleus. Wave 4 is also a pontine wave and represents activity in the lateral lemniscus.

Pontine somatosensory activity is reflected in the N14, which arises from the medial lemniscus; this pathway contains second-order neurons with cell bodies in the nucleus gracilis and nucleus cuneatus that cross to the opposite side in the lower medulla. In clinical recordings, the N14 is less apparent than the N13, which contains components arising from the ipsilateral dorsal horn and cuneate nucleus. Also, in many clinical recordings, the peak referred to as N14 is probably actually the N13 and reflects activity arising in the spinal cord and medulla.

2.5.6.3. Midbrain. No recognizable electrical components of the SEP can be identified as arising from the midbrain. The most prominent component of the BAEP, wave 5, arises in the area of the inferior colliculus. The III-V interpeak latency is thus a measure of conduction between the rostral pontine and midbrain portions of the auditory pathway.

2.5.6.4. Thalamus. Wave 6 of the BAEP is thought to arise in the medial geniculate body of the thalamus. The P15 component of the SEP, recorded to a noncephalic reference, probably arises in the ventral posterolateral nucleus of the thalamus.

2.5.6.5. Thalamocortical Radiations. Wave 7 of the BAEP and P16 and P18 of the SEP are thought to arise from thalamocortical afferents.

Table 2
Purported Generators of Brainstem Auditory-Evoked Potentials

Peak[a]	Generator
I	Acoustic nerve
II	Intracranial acoustic nerve, cochlear nucleus (medulla), or both
III	Superior olive (pons)
IV	Lateral lemniscus (pons)
V	Inferior colliculus (midbrain)
VI	Medial geniculate (thalamus)
VII	Thalomocortical radiations

[a]Listed peaks are positive at the vertex (reproduced with permission from Grundy, 1982).

2.5.6.6. PRIMARY SENSORY CORTEX. Although evidence is not absolutely conclusive, EPs can be recorded that are thought to arise from the primary cortical receiving areas of the somatosensory, auditory, and visual systems. The N20 of the somatosensory cortex probably arises from area 3b in the posterior bank of the central sulcus. The following positivity (P25) is thought to arise from area 1 in the crown of the post central gyrus (Allison et al., 1986). The middle-latency auditory responses probably arise in the primary auditory cortex, areas 41 and 42 of the superior temporal gyrus (Heschl's gyrus). The P100 of the pattern reversal VEP is thought to arise in the primary visual cortex, area 17 (Sokol, 1986). Flash-evoked potentials probably arise from secondary or association areas of the visual cortex, areas 18 and 19. Positron emission tomography studies, however, show strong activation of areas 17, 18, and 19, including both primary and association areas of the visual cortex, with either pattern reversal or flash stimulation (Phelps et al., 1981; Celesia, 1985).

2.5.6.7. ASSOCIATION AREAS OF CORTEX. The P50 of the somatosensory EP may arise from the sensory association cortex, areas 5 and 7 of the parietal lobe. SEP generated by stimulation of a single extremity can be recorded over the ipsilateral sensory cortex 4–8 ms later than the corresponding SEP arising in the somatosensory cortex contralateral to the site of stimulation. The latency difference is probably explained by the time required for conduction through the corpus callosum from one hemisphere to the other. Auditory EPs recorded from the vertex approximately 100–250 ms after stimulus probably represent a widespread activation of frontal cortex (Picton and Hillyard, 1974). Activity recorded from the vertex between 140 and 200 ms after stimulation is nonspecific and can be elicited by either somatosensory, auditory, or visual stimulation (Goff et al., 1977; Hillyard and Picton, 1978; Hillyard et al., 1978).

The P300 potential is a late positive component of the slow vertex potential and can be elicited by somatosensory, visual, or auditory stimulation. A large P300 produced by laser heat stimulation of somatosensory areas has been associated with the subjective sensation of pain (Carmon et al., 1976). The P300 is altered by the subject's psychological state, attention, intentions to respond to stimuli, and discrimination between differing stimuli. This potential relates to the amount of task-relevant information in the stimulus and to whether the subject feels he or she has correctly guessed the significance of the stimulus (Sutton et al., 1967). The P300 probably involves wide areas of the association cortex, including parietal components.

3. Equipment and Techniques

Evoked-potential systems in current use are centered on digital computers that trigger the stimulation, time the acquisition of electrophysiologic data, sum or average the responses to many repeated stimuli, then display the average waveform. Provision must be made for measurement and storage of waveforms. Complete electrophysiologic monitoring systems are available from a number of manufacturers (Table 3). Individual hardware components cost less than "turnkey" systems. Since software development is time-consuming and expensive, most users purchase complete systems that are ready to use or that require minimal program development.

Recommendations for EP capabilities have been provided by the American Electroencephalographic Society (1984). Semiautomatic acquisition and logging of data can be quite helpful in both clinical and experimental settings. With many systems, programs appropriate for a particular application can be developed

Table 3
Manufacturers of Evoked-Potential Monitors[a]

Company	Address
Bio-logic Systems Corp.	Computers in Medicine 425 Huehl Road Northbrook, IL 60062
Cadwell	1021 N. Kellogg Street Kennewick, WA 99336
CNS, Inc.	10349 W. 70th Street Eden Prairie, MN 55344
Interspec	1100 E. Hector Street Conshohocken, PA 19428
Neuroscience	210 Topaz Street Milpitas, CA 95033
Nicolet Biomedical Division	5225-4 Verona Road P.O. Box 4287 Madison, WI 53711-0287
Teca	3 Campus Drive Pleasantville, NY 10570

[a]These manufacturers participated in the workshop Evoked-Potential Monitoring: State of the Art 1985, held September 18–21, 1985, at Lake Buena Vista, Florida, under the sponsorship of the Anesthesiology Alumni Association of Florida, Inc. and the International Anesthesia Research Society, Inc.

with minimal programing by using software provided by the manufacturer.

Devices should be engineered to minimize the possibility of injury to either patient or personnel. In electrically sensitive patients (those with cardiac pacemaker wires or central vascular catheters that could conduct electrical current to the myocardium), leakage of current should be no more than 10 μA to the patient and no more than 100 μA to the case of the instrument.

To record an EP, one must select the particular stimulus to be used and must specify its intensity, duration, frequency, and pattern (regular vs. pseudorandom, for example). The number and location of recording electrodes must be determined for each channel recorded and a grid 1, grid 2, and patient reference or "ground" electrode must be specified. Amplification and filter settings as well as the time base or "sweep time" to be recorded may be selected for all channels or may differ among channels. Analog-to-digital converters should sample the analog signal at a frequency at least twice the highest frequency admitted by the analog filters, and the memory block designated for each waveform must contain enough addresses to include samples of this frequency.

Sensory stimulation may be repeated at rates as low as 0.5 Hz, or slower to as fast as 50 or even 80 Hz. Brainstem auditory-evoked potentials show reproducible increases in latency and decreases in amplitude as stimulus rate increases above 11–15 Hz. Early components of the waveform are progressively attenuated until, at 60–80 Hz, only a small, delayed wave 5 is seen. Somatosensory-evoked cortical potentials show some degradation at interstimulus intervals briefer than 5 s; clinically, however, potentials are usually recorded at stimulus rates between 1 and 10 Hz. Interstimulus intervals of up to 2 s or longer are used when recording cognitive EPs. Pseudorandom interstimulus intervals with actual values ± 25 or 50% of the nominal interstimulus interval help to prevent either habituation of the EP signal or confusion of EP with spontaneous EEG rhythms. Psuedorandom intervals are more important in recording later potentials.

For quality control, the analog signal should be observed visually as data are acquired so that data acquisition can be suspended when the signal is contaminated by artifact. The options on commercially available systems for automatic artifact rejection depend solely on amplitude criteria and therefore allow the inclusion of artifacts not reaching that amplitude. Quality control is enhanced by observation and paper plotting of waveforms as data are acquired. Electromagnetic data storage is essential for ease of subsequent data processing, and detailed measurements of waveforms can be performed after an experiment or study is completed.

The guidelines published by the American Electroencephalographic Society (1984) are specific and detailed with regard to data acquisition of clinical EPs. Some flexibility is required in particular settings, particularly the operating room and intensive care unit. Experimental protocols may dictate their own criteria for data acquisition and processing, but these should be related to generally accepted standards, and reasons for deviation from them should be justifiable. Techniques suggested for intraoperative monitoring of EPs are listed in Tables 4–6.

Table 4
Stimulation and Recording Parameters
for Somatosensory-Evoked Potentials[a]

Parameter	Description
Stimulus	Constant current: 2–20 mA (sensory + motor). Duration: 200 s (150–300 s). Rate: 0.9, 1.9, or 5.2 Hz[b]
Sites[c]	Median nerve at wrist: right (RMN) or left (LMN). Posterior tibial nerve at ankle: right (RPTN) or left (LPTN)
Electrodes	
Stimulating	Subdermal platinum electrode pairs[d]: 3 cm apart, cathode proximal.
Recording	Ag/AgCl cup or disk electrodes applied with collodion and sealed with waterproof tape; impedance <3000 Ω
Recording channels[e]	
With four sites	
1	C3'-EP2 (RMN)
2	C4'-EP1 (LMN)
3	L3S-4 cm rostral or iliac crest; or T12S-4 cm rostral or iliac crest (LPTN, RPTN)
4	PFi (ipsilateral poipliteal fossa)-iliac crest (LPTN, RPTN)
5	C3'-FZ (RMN)
6	C4'-FZ (LMN)
7	C2S-FZ (all)
8	Cz'-FZ (LPTN, RPTN)
Common or ground	Either shoulder or linked ears
RMN and LMN only	
1	C3'-EP2 (RMN)
2	C4'-EP1 (LMN)
3	C2S-FZ (RMN, LMN)
4	C5S-FZ
5	C3'-FZ (RMN)
6	C4'-FZ (LMN)
7	CZ'-FZ
8	C3'-C4'
Common or ground	Either shoulder

(continued)

Table 4
(Continued)

Parameter	Description
Stimulation	
Filters	
Channels 1–4	30–3000 Hz
Channels 5–8	1–1500 Hz
Sweep time	
Channels 1–4	35 or 50 ms
Channels 5–8	140 or 175 ms
Minimum sampling rate for digitization[f]	
Channels 1–4	6000 Hz
Channels 5–8	3000 Hz
Sensitivity	± 50–100 μV full scale ± 200 μV for tense patients with PTN
Repetitions	128–4000

[a]Modified and reproduced with permission from Grundy B. L. (1985) Intraoperative Applications of Evoked Responses, in *Evoked Potential Testing. Clinical Applications* (Owen J. H. and Davis H., eds.) Grune and Stratton, Orlando, Florida.

[b]Avoid harmonics of 60 Hz. For sweep time greater than 175 ms, use only 0.9 or 1.9 Hz; pseudorandom; interstimulus interval ±25% of nominal value.

[c]Only a single site is stimulated during any one average.

[d]Grass Instrument Corp., Quincy, Massachusetts.

[e]When only four channels are available, the montage can be varied according to stimulus site. Stimulus sites for which each channel may be critical are shown in parentheses. With invasive electrodes, replace the usual montage for any two of the first four channels; we prefer bipolar eipdural electrodes above and below the operative site. With noninvasive recordings of lumbar and thoracic spinal cord potentials, 1000 or more repetitions are required; a running average of the four most recent waveforms, averaged from 250 repetitions each, might be useful.

[f]These sampling rates are minimal; oversampling is desirable, particularly in challenging environments such as the operating room or critical care unit.

4. Applications

Since their introduction, sensory EP techniques have come to play important roles in both experimental and clinical settings.

Table 5
Montages for Recording Somatosensory Evoked Potentials

Channel	Two-channel system	Four-channel system
Median nerve		
Right		
1	EP1 or C2S-FZ	EP1-EP2
2	C3'-FZ	C3'-EP2
3	—	C2S-FZ
4	—	C3'-FZ
Left		
1	EP2 or C2S-FZ	CE2-EP1
2	C4'-FZ	C4'-EP1
3	—	C2S-FZ
4	—	C4'-FZ
Right or left		
1	—	EP1-EP2
2	—	C2S-FZ
3	—	C3'-FZ
4	—	C4'-FZ
Posterior tibial nerves		
Noninvasive		
1	C2S-FZ; or lumbar or thoracic spine to iliac crest or rostral spine	PFi-iliac crest
2	CZ'-linked ears or FZ	T12S-4 cm rostral or iliac crest; L3S-4 cm rostral or iliac crest
3		C2S-FZ
4		CZ'-linked ears or FZ
Invasive		
1	Rostral bipolar epidural	Rostral bipolar epidural
2	Caudal bipolar epidural	Caudal bipolar epidural
3		PFi-iliac crest or C2S-FZ
4		CZ'-linked ears or FZ

[a]Modified and reproduced with permission from Grundy B. L. (1985) Intraoperative Applications of Evoked Responses, in *Evoked Potential Testing. Clinical Applications* (Owen J. H. and Davis H., eds.) Grune and Stratton, Orlando, Florida.

Table 6
Stimulation and Recording Parameters for Monitoring Brainstem Auditory-Evoked Potentials[a]

Parameter	Description
Stimulation	
Sites	Ears, left or right
Transducers	Insert earphones[b]
Clicks	Broad-band, filtered, alternating rarefaction/condensation
Intensity	60 dBSL
Duration	100 s
Rate	10.9 Hz in United States, 11.3 Hz in Europe and Japan
Contralateral masking	Wide-band pseudorandom noise, 30 dB lower than stimulus intensity
Recording	
Channels	
1[c]	CZ-A1
2	CAM1-A1[d] or A1-A2
3[c]	CZ-A2
4	EAM2-A2[d] or A2-A1
Common or ground	FZ
Filters	30–3000 Hz
Sampling rate for digitization	10,000–50,000 Hz (minimum: 6000)
Sweep time	15 ms
Sensitivity	±20–50 μV full-scale
Repetitions	2000

[a]Modified and reproduced with permission from Grundy B. L. (1985) Intraoperative Applications of Evoked Responses, in *Evoked Potential Testing. Clinical Applications* (Owen J. H. and Davis H., eds.) Grune and Stratton, Orlando, Florida.

[b]Madsen Electronics, Inc., Buffalo, New York.

[c]If only two channels are available, use CZ-A1 and CZ-A2.

[d]If wave 1 is not well seen in the brainstem auditory-evoked potential.

4.1. Experimental

Early studies of EPs were designed primarily to identify the neural generators of EP waveforms recorded in humans. Some studies were based on invasive recordings in animals or patients.

Others depended on acute or chronic ablative experiments in animals. Perhaps the most important early reports were observational and described clinical pathologic correlations in patients. Although neural generators of SEPs are not yet fully understood, sufficient progress was made to facilitate experiments that use EP recordings to examine acute changes in neurologic function related to drug administration and various physiologic and traumatic insults to the nervous system.

Better methods are becoming available for studies of blood flow, metabolism, and neural transmission and for in vivo imaging of anatomic structures, neurotransmitter receptor sites, and acute or chronic pathophysiologic changes. Techniques such as CT, magnetic resonance imaging (MRI), nuclear magnetic resonance spectroscopy (MRS), positron emission tomography (PET) for examination of brain metabolism, and single photon emission computerized tomography (SPECT) for imaging of specific neurotransmitter receptor sites in vivo open vast new horizons for understanding neurologic function. Correlating EPs with the changes in neuronal function that are detected with these expensive and complex techniques will broaden the range of studies that can make use of the information gained from technically sophisticated neuroradiologic observations. As a result, acute invasive studies of EPs in animals are becoming less important experimentally in establishing how EPs relate to physiologic, pharmacodynamic, and pathophysiologic phenomena that require microsphere injections, neural tissue samples, or nervous system imaging with expensive and complex equipment.

4.2. Clinical

Electrophysiologic techniques in diagnostic neurology now clearly embrace EP recording as well as electroencephalography. New methods for imaging the central nervous system provide much information not available from either clinical neurologic assessment or electrophysiologic recordings; but, with the exception of a few techniques that are still experimental or not generally available, imaging techniques do not show function directly. Thus, EPs do have value as diagnostic tools in clinical neurology, even though some workers have questioned the value of many of the tests being done (Eisen and Cracco, 1983).

The widest clinical application of EPs has been in the diagnosis of multiple sclerosis. Documentation of localized abnormalities at multiple sites in the nervous system, a key criterion for making the diagnosis, can often be demonstrated by evoked-potential abnormalities at one or more sites that are remote from the lesion responsible for a patient's presenting symptoms (Chiappa, 1980). For example, in the patient with ataxia or an isolated lesion in the spinal cord, abnormal VEPs can lead to a diagnosis of multiple sclerosis even in the absence of visual symptoms or signs.

Electrophysiologic recordings have been used to aid diagnosis and prognosis in spinal cord injury (Table 7) (Ertekin et al., 1984; Kaplan, 1983; Larson et al., 1976; Perot, 1972, 1976; Rowed et al., 1978; Rowed, 1982; York et al., 1983; Young, 1982). Reports from the English literature have recently been summarized (Grundy and Friedman, 1987). Although the data obtained so far hold much promise, more study is needed to delineate precisely what the electrophysiologic responses indicate clinically. The noninvasiveness and the precision of these monitoring techniques make such study an important priority for further development of methods to reduce morbidity and mortality after spinal cord injury.

Evoked potentials can be used to assess the functional integrity of specific neural pathways and structures in anesthetized or comatose patients. This is particularly valuable when structures of the central nervous system are at risk, but cannot be adequately evaluated by clinical assessment because the patient is unconscious. The earliest widespread application in the operating room was SEP monitoring of spinal cord function during operative treatment of scoliosis (Nash et al., 1977). Numerous reports appeared of SEP monitoring during neurosurgical procedures on the spine and spinal cord (McCallum and Bennett, 1975; Lueders et al., 1982). Neurosurgical uses of intraoperative SEP monitoring also include localization of the somatosensory and motor cortex (Kelly et al., 1965; Woolsey et al., 1979), identification of nerve trunks (Kaplan et al., 1984), and determination of continuity versus complete disruption during operative treatment of injuries to peripheral nerves or the brachial plexus (Landi et al., 1980; Kline and Judice, 1983), and detection of cerebral ischemia during carotid endarterectomy (Moorthy et al., 1982) or operations on intracranial aneurysms (Friedman et al., 1987). Although it is clear that SEPs are transmitted primarily through the dorsal columns of the spinal cord, many observations suggest that even anterior vascular insults such as those sometimes seen during occlusion of the thoracic aorta are at least transiently reflected in altered or obliterated SEPs (Kaplan et al., 1986).

Brainstem auditory EPs are monitored during operations in the posterior cranial fossa, particularly those in the cerebellopontine angle, which may jeopardize the eighth cranial nerve or its blood supply (Grundy, 1982; Friedman et al., 1985). Visual EPs have been monitored during operations on the pituitary gland or in the anterior cranial fossa, but VEP monitoring has been less widely accepted than SEP or BAEP monitoring because of difficulties with appropriate stimulation (Grundy, 1985) and the exquisite sensitivity of VEP to anesthetic agents (Uhl et al., 1980). Combinations of trigeminal and brainstem auditory, visual, and motor EPs can be used to monitor multiple cranial nerves during neurosurgical operations on the base of the skull (Sekhar and Moller, 1986). Transcranial stimulation of the motor cortex with recording of the electrical potentials evoked in motor nerve or muscle is now being introduced into clinical practice for direct monitoring of motor pathways (Levy et al., 1984).

Some controversy has surrounded so-called "false-negative" results of intraoperative EP monitoring. These reports are very few compared with those demonstrating totally satisfactory experiences with the monitoring of EPs, but are clearly of great concern to all physicians working in this area. Most "false-negative" results seem related to failure to monitor during the insult that produced neurologic injury or misinterpretation of waveforms (Friedman and Grundy, 1987). The current introduction of clinical EP monitoring in the operating room resembles in many respects the introduction of other monitoring techniques for surgical patients in years past. Both blood pressure monitoring (Beecher, 1940) and electrocardiography in the operating room were viewed with skepticism when first introduced, yet these techniques are part of the standard of care today. Evoked-potential monitoring now has a clear role in the clinical armamentarium of anesthesiologists and surgeons (Friedman and Grundy, 1987) and likely will become accepted as a standard of care for certain high-risk neurologic, orthopedic, vascular, and cardiac operations within the near future.

5. Conclusions

Sensory recordings are an important technique in the armamentarium of the neuroscientist. These electrophysiologic signals help relate structure to function in the nervous system and

Table 7
Correlation of Outcome with Sensory-Evoked Potentials (SEPs) Recorded from the Scalp of Patients with Acute Spinal Cord Injury[a]

Study	Insult	Subjects, *n*	Time after injury	Sensory-evoked potential recordings: Nerve stimulated	Findings, *n*	Correlation with clinical neurologic assessment
Perot (1972)					SEP present in:	No SEPS were transmitted past complete cord lesions. SEPs transmitted past incomplete lesions were abnormal in shape and had prolonged latencies
	Cervical					
	Complete	10	—	Median	1 in 5	
				Common peroneal	0 in 8	
				Sural	0 in 5	
	Incomplete	18		Median	13 in 13	
				Common peroneal	10 in 18	
				Sural	4 in 10	

No deficit	5		Median	3 in 3
			Common peroneal	3 in 4
			Sural	1 in 1
Thoracic				
Complete	6		Median	5 in 5
			Common peroneal	0 in 6
			Sural	0 in 5
Incomplete	2		Common peroneal	1 in 1
Lumbar				
Complete	2		Median	2 in 2
			Common peroneal	0 in 2
			Sural	0 in 2
Incomplete	4		Median	3 in 3
			Common peroneal	1 in 4
			Sural	0 in 2

	Assessment of outcome				
Study	Time after injury	Technique	Findings	Correlation with initial SEP	Comments
Perot (1972)	Not specified	Clinical assessment	Waveform and latencies returned toward normal with clinical improvement	Not stated	Recordings done at bedside in ward or ICU. Not all nerves were stimulated in each patient Transient abnormalities in SEPs evoked by median

(continued)

Table 7 *(continued)*

Study	Assessment of outcome: Time after injury	Technique	Findings	Correlation with initial SEP	Comments
					nerve stimulation were seen with injuries of the thoracic or lumbar cord, suggesting transient ascending myelopathy

Study	Insult	Subjects, *n*	Time after injury	Sensory-evoked potential recordings: Nerve stimulated	Findings, *n*	Correlation with clinical neurologic assessment
Perot (1976)	Cervical				SEP present in:	
	Complete	20	Not specified	Median	6 in 15	Six patients with low cervical complete lesions had SEPs to medial nerve stimulation. In patients with
				Ulnar	0 in 10	
				Common peroneal	0 in 15	
				Bilateral common peroneal	0 in 9	

		Sural	0 in 10	complete cervical lesions, no SEPs were elicited by stimulation of ulnar, common peroneal, bilateral common peroneal, or sural nerve.
Incomplete	32	Median	25 in 26	
		Ulnar	9 in 16	
		Common peroneal	15 in 28	
		Bilateral common peroneal	5 in 9	
		Sural	6 in 16	
Thoracic				
Complete	9	Median	8 in 8	Good correlation with clinical assessment was found in patients with thoracic lesions.
		Ulnar	4 in 9	
		Common peroneal	0 in 8	
		Bilateral common peroneal	0 in 4	
		Sural	0 in 7	
Incomplete	3	Median	1 in 1	
		Ulnar	1 in 1	
		Common peroneal	1 in 3	
		Bilateral common peroneal	0 in 1	
Lumbar				
Complete	1		Not spec.	
Incomplete	5		Not spec.	

(continued)

Table 7 *(continued)*

	Assessment of outcome				
Study	Time after injury	Technique	Findings	Correlation with initial SEP	Comments
Perot (1976)	Variable	Clinical SEPs Pathology			Patients with low spinal injuries had marked abnormalities that improved over days or weeks. No patient with a complete cervical, thoracic, or lumbar cord injury responded to stimulation of common peroneal or sural nerve. Need to stimulate each of several nerves to best characterize a lesion.

				Sensory-evoked potential recordings		
Study	Insult	Subjects, n	Time after injury	Nerve stimulated	Findings, n	Correlation with clinical neurologic assessment
Larson et al. (1976)	Trauma	28	<1 d to >1 yr, before and after operation	Peroneal	Preoperative SEP Normal/n=6 Abnormal/n=22 Postoperative SEP in eight patients with abnormal preoperative SEP Normal, 6/8 Abnormal, 2/8 Postoperative SEP in nine patients with absent preoperative SEP SEP present, 4/9 SEP absent, 5/9	Perception of joint rotation Normal in 6/6 Abnormal in 20/22

(continued)

Table 7 *(continued)*

	Assessment of outcome				
Study	Time after injury	Technique	Findings	Correlation with initial SEP	Comments
Larson et al. (1976)		Clinical and SEP		No EP before or after operation. Clinical improvement, 1 in 5. SEP and clinical status returned to normal, 3 in 5. Postoperative normalization of SEPs that were abnormal before operation.	

						Improved perception of joint rotation and overall neurologic status, 6 in 6. No postoperative normalization of SEPs that were abnormal before operation. Recovered clinically, 1 in 2.

Study	Insult	Subjects, *n*	Time after injury	Sensory-evoked potential recordings: Nerve stimulated	Findings, *n*	Correlation with clinical neurologic assessment
Rowed et al. (1978)	Trauma	38	As early as possible usually within 24 hr; then	Median Ulnar Posterior tibial Bilateral posterior tibial		No response when stimulated below level of complete injury. With complete le-

(continued)

Table 7 *(continued)*

Study	Insult	Subjects, *n*	Time after injury	Sensory-evoked potential recordings: Nerve stimulated	Findings, *n*	Correlation with clinical neurologic assessment
			daily for first wk and at least weekly for 1 mo			sions, clinical findings correlated closely.
	Cervical					
	Complete	7		Median	Absent	
				Ulnar	Absent	
				Posterior tibial	Absent	
Rowed et al. (1978)	Thoracic					
	Complete	8		Median	Normal, 6 in 7 Abnormal, 1 in 7	
				Ulnar	Normal, 5 in 6 Abnormal, 1 in 6	
				Posterior tibial	Absent, 8 in 8	

Study	Insult	Subjects, *n*	Time after injury	Sensory-evoked potential recordings		
				Nerve stimulated	Findings, *n*	Correlation with clinical neurologic assessment
	Cervical Incomplete	15		Median	Normal, 4 in 13	
					Abnormal, 9 in 13	
				Ulnar	Normal, 1 in 3	
					Abnormal, 9 in 13	
					Absent, 3 in 13	
				Posterior tibial	Normal, 1 in 14	
					Abnormal, 11 in 14	
					Absent, 2 in 14	
	Thoracic Incomplete	4		Median	Normal, 1 in 1	
				Ulnar	Normal, 1 in 1	
				Posterior tibial	Normal, 1 in 3	
					Abnormal, 1 in 3	
					Absent, 1 in 3	
	Lumbar Incomplete	4		Median	Normal, 4 in 4	
				Ulnar	Normal, 4 in 4	
				Posterior tibial	Normal, 1 in 4	
					Abnormal, 3 in 4	

(continued)

Table 7 *(continued)*

Study	Assessment of outcome				Comments
	Time after injury	Technique	Findings	Correlation with initial SEP	
Rowed et al. (1978)	24 h to 1 mo	Clinical and SEP		All patients in whom SEP was present first day, or returned in first week and progressively improved, had substantial clinical recovery.	Return of recognizable SEP in all cases preceded substantial clinical recovery.

Study	Insult	Subjects, *n*	Time after injury	Sensory-evoked potential recordings		
				Nerve stimulated	Findings, *n*	Correlation with clinical neurologic assessment
Rowed (1982)	Cervical trauma	71	As soon as possible,	Median, ulnar, posterior tibial, and	SEP grading: 0, absent; 1, gross-	Clinical grading: 1, complete motor

majority in less than 24 h, then daily for first week and weekly for at least first mo	bilateral posterior tibial were all done in each subject	ly abnormal; 2, slightly abnormal; 3, normal	and sensory loss; 10, no deficit.
			Strong correlation between initial SEP and clinical grades.

	Assessment of outcome				
Study	Time after injury	Technique	Findings	Correlation with initial SEP	Comments
Rowed (1982)	Variable, range not specified	Clinical and mean time to walking with a cane	27 patients walked with a cane, 39–152 d after injury	Mean time to walking: with SEP absent, 152 d ($n = 1$); with SEP grossly abnormal, 129 d ($n = 7$); with SEP slightly abnormal, 50	When weighting factors were assigned to SEP measurements, initial SEP correlated strongly with initial and with final neurological grade. Predictive value of the SEP model had not yet been tested prospectively.

(continued)

Table 7 *(continued)*

Study	Assessment of outcome: Time after injury	Technique	Findings	Correlation with initial SEP	Comments
				d ($n = 11$); with SEP normal, 39 d ($n = 8$)	

Study	Insult	Subjects, n	Time after injury	Sensory-evoked potential recordings: Nerve stimulated	Findings, n	Correlation with clinical neurologic assessment
Young (1982)	Cervical trauma Neurologically complete lesions	37 31	Within 48 h, 1 wk, 3 wk	Median Posterior tibial Bilateral posterior tibial	SEP graded: 0, absent; 1, present, but markedly decreased in amplitude; 2, present, but abnormal in waveform; 3, normal	Detailed neuroscoring system. Two patients had clinically complete lesions, but definite SEP. Initial SEP and neuroscores had significant linear correlation.

	Assessment of outcome				
Study	Time after injury	Technique	Findings	Correlation with initial SEP	Comments
Young (1982)	6 wk (also sometimes 3–6 mo)	Clinical and SEP	Median nerve, SEP clearly improved over time. Posterior tibial nerve, SEP improved only slightly.	6 in 21 patients with neurologically complete cervical lesions had PTN-SEP on admission, but only 2 in 6 retained SEP 3–6 mo later. Admission SEP scores of 11 in 24 (almost 50%) correctly predicted neurological scores 6 wk later.	In some patients with preserved SEP to stimulation of posterior tibial nerve, stimulation of bilateral posterior tibial nerves suppressed SEP, whereas motor responses remained the same.

(continued)

Table 7 *(continued)*

Study	Insult	Subjects, *n*	Time after injury	Sensory-evoked potential recordings: Nerve stimulated	Findings, *n*	Correlation with clinical neurologic assessment
Kaplan (1983)	Cervical					
	Complete	10	Not stated	Pudendal	Absent, 10 in 10	Clinically complete lesions
	Incomplete and central cord syndromes	15	Not stated	Pudendal	Normal, 15 in 15	Normal bulbocavernous and micturation reflexes; normal cystourethrography and cystometrography
	Thoracic or lumbar with paraplegia					
	Complete lesions	10	Not stated	Bladder[b]	Absent, 10 in 10	Clinically complete lesions, 10 in 10

	Incomplete lesions	10	≥3 mo	Bladder	Present, 10 in 10	Normal bulbocavernous or anocutaneous reflexes and hyperreflexic neurogenic bladders, 10 in 10

Study	Assessment of outcome: Time after injury	Technique	Findings	Correlation with initial SEP	Comments
Kaplan (1983)					SEPs upon stimulation of pelvic or pudendal nerves were sensitive indicators of incomplete cord lesions.

Study	Insult	Subjects, n	Time after injury	Sensory-evoked potential recordings: Nerve stimulated	Findings, n	Correlation with clinical neurologic assessment
York et al. (1983)		71	Immediately upon admission	Deep peroneal (foot) or posterior tibial (ankle); 28–64 receptions per average		

(continued)

Table 7 *(continued)*

Study	Insult	Subjects, n	Time after injury	Sensory-evoked potential recordings: Nerve stimulated	Findings, n	Correlation with clinical neurologic assessment
	Complete lesions	28			SEP absent, 12 in 28;	Good
					SEP present, 16 in 28;	Poor
	Incomplete lesions	43			SEP present	

Study	Assessment of outcome: Time after injury	Technique	Findings	Correlation with initial SEP	Comments
York et al. (1983)	Weeks to months	Clinical and SEP	No change in SEP findings with complete or incomplete lesions	Complete lesions (n, 28): no clinical recovery of cord function even when SEP was present.	Nonstandard SEP techniques; SEP excessively noisy and not easy to evaluate; no explanation for preservation of SEP with no sensory function

					Incomplete lesions (n, 43): 16 had some sensory return, six had some motor and sensory return, three died, 18 were lost to followup.	

Study	Insult	Subjects, n	Time after injury	Sensory-evoked potential recordings: Nerve stimulated	Findings, n	Correlation with clinical neurologic assessment
Ertekin et al. 1984	Trauma or compression by tumor or disk	52		Intrathecal T10-11 or T11-12 vertebral level		
	Cauda equina and/or conus medullaris trauma	15	5–75 d (n, 5); 1–18 yr (n, 10)	Above injury	Present, 15 in 15; prolonged latencies with injury <2 yr old and normal with injury >2 yr old (n, 6)	Varied from impotence only or urinary incontinence to flaccid paraplegia.

(continued)

Table 7 *(continued)*

				Sensory-evoked potential recordings		
Study	Insult	Subjects, *n*	Time after injury	Nerve stimulated	Findings, *n*	Correlation with clinical neurologic assessment
Ertekin et al. (1984)	Compressive lesions Cauda equina by tumor (*n*, 6) or midline disc (*n*, 7)	13	Before operation	Above lesion	Present, 13 in 13; prolonged latencies	Not stated; presumably similar to above
	Traumatic or compressive lesions of lumbar cord: trauma (*n*, 17), local arachnoiditis (*n*, 3), tumor (*n*, 4)	24	Within 2 mo of trauma (n, 7) or later (*n*, 17)	At or near lesion	Absent 6 in 24 Present, 18 in 24 Prolonged latencies and decreased amplitudes	Pyramidal signs and severe paraplegia

Study	Assessment of outcome				Comments
	Time after injury	Technique	Findings	Correlation with initial SEP	
Ertekin et al. (1984)					Conduction disturbances were greater in traumatic lesions than in compressive lesions. Amplitudes of early components were greater with lateral than with midline stimulation, perhaps because of cortical potentials elicited by antidromic stimulation of the pyramidal tracts. Latency increases during stimulation above lesion probably reflect degeneration of afferent tracts, which may be regenerated by 2 yr after injury when latencies are normal.

[a]Reproduced with permission from Grundy B. L. Electrophysiological evaluation of the patient with acute spinal cord injury. *Crit. Care Clin.*, in press. Abbreviation: Spec, specified.

[b]Bladder was well stimulated by a fine-wire electrode via a Foley catheter.

directly reflect the function of neural pathways and neural structures during various physiologic, pharmacodynamic, and pathophysiologic phenomena. Although much work remains to be done in understanding the detailed anatomy and physiology underlying the generation of these electrophysiologic signals, the present state of the art permits useful applications of these techniques in both laboratory and clinical investigations of the nervous system. As correlations between EPs and localized metabolic states, neurotransmitter function, and various kinds of pathophysiologic dysfunction are established, EP monitoring during acute and chronic neurophysiologic experiments will become even more valuable.

References

Allison T., Wood C. C., and McCarthy G. (1986) The Central Nervous System, in *Psychophysiology* (Coles M. G. H., Donchin E., and Porges S. W., eds.) Guilford, New York.

American Electroencephalographic Society (1984) Guidelines for clinical evoked potential studies. *J. Clin. Neurophysiol.* **1,** 3–53.

Armington J. C. (1986) Electroretinography, in *Electrodiagnosis in Clinical Neurology,* 2nd Ed. (Aminoff M. C., ed.) Churchill Livingstone, New York.

Aunon J. I. (1978) Computer techniques for the processing of evoked potentials. *Computer Programs Biomed.* **8,** 243–255.

Barajas J. J. (1985) Brainstem response audiometry as subjective and objective test for neurological diagnosis. *Scand. Audiol.* **14,** 57–62.

Barker A. T., Jalinous R., and Freeston I. L. (1985) Non-invasive magnetic stimulation of human motor cortex (letter). *Lancet* **1,** 1106–1107.

Beecher H. K. (1940) The first anesthesia records (Codman, Cushing). *Surg. Gynecol. Obstet.* **71,** 689–693.

Bennett M. H. and Jannetta P. J. (1980) Trigeminal evoked potentials in humans. *Electroencephalogr. Clin. Neurophysiol.* **48,** 517–526.

Brown R. H. and Nash C. L. Jr. (1979) Current status of spinal cord monitoring. *Spine* **4,** 466–470.

Burke D., Skuse N. F., and Lethlean A. K. (1980) Cutaneous and muscle afferent components of the cerebral potential evoked by electrical stimulation of human peripheral nerves. *Electroencephalogr. Clin. Neurophysiol.* **51,** 579–588.

Carmon A., Mor J., and Goldberg J. (1976) Evoked cerebral responses to noxious thermal stimuli in humans. *Exp. Brain. Res.* **25,** 103–107.

Celesia G. G. (1982) Steady-state and transient visual evoked potentials in clinical practice. *Ann. NY Acad. Sci.* **388,** 290–305.

Celesia G. G. (1985) Visual Evoked Responses, in *Evoked Potential Testing: Clinical Applications* (Owen J. H. and Davis H., eds.) Grune and Stratton, Orlando.

Chiappa K. H. (1980) Pattern shift visual, brainstem auditory, and short-latency somatosensory evoked potentials in multiple sclerosis. *Neurology* **30,** 110–123.

Chiappa K. H. and Ropper A. H. (1982a) 1. Evoked potentials in clinical medicine. *N. Eng. J. Med.* **306,** 1140–1150.

Chiappa K. H. and Ropper A. H. (1982b) 2. Evoked potentials in clinical medicine. *N. Eng. J. Med.* **306,** 1205–1211.

Childers D. G., Fischler I. S., Boaz T. L., Perry N. W., Jr., and Arroyo A. A. (1986) Multichannel, single trial event related potential classification. *I.E.E.E. Trans. Biomed. Eng.* **BME-33,** 1069.

Cooper R., Osselton J. W., and Shaw J. C. (1980) *EEG Technology,* 3rd Ed. Butterworths, London.

Cracco J. B., Cracco R. Q., and Stolove R. (1979) Spinal evoked potential in man: A maturational study. *Electroencephalogr. Clin. Neurophysiol.* **46,** 58–64.

Daube J. R. (1986) Nerve Conduction Studies, in *Electrodiagnosis in Clinical Neurology,* 2nd Ed. (Aminoff M. J., ed.) Churchill Livingstone, New York.

Dawson G. D. (1947) Cerebral responses to electrical stimulation of peripheral nerve in man. *J. Neurol. Neurosurg. Psychiat.* **10,** 137–140.

Derendorf H., Drehsen G., and Rohdewald P. (1982) Cortical evoked potentials and saliva levels as basis for the comparison of pure analgesic to analgesic combinations. *Pharmacology* **25,** 227–236.

deWeerd J. P. C. (1981) A posteriori time-varying filtering of averaged evoked potentials. I. Introduction and conceptual basis. *Biol. Cybern.* **41,** 211–222.

deWeerd J. P. C. and Kap J. I. (1981) A posteriori time-varying filtering of averaged evoked potentials. II. Mathematical and computational aspects. *Biol. Cybern.* **41,** 223–234.

deWeerd J. P. C. and Martens W. L. J. (1978) Theory and practice of a posteriori "Wiener" filtering of average evoked potentials. *Biol. Cybern.* **30,** 81–94.

Don M., Elberling M. D. C., and Waring M. (1984) Objective detection of averaged auditory brainstem responses. *Scand. Audiol.* **13,** 219–228.

Eisen A. and Aminoff M. J. (1986) Somatosensory evoked potentials, in *Electrodiagnosis in Clinical Neurology,* 2nd Ed. (Aminoff M. J., ed.) Churchill Livingstone, New York.

Eisen A. and Cracco R. Q. (1983) Overuse of evoked potentials: Caution. *Neurology* **33,** 618–621.

Ertekin C., Sarica Y., and Uckardesler L. (1984) Somatosensory cerebral potentials evoked by stimulation of the lumbo-sacral spinal cord in normal subjects and in patients with conus medullaris and cauda equina lesions. *Electroencephalogr. Clin. Neurophysiol.* **59,** 57–66.

Friedman W. A. and Grundy B. L. (1987) Monitoring of sensory evoked potentials is highly reliable and helpful in the operating room. *J. Clin. Monit.* **3,** 38–45.

Friedman W. A., Kaplan B. J., Day A. L., Sypert G. W., and Curran M. T. (1987) Evoked potential monitoring during aneurysm surgery—observations after 50 cases. *Neurosurgery* **20,** 678–687.

Friedman W. A., Kaplan B. J., Gravenstein D., and Rhoton A. L., Jr. (1985) Intraoperative brain-stem auditory evoked potentials during posterior fossa microvascular decompression. *J. Neurosurg.* **62,** 552–557.

Galbraith G. C. (1984) Latency compensation analysis of the auditory brain-stem evoked response. *Electroencephalogr. Clin. Neurophysiol.* **58,** 333–342.

Gasser T., Mocks J., Kohler W., and de Weerd J. P. C. (1986) Performance and measures of performance for estimators of brain potentials using real data. *I.E.E.E. Trans. Biomed. Eng.* **BME-33,** 949.

Gasser T., Mocks J., and Verleger R. (1983) Selavco: A method to deal with trial-to-trial variability of evoked potentials. *Electroencephalogr. Clin. Neurophysiol.* **55,** 717–723.

Gevins A. S. (1986) Quantitative Aspects of EEG and Evoked Potentials, in *Electrodiagnosis in Clinical Neurology,* 2nd Ed. (Aminoff M. J., ed.) Churchill Livingstone, New York.

Goff G. D., Matsumiya Y., Allison T., and Goff W. R. (1977) The scalp topography of human somatosensory and auditory evoked potentials. *Electroencephalogr. Clin. Neurophysiol.* **42,** 57–76.

Goodin D. S., Squires K. C., and Starr A. (1978) Long latency event-related components of the auditory evoked potential in dementia. *Brain* **101,** 635–648.

Gorga M. P., Worthington D. W., Reiland J. K., Beauchaine K. A., and Goldgar D. E. (1985) Some comparisons between auditory brain stem response thresholds, latencies, and the pure-tone audiogram. *Ear Hearing* **6,** 105–112.

Greenberg R. P. and Ducker T. B. (1982) Evoked potentials in the clinical neurosciences. *J. Neurosurg.* **56,** 1–18.

Grundy B. L. (1983) Intraoperative monitoring of sensory evoked potentials. *Anesthesiology* **58,** 72–87.

Grundy B. L. (1982) Monitoring of sensory evoked potentials during neurosurgical operations: Methods and applications. *Neurosurgery* **11,** 556–575.

Grundy B. L. (1985) Evoked Potential Monitoring, in *Monitoring in Anesthesia and Critical Care Medicine* (Blitt C. D., ed.) Churchill Livingstone, New York.

Grundy B. L. and Friedman W. (1987) Electrophysiological evaluation of the patient with acute spinal cord injury. *Crit. Care Clin.* **3,** 519–548.

Grundy B. L., Jannetta P. J., Procopio P. T., Lina A., Boston J. R., and Doyle E. (1982) Intraoperative monitoring of brain-stem auditory evoked potentials. *J. Neurosurg.* **57,** 674–681.

Grundy B. L., Nash C. L., and Brown R. H. (1981a) Arterial pressure manipulation alters spinal cord function during correction of scoliosis. *Anesthesiology* **54,** 249–253.

Grundy B. L., Heros R. C., Tung A. S., and Doyle E. (1981b) Intraoperative hypoxia detected by evoked potential monitoring. *Anesth. Analg* **60,** 437–439.

Grundy B. L., Lina A., Procopio P. T., and Jannetta P. J. (1981c) Reversible evoked potential changes with retraction of the eighth cranial nerve. *Anesth. Analg.* **60,** 835–838.

Hacke W. (1985) Neuromonitoring during interventional neuroradiology. *Cent. Nerv. Sys. Trauma* **2,** 123–136.

Harding G. F. A. and Rubinstein M. P. (1980) The scalp topography of the human visually evoked subcortical potential. *Invest. Ophthalmol. Visual Sci.* **19,** 318–321.

Harner S. G., Daube J. R., and Ebersold M. J. (1986) Electrophysiologic monitoring of facial nerve during temporal bone surgery. *Laryngoscope* **96,** 65–69.

Hillyard S. A. and Picton T. W. (1978) Event Related Brain Potentials and Selective Information Processing, in *Cognitive Components in Cerebral Event-Related Potentials and Selective Attention. Progress In Clinical Neurophysiology,* vol. 6 (Desmedt J. E., ed.) Karger, Basel.

Hillyard S. A., Picton T. W., and Regan D. (1978) Sensation, Perception and Attention: Analysis Using ERPs, in *Event-Related Brain Potentials in Man* (Callaway E., Tueting P., and Koslow S. H., eds.) Academic, New York.

Hoke M., Ross B., Wickesberg R., and Lutkenhoner B. (1984) Weighted averaging—theory and application to electric response audiometry. *Electroencephalogr. Clin. Neurophysiol.* **57,** 484–489.

Jones S. J. (1977) Short latency potentials recorded from the neck and scalp following median nerve stimulation in man. *Electroencephalogr. Clin. Neurophysiol.* **43,** 853–863.

Kaplan P. E. (1983) Somatosensory evoked responses obtained after stimulation of the pelvic and pudendal nerves. *Electromyogr. Clin. Neurophysiol.* **23,** 99–102.

Kaplan B. J., Friedman W. A., Alexander J. A., and Hampson S. R. (1986) Somatosensory evoked potential monitoring of spinal cord ischemia during aortic operations. *Neurosurgery* **19,** 82–90.

Kaplan B. J., Gravenstein D., and Friedman W. A. (1984) Intraoperative electrophysiology in treatment of peripheral nerve injuries. *J. Florida Med. Assoc.* **71,** 400–403.

Kelly D. L., Jr., Goldring S., and O'Leary J. L. (1965) Averaged evoked somatosensory responses from exposed cortex of man. *Arch. Neurol.* **13,** 1–9.

Kline D. G. and Judice D. J. (1983) Operative management of selected brachial plexus lesions. *J. Neurosurg.* **58,** 631–649.

Krieger D. and Larimore W. (1986) Automatic enhancement of single evoked potentials. *Electroencephalogr. Clin. Neurophysiol.* **64,** 568–572.

Landi A., Copeland S. A., Wynn Parry C. B., and Jones S. J. (1980) The role of somatosensory evoked potentials and nerve conduction studies in the surgical management of brachial plexus injuries. *J. Bone Joint Surg.* **62B,** 492–496.

Larson S. J., Holst R. A., Hemmy D. E., and Sances A. Jr. (1976) Lateral extracavitary approach to traumatic lesions of the thoracic and lumbar spine. *J. Neurosurg.* **45,** 628–637.

Lesser R. P., Koehle R., and Lueders H. (1979) Effect of stimulus intensity on short latency somatosensory evoked potentials. *Electroencephalogr. Clin. Neurophysiol.* **47,** 377–382.

Levy W. J. (1983) Spinal evoked potentials from the motor tracts. *J. Neurosurgery* **58,** 38–44.

Levy W. J. and York D. H. (1983) Evoked potentials from the motor tract in humans. *Neurosurgery* **12,** 422–429.

Levy W. J., York D. H., McCaffrey M., and Tanzer F. (1984) Motor evoked potentials from transcranial stimulation of the motor cortex in humans. *Neurosurgery* **15,** 287–302.

Lueders H., Gurd A., Hahn J., Andrish J., Weiker G., and Klem G. (1982) A new technique for intraoperative monitoring of spinal cord function: Multichannel recording of spinal cord and subcortical evoked potentials. *Spine* **7,** 110–115.

Mackey-Hargadine J. R. and Hall J. W., III (1985) Sensory evoked responses in head injury. *Cent. Nerv. Sys. Trauma* **2,** 187–206.

McCallum J. E. and Bennett M. H. (1975) Electrophysiologic monitoring of spinal cord function during intraspinal surgery. *Surg. Forum* **26,** 469–471.

Macon J. B., Poletti C. E., Sweet W. H., and Zervas N. T. (1980) Spinal cord conduction velocity measurement during laminectomy. *Surg. Forum* **31,** 453–455.

Merton P. A., Hill D. K., Morton H. B., and Marsden C. D. (1982) Scope of a technique for electrical stimulation of human brain, spinal cord, and muscle. *Lancet* **2,** 597–600.

Møller A. R. and Jannetta P. J. (1983) Monitoring auditory functions during cranial nerve microvascular decompression operations by direct recording from the eighth nerve. *J. Neurosurg.* **59,** 493–499.

Moorthy S. S., Markand O. N., Dilley R. S., McCammon R. L., and Warren C. H., Jr (1982) Somatosensory-evoked responses during carotid endarterectomy. *Anesth. Analg.* **61,** 879–883.

Moser J. M. and Aunon J. I. (1986) Classification and detection of single evoked brain potentials using time-frequency amplitude features. *I.E.E.E. Trans. Biomed. Eng.* **BME-33,** 1096.

Nagelkerke N. J. D., de Weerd J. P. C., and Strackee J. (1983) Some criteria for the estimation of evoked potentials. *Biol. Cybern.* **48,** 27–33.

Nash C. L. Jr., Lorig R. A., Schatzinger L. A., and Brown R. H. (1977) Spinal cord monitoring during operative treatment of the spine. *Clin. Orthop.* **126,** 100–105.

Nordwall A., Axelgaard J., Harada Y., Valencia P., McNeal D. R., and Brown J. C. (1979) Spinal cord monitoring using evoked potentials recorded from feline vertebral bone. *Spine* **4,** 486–494.

Peregrin J. and Valach M. (1981) Averaging, selective averaging and latency-corrected averaging. *Pflugers. Arch.* **391,** 154–158.

Perot P. L. (1972) The clinical use of somatosensory evoked potentials in spinal cord injury. *Clin. Neurosurg.* **20,** 367–381.

Perot P. L. Jr. (1976) Somatosensory Evoked Potentials in the Evaluation of Patients with Spinal Cord Injury, in *Current Controversies* (Morley T. P., ed.) Saunders, Philadelphia.

Phelps M. E., Mazziotta J. C., Kuhl D. E., Nuwer M., Packwood J., Metter J., and Engel J. Jr. (1981) Tomographic mapping of human cerebral metabolism: Visual stimulation and deprivation. *Neurology* **31,** 517–529.

Picton T. W. and Hillyard S. A. (1974) Human auditory evoked potentials. II. Effects of attention. *Electroencephalogr. Clin. Neurophysiol.* **36,** 191–200.

Pratt H. and Starr A. (1981) Mechanically and electrically evoked somatosensory potentials in humans: Scalp and neck distributions of short latency components. *Electroencephalogr. Clin. Neurophysiol.* **51,** 138–147.

Reite M., Zimmerman J. T., and Zimmerman J. E. (1982) MEG and EEG auditory responses to tone, click and white noise stimuli. *Electroencephalogr. Clin. Neurophysiol.* **53,** 643–651.

Rowed D. W. (1982) Value of Somatosensory Evoked Potentials for Prognosis in Partial Cord Injuries, in *Early Management of Acute Spinal Cord Injury* (Tator C. H., ed.) Raven, New York.

Rowed D. W., McLean J. A. G., and Tator C. H. (1978) Somatosensory evoked potentials in acute spinal cord injury: Prognostic value. *Surg. Neurol.* **9,** 203–210.

Sekhar L. N. and Møller A. R. (1986) Operative management of tumors involving the cavernous sinus. *J. Neurosurg.* **64,** 879–889.

Sgro J. A., Emerson R. G., and Pedley T. A. (1985) Real-time reconstruction of evoked potentials using a new two-dimensional filter method. *Electroencephalogr. Clin. Neurophysiol.* **62,** 372–380.

Sjøntoft E. (1980) A deconvolution procedure for use in extracting information in average evoked response EEG signals. *IEEE Trans. Biomed. Engineer* **BME27,** 227–230.

Sokol S. (1986) Visual Evoked Potentials, in *Electrodiagnosis in Clinical Neurology* (Aminoff M. J., ed.) Churchill Livingstone, New York.

Stockard J. J., Stockard J. E., and Sharbrough F. W. (1986) Brainstem Auditory Evoked Potentials in Neurology: Methodology, Interpretation, and Clinical Application, in *Electrodiagnosis in Clinical Neurology* (Aminoff M. J., ed.) Churchill Livingstone, New York.

Stockard J. E., Stockard J. J., Westmoreland B. F., and Corfits J. L. (1979) Brainstem auditory-evoked responses: Normal variation as a function of stimulus and subject characteristics. *Arch. Neurol.* **36,** 823–831.

Stohr P. E. and Goldring S. (1969) Origin of somatosensory evoked scalp responses in man. *J. Neurosurg.* **31,** 117–127.

Sutton S., Tueting P., and Zubin J. (1967) Information delivery and the sensory evoked potential. *Science* **155,** 1437–1439.

Tamaki T. (1977) Clinical benefits of ESP. *Seikagaku* **29,** 681–689.

Uhl R. R., Squires K. C., Bruce D. L., and Starr A. (1980) Effect of halothane anesthesia on the human cortical visual evoked response. *Anesthesiology* **53,** 273–276.

Walter D. O. (1969) A posteriori "Wiener filtering" of average evoked responses. *Electroencephalogr. Clin. Neurophysiol. (suppl.)* **27,** 61–70.

Westerkamp J. J. and Aunon J. I. (1987) Optimum multielectrode *a posteriori* estimates of single-response evoked potentials. *I.E.E.E. Trans. Biomed. Eng.* **BME-34,** 13.

Wiener N. (1949) *Extrapolation, Interpolation and Smoothing of Stationary Time Series,* John Wiley, New York.

Woody C. D. (1967) Characterization of an adaptive filter for the analysis of variable latency neuroelectric signals. *Med. Biol. Eng.* **5,** 539–553.

Woolsey C. N., Erickson T. C., and Gilson W. E. (1979) Localization in somatic sensory and motor areas of human cerebral cortex as determined by direct recording of evoked potentials and electrical stimulation. *J. Neurosurg.* **51,** 476–506.

York D. H., Watts C., Raffensberger M., Spagnolia T., and Joyce C. (1983) Utilization of somatosensory evoked cortical potentials in spinal cord injury. Prognostic limitations. *Spine* **8,** 832–839.

Young W. (1982) Correlation of Somatosensory Evoked Potentials and Neurological Findings in Spinal Cord Injury, in *Early Management of Acute Spinal Cord Injury* (Tator C. H., ed.) Raven, New York.

INDEX

N-acetylaspartate (NAA), 247, 249, 250, 258
Adenosine, 77
Adenosine diphosphate (ADP), 235–237
Adenosine triphosphate (ATP), 235–238, 257, 299
ADP (*see* Adenosine diphosphate)
Alanine, 249, 252
Alkyl aryl thiourea compounds, 116
Alpha-chloralose, 323
γ-Aminobutyric acid (GABA), 248, 249, 252
Angular frequency, 269
Annihilation photons, 181, 182
Antipyrine, 113–116, 119, 120, 122
A posteriori Wiener filtering (APWF), 389
APWF (*see* A posteriori Wiener filtering)
Arachnoid membrane, 42
Arterial hypertension, 59, 79, 85
Arterial time-activity curve, 201
Artificial CSF, 39
Ascorbic acid, 31
Aspartate, 248, 250
Aspartate amino transferase, 250
Atomic flame emission spectrophotometry, 312
ATP (*see* Adenosine triphosphate)
ATT-E (*see* Attenuating the signal)
Attenuating the signal (ATT-E), 389
Auditory EPs, 396, 402
Autoradiograms, 154
Autoradiography, 22, 107–170
  image analysis, 166–169
  local cerebral blood flow, 108–131
  local cerebral glucose utilization, 131–146
  multiple-label autoradiography, 155–166
  resolution and tritiated autoradiography, 146–155
Autoregulation, 42, 80, 81
Avidin-Biotin, 346

BAEPs (*see* Brainstem auditory EPs)
Bicarbonate (HCO_3^-), 300
Bicuculline, 238, 239
Blood–brain barrier, 50, 81
BPTU (*see* 1-Butyl-3-phenothiourea)
^{75}Br (*see* Bromine-75)
Brain–blood partition coefficient, 197–199, 201, 202
Brainstem auditory EPs (BAEPs), 382, 387, 390, 391, 393, 394, 405
Brain tissue, 347
Brain uptake index (BUI), 121, 122
Bromine-75 (^{75}Br), 187
Buffers, 368
BUI (*see* Brain uptake index)
1-Butyl-3-phenothiourea (BPTU), 116, 122

^{11}C (*see* Carbon-11)
Carbon-11 (^{11}C), 187, 193, 196, 210, 211

$^{11}C-CO_2$, 213
^{13}C, 247
^{45}Ca, 315
Ca^{2+} (*see* Calcium)
Calcium (Ca^{2+}), 300, 302, 304, 307, 315, 318, 322, 325–328
Ca_e, 325–328
Ca_1, 304
Calcium channel blockers, 84
Carbon monoxide, 193, 196
Catecholamines, 31
CBF (*see* Cerebral blood flow)
CBV (*see* Cerebral blood volume)
Cerebral arteries, 352, 355, 363
Cerebral blood flow (CBF) (*see also* Measurement of cerebral blood flow using diffusible gases and Regional cerebral blood flow), 1–32, 72–75, 77, 78, 80, 81, 91, 117, 195, 204, 205, 217, 218, 299, 301, 302
Cerebral blood volume (CBV), 79, 195, 217
Cerebral cortex, 323, 356
Cerebral infarction, 217–219, 288
Cerebral ions
  measurement of 299–329
    general methods, 305–315
    ion-selective electrodes in studies of focal ischemia, 322–328
    ions in normal cerebral physiology, 300–305
    making and calibrating an extracellular ion-selective microelectrode, 315–322
Cerebral metabolic rate for glucose (CMRGlu) (*see also* Regional cerebral metabolic rate for glucose), 207, 209, 210–213, 217
Cerebral metabolic rate for oxygen ($CMRO_2$) (*see also* Regional cerebral metabolic rate for oxygen), 217
Cerebral oxygen extraction fraction (*see also* Regional oxygen extraction fraction), 195, 204, 215, 217
Cerebral perfusion pressure, 82
Cerebral spinal fluid (CSF), 306, 319
Cerebral vessels, 347, 361
Cervical sympathetic stimulation, 80, 86
Chloride (Cl^-), 300
Choline, 242, 247
Chromalum gelatin, 369
Closed cranial window, 43
$CMRO_2$ (*see* Cerebral metabolic rate for oxygen)
Coincidence circuit, 181, 182
Coincidence resolving time, 181, 184
Colloidal Gold, 347
Compartmental model, 191
Complex partial seizures, 220
Compton scatter, 18, 19
Computed tomography (CT), 234, 380, 403
Computerized transmission tomography, 143
Contrast, 285, 288
Coordinate frame, 269
Cranial window, 40, 72
Cranial window technique, 38
Craniovertebral junction, 291
Creatine, 247, 248, 251
Creatine kinase, 239, 240
CSF (*see* Cerebrospinal fluid)
CSF composition, 42
CT (*see* Computed tomography)
CT scanning, 21, 22
Cyclotron, 189

DANTE (delays alternating with nutations for tailored excitation), 255
DC_e (*see* Extracellular steady-state potential)
Decibel hearing levels (dBHLs), 377
^{18}F-2-deoxyglucose (FDG), 209
Deoxyglucose (DG), 132–142, 144, 145, 147, 150, 151, 153, 154, 157, 158, 161–164, 170
Deoxyglucose method, 206, 208
Deoxyglucose 6-phosphate (DG-6-P), 146, 206, 207, 209
^{18}F-labeled deoxymannose, 210
2-DFT (*see* Two-dimensional Fourier transform)
DG (*see* Deoxyglucose)
Diffusible tracers, 112
5,5-Dimethyl-2,4-oxazolidinedione (^{14}C-DMO) autoradiography, 314
Dirac δ function, 128, 129
Distributed model, 191, 192
DMO (see 5,5-Dimethyl-2,4-oxazolidinedione)
Down's syndrome, 387
Dual photon tomography, 20
Dye microangiography, 49

EEG (*see* Electroencephalogram)
Electroencephalogram (EEG), 299, 383–386, 388, 398
Electromagnetic field B_1, 268, 269
Electromagnetic irradiation, 267, 268
Electromotive force (EMF), 309, 310
Electron probe X-ray microanalysis, 312
Electroretinography (ERG), 378, 390–392
EMF (*see* Electromotive force)
Encoding gradients, 275
EPs (*see* Sensory-evoked potentials)
ERG (*see* Electroretinography)
ERPs (*see* Event-related potentials)
Ethanolamine, 242
Event-related potentials (ERPs), 388
Extracellular steady-state potential (DC_e), 327

Faraday's law, 24
Faraday's number, 24
Far-field potentials, 382
FDG (*see* ^{18}F-2-Deoxyglucose)
Fick equation, 6, 9, 11
Fick principle, 2, 109, 197
Fick's first law of diffusion, 24
FID (*see* Free induction decay)
Fixatives, 368
Flow velocity, 66, 68
Fluorescence, 345
Fluorescence angiography, 37, 44
Fluorescence microscope, 352
^{18}Fluorine (^{18}F), 187
Fluorodeoxyglucose, 140, 165, 209
^{18}F-fluoromethane, 203
Fourier transform, 273
Free induction decay (FID), 277
Frequency encoding, 274
Functional activation studies, 220

^{68}Ga (*see* Gallium-68)
GABA (*see* γ-Aminobutyric acid)
Gadolinium (Gd^{3+}) ion, 284
Gallium-68 (^{68}Ga), 187
Gd^{3+} (*see* Gadolinium ion)
GD DTPA, 285
Glasgow analysis, 13
Glasgow slope technique, 12
Glass electrodes, 305

Glucose, 131, 133–136, 138, 141–146, 150, 153, 158, 163, 206, 212, 243–245, 256, 257
^{11}C-Glucose, 212
Glucose metabolism, 133, 206
Glucose-6-phosphatase, 138, 142, 145, 146
Glutamate, 245, 247, 250, 252, 253, 256, 257
Glutamine, 245, 251
Gluteraldehyde, 153
Glycerol 3-phosphorylcholine, 239
Glycerol 3-phosphorylethanolamine, 239

H^+ (*see* Proton)
HCO_3^- (*see* Bicarbonate)
Hemiparkinsonism, 218
Hemoglobin, 31
Henderson-Hasselbalch equation, 239
Hexokinase, 134
5-HT (*see* 5-Hydroxytryptamine)
Hydrogen administration, 26
Hydrogen clearance methods, 23
Hydrogen polarography, 24
5-Hydroxytryptamine (5-HT), 347, 349, 350, 356, 358–361
Hypercapnia, 78, 79, 84, 86
Hyperglycemia, 140
Hypoglycemia, 140, 251, 305

IAP (*see* Iodoantipyrine)
ICP (*see* Intracranial pressure)
IER (*see* Ion exchange resin)
Image intensity, 213
Image noise, 184
Image resolution, 184
Immunoblots, 345
Immunocytochemistry procedures, 343–369
  immunocytochemistry of selected tissues, 352–365
  immunological basis, 344, 345
  protocols, 366–369
  types of methods, 345–352
Immunolabeling, 351
Immunoprecipitation, 345
Inactivation, 345
Indicator fractionation technique, 122
Indirect immunofluorescence, 346, 347
Inhalation technique, 8
Inorganic phosphate (Pi), 235, 237–239, 257
Insulin, 251
Interactive computer programs, 214
Intracranial hemorrhage, 289
Intracranial pressure (ICP), 39, 92, 93
Intraluminal pressure, 74, 75
Intravital microscopy, 37, 44, 47
Inulin, 120
Invasive recording, 382
Inversion recovery, 281, 286
Iodoantipyrine (IAP), 114–116, 122, 123, 126, 141, 157, 158, 161–165
Ion exchange resin (IER), 307, 309, 310, 312, 316, 322
Ion-selective electrodes (ISEs), 305, 308, 310, 316–318, 320–324, 329
Ischemia, 323, 326
Ischemic penumbra, 299
ISEs (*see* Ion-selective electrodes)

K^+ (*see* Potassium)
K^+-ATPase, 304
Ke, 300–303, 325–327
Ketanserin, 97

Kety-Schmidt technique, 4, 5, 107, 117, 119
Kinematography, 37

Lactate, 239, 243, 247, 249, 250, 252, 254
Larmor frequency, 268, 271, 272, 285
LC (*see* Lumped constant)
lCBF (*see* Local cerebral blood flow)
lCGU (*see* Local cerebral glucose utilization)
LEDs (*see* Light-emitting diodes)
Light-emitting diodes (LEDs), 378, 379
Liquid membrane electrodes, 306
LMN (*see* Left median nerve)
Local cerebral blood flow (lCBF), 108, 111, 112, 114, 163
Local cerebral glucose utilization (lCGU), 131, 141, 142, 146, 158, 163
Longitudinal relaxation (T_1), 276
Lumped constant (LC), 207–209, 211, 212

Magnesium (Mg^{2+}), 300, 312
Magnetic field gradient, 271–273
Magnetic resonance imaging (MRI) (*see* also Nuclear magnetic resonance spectroscopy), 403
Manganese (Mn^{2+}), 326
Mathematical models, 191
Median nerve at wrist: right (RMN) or left (LMN), 399
3-0-Methylglucose, 144, 212
Mg^{2+} (*see* Magnesium)
Michaelis-Menten constants, 132, 138
Michaelis-Menten equation, 134, 136
Microscopes and microscopical techniques, 43
Mn^{2+} (*see* Manganese)
Motional rate, 284
MRI (*see* Magnetic resonance imaging and Nuclear magnetic resonance)
Multiple sclerosis, 290
Myelination, 290, 291

^{13}N (*see* Nitrogen-13)
N12b, 393
N13b, 393
N13, 394
N14, 394
N20, 395
Na^+ (*see* Sodium)
NAA (*see* *N*-Acetylaspartatate)
NAD (*see* Nicotinamide adenine dinucleotide)
NADH (*see* Nicotinamide adenine dinucleotide, reduced)
NDP, 251
Near-field potentials, 382
Nernst equation, 309
Net magnetic moment (*see* Nuclear magnetic moment)
Neuropeptide Y (NPY), 347, 348, 350, 352, 354, 356, 358
Nicotinamide adenine dinucleotide (NAD), 236
Nicotinamide adenine dinucleotide, reduced (NADH), 300
Nifedipine, 85, 95
Nimodipine, 84, 94, 96
Nitrogren-13 (^{13}N), 187
Nitrous oxide method, 1, 179
NMR (*see* Nuclear magnetic resonance spectroscopy)
Noninvasive recording, 382
Noradrenaline, 77, 81
Normal anatomy, 285

NPY (*see* Neuropeptide Y)
NTP (*see* Nucleotide triphosphate)
Nuclear gyromagnetic ratio, 267
Nuclear magnetic moment (*see also* Net magnetic moment), 266, 268, 271, 273
Nuclear magnetic resonance (NMR) spectroscopy, 233–258, 265, 294, 314, 403
Nuclear relaxation, 275, 277
Nuclear spin quantum number, 187, 193, 196
Nucleotide triphosphate (NTP), 251

^{15}O (*see* Oxygen-15)
^{15}O-labeled CO_2 ($C^{15}O_2$), 198, 204, 205
^{15}O-labeled water ($H_2{}^{15}O$), 196, 200, 201, 205
Obrist method, 11, 12
Open cranial window, 43
Oxygen-15 (^{15}O), 1, 196, 198, 200, 201, 203–205

^{31}P (*see* Phosphorus), 257
P25, 395
P300 potential, 396
$PaCO_2$ (*see* Partial pressure of carbon dioxide)
PAP (*see* Peroxidase-antiperoxidase)
Paramagnetic agents, 283
Partial pressure of carbon dioxide ($PaCO_2$), 77, 78
PCr (*see* Phosphocreatinine), 247, 249
Perfusion pressure, 43
Peroxidase-antiperoxidase (PAP) complex, 345
PET (*see* Positron emission tomography)
pH, 77, 82
Phase encoding, 274, 278
Phosphatase activity, 145
Phosphate-buffered saline, 369
Phosphocreatine (PCr), 235, 236, 238–242, 247, 249, 257
Phosphodiesters, 258
Phosphoethanolamine, 236
Phospholipases, 304
Phosphomonodiesters, 258
Phosphorus, 235, 236
Photon attenuation, 182
Pi (*see* Inorganic phosphate)
Pial arterioles, 356, 359
Pial blood volume, 55
Pial vessel hemodynamics, 37–97
  analysis of relevance of techniques, 72–83
  direct investigation methods, 38–72
  direct pharmacological cerebrovascular effects, 84–91
Picric acid-formaldehyde, 368
Poisson statistical distribution, 184
Polytetrafluorethylene (PTFE), 317
Polyvinyl chloride (PVC), 307, 308
Positron, 180
Positron emission tomography (PET), 1, 179–221, 314, 403
  analysis of PET data, 213–217
  applications, 217–221
  cerebral blood flow, 196–203
  cerebral blood volume, 193–196
  cerebral oxygen metabolism, 203–206
  detection systems, 180–187
  glucose metabolism, 206–213
  radiotracers, 187–191
  tracer techniques, 191–193

Positron-emitting radionuclides, 187, 191
Posterior tibial nerve (PTN), 399, 419
Potassium (K^+), 83, 300, 302, 307, 312, 318, 321, 322
Precessional motion, 269
Precipitation reactions, 344
Proton (H^+), 300, 302
PTFE (*see* Polytetrafluoroethylene)
PTN (*see* Posterior tibial nerve)
PVC (*see* Polyvinyl chloride)

Radiochemical labeling, 189
Radiofrequency irradiation, 271
Radiolabeled metabolites, 192
Random coincidences, 184
Rate constants, 207–209, 211
^{82}Rb (*see* Rubidium-82)
rCBF (*see* Regional cerebral blood flow)
rCMRGlu (*see* Regional cerebral metabolic rate for glucose)
$rCMRO_2$ (*see* Regional cerebral oxygen metabolism)
Regional cerebral blood flow (rCBF), 204, 205, 218
Regional cerebral blood volume (rCBV), 193, 196, 205
Regional cerebral metabolic rate for glucose (rCMRGlu), 207, 211–213
Regional cerebral oxygen metabolism ($rCMRO_2$), 203–205, 218
Regional oxygen extraction fraction (rOEF), 204, 205, 217
Resolution, 189
Retina, 347, 356, 358, 361, 362, 364
Risberg's algorithm, 14
RMN (*see* Right median nerve)
rOEF (*see* Regional oxygen extraction fraction)
Rubidium-82, 187

Saturation recovery, 279, 280
Sausage strings, 80
Scan time, 192
Scintillation crystals, 183
SE (*see* Spin-echo)
Sections, 351
Senile dementia, 218
Sensory-evoked potentials (EPs), 375–426
  applications, 400–405
  classification of evoked potentials, 375–396
  equipment and techniques, 396–400
SEPs (*see* Somatosensory-evoked potentials)
Short-lived radionuclides, 20
Single photon emission computerized tomography (SPECT), 403
Single photon tomography, 18
Slice definition, 273
Sodium (Na^+), 300, 302, 304, 307, 312
Sodium imaging, 291
Solid membrane electrodes, 307
Solid-phase binding, 345
Somatosensory-evoked potentials (SEPs), 376, 379, 381, 383, 384, 390, 393–395, 403–406, 411–413, 416–419, 421, 422
Sorenson buffer, 368
Sound pressure level (SPL), 377
SP (*see* Substance P)
Spatial coordinates, 274
Spatial encoding, 271
Spatial resolution, 186
Specific activity, 189

SPECT (*see* Single photon emission computerized tomography)
Spinal cord, 291
Spin angular momentum, 266
Spin-echo imaging, 282
Spin-echo (SE) sequences, 282
SPL (*see* Sound pressure level)
Starling resistor, 43
Starling resistor phenomenon, 83
Static magnetic field B_0, 267, 268
Steady-state rCBF method, 199
Steady-state $rCMRO_2$ method, 204
Stereo zoom microscope, 44
Substance P (SP), 348, 350, 351, 354, 358, 361, 363
Superfusion, 39, 42
Sympathetic stimulation, 77, 79, 89

T_1 (*see* Longitudinal relaxation)
T_2 (*see* Transverse relaxation)
Taurine, 252
Television reflectometry, 72
Three compartment model, 206, 207
Time-varying filtering (TVF), 389
Tissue discrimination, 275
Tracer kinetic model, 192
Transmural pressure, 43
Transverse relaxation (T_2), 276
Tricarboxylic acid cycle, 250
Trifluoroiodomethane, 115, 122
Tumors 289, 290
TVF (*see* Time-varying filtering)
Two compartment model, 205
Two-dimensional Fourier transform technique (2-DFT), 274, 275

Umbelliferone fluorescence, 314

Vasoactive intestinal peptide (VIP), 349–351, 354, 358, 363, 364
VEPs (*see* Visually evoked potentials)
Video angiometry, 72
Video densitometry, 68, 72
VIP (*see* Vasoactive intestinal peptide)
Visually evoked potentials (VEPs), 379, 382, 390, 391, 395, 405
Volume averaging, 187
Volume flow, 66–68

Water concentration, 277, 278, 283, 286, 287
Whole mounts, 351, 367

Xenon, 1, 4–6, 8–12, 14–20, 22, 23, 116, 118, 119, 179
  Xenon-123, 179
  Xenon-127, 19
  Xenon-133, 1, 4–6, 8–12, 14–19, 23, 116, 118
Xenon tomographic method, 22
X-ray absorption, 21
X-ray computerized tomography scanning, 1, 17

GPSR Compliance

The European Union's (EU) General Product Safety Regulation (GPSR) is a set of rules that requires consumer products to be safe and our obligations to ensure this.

If you have any concerns about our products, you can contact us on ProductSafety@springernature.com

In case Publisher is established outside the EU, the EU authorized representative is:

Springer Nature Customer Service Center GmbH
Europaplatz 3
69115 Heidelberg, Germany

Batch number: 10837688

Printed by Printforce, the Netherlands